Van de Voorde (Ed.)
Nanoscience and Nanotechnology

Also of Interest

Chemistry of Carbon Nanostructures.
Muellen, Feng (Eds.); 2017
ISBN 978-3-11-028450-8, e-ISBN 978-3-11-028464-5

Nano-Safety.
What We Need to Know to Protect Workers
Fazarro, Trybula, Tate, Hanks (Eds.); 2017
ISBN 978-3-11-037375-2, e-ISBN 978-3-11-037376-9

Biomimetic Nanotechnology.
Mueller; 2017
ISBN 978-3-11-037914-3, e-ISBN 978-3-11-037916-7

Intelligent Materials and Structures.
Abramovich; 2016
ISBN 978-3-11-033801-0, e-ISBN 978-3-11-033802-7

Nanotechnology Reviews.
Challa Kumar (Editor-in-Chief)
ISSN 2191-9089, e-ISSN 2191-9097

Nanoscience and Nanotechnology

Advances and Developments in Nano-sized Materials

Edited by
Marcel Van de Voorde

DE GRUYTER

Editor
Prof. Dr. ing. ir. Dr h.c. Marcel Van de Voorde
Professor at Delft University of Technology (The Netherlands)
and
Member of the Science Council of the French Senate and National Assembly, Paris
Rue du Rhodania 5
BRISTOL A, Appartment 31
3963 Crans-Montana
Switzerland

ISBN 978-3-11-054720-7
e-ISBN (PDF) 978-3-11-054722-1
e-ISBN (EPUB) 978-3-11-054729-0

Library of Congress Control Number: 2018943370

Bibliographic information published by the Deutsche Nationalbibliothek
The Deutsche Nationalbibliothek lists this publication in the Deutsche Nationalbibliografie; detailed bibliographic data are available on the Internet at http://dnb.dnb.de.

Typesetting: Integra Software Services Pvt. Ltd.
Printing and binding: CPI books GmbH, Leck
Cover image: cybrain/iStock/Getty Images

www.degruyter.com

Acknowledgments

The editor is deeply indebted to Professor Michael FITZPATRICK, Executive Dean, Faculty of Engineering, Environment and Computing and Lloyd's Register Foundation Chair in Structural Integrity and Systems Performance, Coventry University (UK), for his work throughout the book in technical and academic editing to ensure a consistent style and accessibility of the book for a nonexpert reader, as well as for numerous technical contributions.

https://doi.org/10.1515/9783110547221-201

Contents

List of Contributors

Livio Baldi
baldi.livio@gmail.com

Cecilia Bartolucci
National Research Council of Italy
Piazzale Aldo Moro 7
00185 Roma RM
Italy
cecilia.bartolucci@cnr.it

Bikramjit Basu
Materials Research Centre
Indian Institute of Science (IISc)
Bangalore-560012
India
bikram.iisc@gmail.com

Kanishka Biswas
CSIR
New Chemistry Unit & International Centre for Materials Science
Jawaharlal Nehru Centre
Bangalore-560064
India
kanishka@jncasr.ac.in

Gerrit Borchard
Section des Sciences Pharmaceutiques
Biopharmacie
Université de Genève
Rue Michel-Servet 1
1211 Geneva 4
Switzerland
Gerrit.Borchard@unige.ch

Urs Braegger
Universität Bern
Klinik für Rekonstruktive Zahnmedizin und Gerodontologie
Freiburgstrasse 7
3010 Bern
Switzerland
urs.braegger@zmk.unibe.ch

Thomas H. Brock
Berufsgenossenschaft Rohstoffe und chemische Industrie
BGRCI
Kurfürsten-Anlage 62
69115 Heidelberg
Germany
Thomas.Brock@bgrci.de

Atasi Dan
Materials Research Centre
Indian Institute of Science (IISc)
Bangalore-560012
India
atasi.lbc@gmail.com

Tiziana Di Francesco
Section des sciences pharmaceutiques
Biopharmacie
Université de Genève
Rue Michel-Servet 1
1211 Geneva 4
Switzerland
Tiziana.DiFrancesco@unige.ch

Ioana Fechete
Université de Strasbourg
ICPEES
UMR 7515 CNRS
25 rue Becquerel
67087 Strasbourg CEDEX 2
France
ifechete@unistra.fr

Hans-Jörg Fecht
Abt. Werkstoffe der Elektrotechnik
Universität Ulm
Albert-Einstein-Allee 47
89081 Ulm
Germany
hans.fecht@uni-ulm.de

https://doi.org/10.1515/9783110547221-202

Peter Gluche
GFD Gesellschaft für Diamantprodukte mbH
Lise-Meitner-Str. 13
89081 Ulm
Germany
peter.gluche@gfd-diamond.com

Petra Göring
SmartMembranes GmbH
Heinrich-Damerow-Str. 4
06120 Halle
Germany
petra.goering@smartmembranes.de

Oluwatosin Ijabadeniyi
Durban University of Technology
Department of Biotechnology and Food Technology
Faculty of Applied Sciences
PO Box 1334
Durban 4000
South Africa
oluwatosini@dut.ac.za

Abosede Ijabadeniyi
Durban University of Technology
Department of Marketing and Retail Management
Durban 4001
South Africa
bosede55@yahoo.com

Colin Johnston
Department of Materials
Oxford University
Begbroke Science Park
Sandy Lane
Yarnton
Oxford OX5 1PF
United Kingdom
colin.johnston@materials.ox.ac.uk

Karolina Jurczyk
Universität Bern
Klinik für Rekonstruktive Zahnmedizin und Gerodontologie
Freiburgstrasse 7
3010 Bern
Switzerland
karolina.jurczyk@zmk.unibe.ch

Mieczyslaw Jurczyk
Department of Functional Nanomaterials
Poznán University of Technology
ul. Jana Pawła II 24
60-965 Poznán
Poland
mieczyslaw.jurczyk@put.poznan.pl

Fanny Knorr
Charité – Universitätsmedizin Berlin
Klinik für Dermatologie, Venerologieund Allergologie
Charitéplatz 1
10117 Berlin
Germany
fanny.knorr@charite.de

Jürgen Lademann
Charité – Universitätsmedizin Berlin
Klinik für Dermatologie, Venerologie und Allergologie
Charitéplatz 1
10117 Berlin
Germany
juergen.lademann@charite.de

Monika Lelonek
SmartMembranes GmbH
Heinrich-Damerow-Str. 4
06120 Halle
Germany
monika.lelonek@smartmembranes.de

Silke Lohan
Charité – Universitätsmedizin Berlin
Klinik für Dermatologie, Venerologie und Allergologie
Charitéplatz 1
10117 Berlin
Germany
silke.lohan@charite.de

Martina Meinke
Charité – Universitätsmedizin Berlin
Klinik für Dermatologie, Venerologie
und Allergologie
Charitéplatz 1
10117 Berlin
Germany
martina.meinke@charite.de

Bert Müller
Universität Basel
Gewerbestrasse 14
4123 Allschwil, Basel
Switzerland
bert.mueller@unibas.ch

Urszula Narkiewicz
West Pomeranian University of Technology
Institute of Chemical and Environment
Engineering
Aleja Piastów 17
70-310 Szczecin
Poland
urszula.narkiewicz@zut.edu.pl

Alexa Patzelt
Charité – Universitätsmedizin Berlin
Klinik für Dermatologie, Venerologie
und Allergologie
Charitéplatz 1
10117 Berlin
Germany
alexa.patzelt@charite.de

Baldev Raj†
National Institute of Advanced Studies
Indian Institute of Science Campus
Bangalore
Karnataka-560012
India
baldev.dr@gmail.com

Mihail C. Roco
National Science Foundation
Alexandria VA 22314
USA
mroco@nsf.gov

Subhajit Roychowdhury
CSIR
New Chemistry Unit & International Centre for
Materials Science
Jawaharlal Nehru Centre
Bangalore-560064
India
subhajit@jncasr.ac.in

Jacques Védrine
Université P. & M. Curie-Paris 06
Sorbonne Universités
Laboratoire de Réactivité de Surface
UMR-CNRS 7197
4 Place Jussieu
75252 Paris
France
jacques.vedrine@upmc.fr

Ernst Wagner
Ludwig-Maximilians-Universität München
Fakultät für Chemie und Pharmazie
Butenandtstr. 5-13
81377 München
Germany
ernst.wagner@cup.uni-muenchen.de

Matthias Werner
NMTC
Reichsstr. 6
14052 Berlin
Germany
werner@nmtc.de

Wolfgang Wondrak
E-Motor Development and Power Electronics
Daimler AG
Hanns-Klemm-Str. 45
71034 Böblingen
Germany
wolfgang.wondrak@daimler.com

Foreword: Nanotechnology divergence

Mihail C. Roco
National Science Foundation and U.S. National Nanotechnology Initiative

The book illustrates the new phase of nanotechnology development with inspiring topics. This phase is characterized by the integration of knowledge at the nanoscale and the creation of larger nanostructure-based material architectures, devices, and systems for fundamentally new products. Equally important, the nanotechnology field *diverges* (expands and branches out) from basic knowledge of control of matter at the nanoscale to novel science platforms and production paradigms.

In the first decade after 2000, the focus of nanotechnology development was on the discovery of new phenomena and the convergence (deep integration for a common goal) of disciplines to reach understanding and control of matter at the nanoscale. It has been a worldwide scientific revolution with balanced contributions from Europe, Asia, and the Americas. After 2010, the focus has shifted to integration, invention, and innovation for technology development and applications. According to Lux Research, global production of products and services incorporating nanotechnology as a condition for their competitiveness reached $1 trillion in 2013 – two years earlier than estimations made in 2000 (see Roco and Bainbridge, "Societal Implications of Nanoscience and Nanotechnology", Kluwer Academic Publishers (now Springer), Dordrecht, 2001, 370p.).

After about 2015, increased attention has been on divergence of nanotechnology field into new topics in knowledge, technology, and applications, by using as a foundation the convergent concepts and methods of nanotechnology already developed. Several topics of these are described in this book. Part III is about advanced industrial applications, including special nanomaterials such as diamond, catalysts, membranes, for energy application, and in automotive industry. Part II is about information and telecommunication including emerging areas such as nanophotonics, quantum computing, and spintronics. Part I is about key areas of health care such as targeted drugs, pharmaceutics, dental implants, food and cosmetics; finally Part IV treats occupational safety and Part V gives an outlook on the future of the field.

In the last few years, the growth of nanotechnology applications seems only to have accelerated via convergence with other emerging technologies such as biotechnology, information technology, cognitive sciences, and artificial intelligence (www.wtec.org/NBIC2/), followed by divergence into new competencies and fields of applications. This book supports this trend through the selected topics. It is noted that the average annual growth rate, even if not uniform, of combined research investments, inventions, and revenues from nanotechnology is about 25% since 2000.

The contributions in this book bring together several of the most active and visionary researches. This book offers a diverse collection of opinions and ideas on the

https://doi.org/10.1515/9783110547221-203

state of the art and perspectives for the future assembled by Professor Marcel Van de Voorde, who continues his leadership role in advancing nanoscale science and technology with this book. Readers, whether enthusiasts or experts in nanotechnology, will embark on a thought-provoking exploration of nanotechnology today and with a hint for tomorrow.

Arlington, Virginia
30 August 2017

Preface

Nanoscience and nanotechnology have gained recognition worldwide as exciting themes in science and engineering, and particularly for their potential industrial applications. Universities have implemented dedicated courses for students; government agencies have launched many research and development programs; and industries have invested in research and generated patents for a wide range of innovative procedures and products. It is expected that nanoscience and nanotechnology will transform the future, drastically influencing our daily life in health care, communication, energy generation and storage, agriculture and food technology, mobility including vehicle safety, and personal security by providing controlled access to data and services.

These revolutionary developments come with concerns over potential safety issues. Therefore, it is of vital importance that scientists, industrialists, and governments also master the safety and ethical aspects of the application of nanotechnology. More research on the environmental impacts of nanoparticles as well as on their toxicity, epidemiology, persistence, and bioaccumulation is essential.

Because of the success of nanotechnology to date, numerous journal articles and books on the topic have been published, but they are often focused on specialized disciplines. This book is unique. It focuses on the synergy between technologies, such as when nanotechnology is applied in medicine and electronics. Attention is also given to advanced and innovative applications, including quantum computing, and the synergies of nanoelectronics and photonics. The innovations and the potential for future applications over a broad spectrum of fields are highlighted.

The book gives (see contents layout for details)

- an overview of the recently developed nanosynthesis methods and transfer routes from the laboratory to industrial scale;
- the spectrum of innovations and applications in human health and health care: nanomedicine, nanopharmacy, nanodentistry, nanocosmetics, and nanofood sciences and technologies;
- advanced innovations in information and communications technologies, in nanoelectronics, nanophotonics, spintronics, and quantum computing;
- the industrial applications of nanoscience and nanotechnology in future energy sources, in modern transportation, in clean chemical industries, on topics related to environmental issues, for example, water and air purification, and finally, in the watch industry and fine mechanics; and
- the guidelines for safe handling of nanomaterials and the protection of the consumer and the environment.

This book is a perfect guide for newcomers to support them in becoming quickly aware of the state of the art in nanoscience and nanotechnology. It will provide a deep insight into the most recent developments and innovations. The book is addressed to

https://doi.org/10.1515/9783110547221-204

a wide spectrum of readers and connects physicists, chemists, and materials scientists with biologists, dentists, and clinicians across the entire field. It is also valuable for researchers, engineers, industrialists, government agencies and consumer organizations. The book is especially recommended as a handbook for students and a reference guide for industrialists.

All authors are authorities in their field working in Europe, the United States, and India. Dr. Mike ROCO, Senior Advisor for Science and Engineering at the NSF and one of the great experts in the field, has provided the Foreword, demonstrating his support for the book. Dr. Roco initiated the first U.S. program devoted to nanoscale science and engineering at the National Science Foundation in 1991. He introduced the National Nanotechnology Initiative at the White House in 1999. He is the founding chair of the U.S. National Science and Technology Council's subcommittee on Nanoscale Science, Engineering, and Technology and the Senior Advisor for Science and Engineering at the National Science Foundation. He is the editor-in-chief of the *Journal of Nanoparticle Research.*

Marcel Van de Voorde
July 2017

Part I: Nanotechnology Innovations for Health

Bert Müller

1 Nanomedicine at a glance

1.1 Introduction

Medical doctors can efficiently diagnose and treat patients because of the broad experience available. Generally, the medical experts have hardly any deep understanding of nanoscience and nanotechnology. In some cases, however, they apply nanotechnology-based drugs, implants, and devices to reach the envisioned success. In our everyday life, we do make similar experiences, as we are using toothpaste and sun protection; both rely on nanometer-sized ingredients, without realizing that nanomedicine is employed. Nanomedicine, however, is a term often related to some danger for our health. Although there is no proof, people believe that nanotechnology could become harmful and the application should be better avoided. This chapter, therefore, summarizes selected examples, which elucidate the potential of nanoscience and nanotechnology for the treatment of common diseases, including caries, incontinence, and cardiovascular diseases.

Here, nanomedicine is defined as the science and technology of diagnosing, treating, and preventing diseases and traumatic injuries; of relieving pain; and of preserving and improving health using the nanometer-sized components [1]. Boisseau and Loubaton have introduced the term nanotechnology-enabled medicine, which might be the better choice compared to the simple word nanomedicine [2]. Recent publications prefer the term nanoscience and nanotechnology for human health, which covers the entire interdisciplinary field without pronouncing medicine [3].

1.2 Nanoscience and Nanotechnology for Oral Health

It should be noted that the subject has been covered in previous books [4–6]. Therefore, the following part is based on the knowledge already available.

1.2.1 Natural Nanomaterials Within the Oral Cavity

The human tissues consist of anisotropic and hierarchically ordered nanostructures. For example, the crowns of the 32 teeth embrace the hardest tissue of the human body – the enamel. It consists of nanometer-sized, ordered hydroxyapatite crystals. This natural material is about three times tougher than the geological hydroxyapatite

https://doi.org/10.1515/9783110547221-001

and much less brittle than the hydroxyapatite sintered at elevated temperatures [7]. The unique mechanical properties stem from the organization of the crystallites in an ordered fibrous continuum in three-dimensional space, where the needle-like crystallites are aligned to form microscopic bundles or rods, sheeted by organic material. These are in return oriented in specific directions depending on their location within the crown. The design is not only restricted to the enamel but also includes the dentin located below the enamel, which is composed of a mineralized collagenous matrix more akin to bone. The softer dentin counteracts the enamel's high brittleness. The result is a high-performance composite structure that maintains its functionality – mastication – over decades under heavy cyclical loads and adverse chemical conditions [8]. Where these two materials meet, they gradually merge in a complex interplay of highly mineralized and collagen-rich regions [9]. Although many research activities have been devoted to understand the impact of the dentin–enamel junction on the mechanical properties of the crown, the role of this complex interface in the tooth's functionality is still under discussion [9, 10]. Especially the organization on the nanometer level seems to play a critical role [11]. Whereas the hydroxyapatite crystals in the enamel are oriented toward the crown surface and the dentin–enamel junction, in the dentin they are oriented parallel to the junction. The orientation of the enamel's crystallites at the crown surface permits an effective way to remineralize the top layer of the crown after the food-related demineralization phase. Such a mineralization cycle occurs for several times during day and, if balanced, the natural teeth last healthy for several decades. An imbalance toward the acidic conditions within the oral cavity, however, leads to destruction within short periods of time. Currently, no engineering process to biomimetically repair or ex vivo recreate the human crowns has been identified. Nevertheless, the state-of-the-art nanoimaging allows for the identification of the design rules to build bioinspired dental fillings [12].

1.2.2 Dental Fillings

The restoration of teeth has been significantly improved during the last century. Many of us do remember the gold crowns and the amalgam fillings, which have been replaced during the twenty-first century by zirconia crowns and polymer-based composites, respectively. Although the restoration materials and related procedures for crown repair have been steadily improved, their life span is limited and does not reach the level of the natural tissue [13, 14]. This means that after about two decades the fillings have to be replaced by larger ones or even by inlays or artificial crowns. The treatment of the root canal is usually the next step, before posts and dental implants become necessary. The costs involved are significant, and alternatives are desirable. One possible approach consists in the development of anisotropic restoration materials that mimic the complex ultrastructure of human teeth [12]. Ideally, such materials will improve bonding to the tooth material as well as better match physical properties, including Young's and shear moduli as well as the thermal expansion of the tooth's components, providing longer life span of the restoration.

1.2.3 Dental Implants

Dental implants are inserted into the jaw and form a stable interface with the surrounding bony tissue. This part of the implant's surface is made rough by means of sand-blasting and etching to reach osseointegration. Here, the roughness on the micro- and nanometer scales is equally important. It has been demonstrated that the nanostructures are especially vital to avoid the inflammatory reactions [15–17].

The other part of the implant is usually smooth and serves for the artificial crown fixation on top. The only current main challenge is the formation of a satisfactory interface between the crown and the soft tissue of the gingiva.

In general, the currently available dental implants remain stable with a success rate close to 100%, and further breakthroughs in nanostructured implants are not expected any more.

Before placing an implant, it is necessary to check for the bone volume and quality. In the case of insufficient bone availability, the bone has to be augmented to guarantee the stability of the implant. A variety of calcium phosphate-based bone graft materials of well-established suppliers are on the market [18], which are used to strengthen the jawbone within months and subsequently allow for a proper fixation of the implant. These calcium phosphate phases should be upgraded to accelerate the jawbone formation.

1.2.4 Challenges of Nanotechnology in Oral Health

Depending on the pH value within the oral cavity, cyclic de- and remineralization of the enamel surface region takes place through diffusion of ions, maintaining tooth crowns in an intact state (see Figure 1.1). If a disequilibrium between the two processes occurs in favor of demineralization, tooth decay occurs. Conversely, artificially supporting the remineralization process can result in tooth repair. The biomimetic repair of the damaged crowns, however, is hardly solved and understood. Ion delivery is a concentration-mediated dynamic process, and such a material flow is more evasive to classical pharmaceutical approaches. In everyday life, we use a more or less nanotechnology-based toothpaste together with a more or less sophisticated brush mainly to clean the crowns usually twice a day. The toothpaste often also provides nanometer-sized species, which promote the remineralization of the crown's surface, and thus, the regeneration processes. Therefore, the small damages that are optically

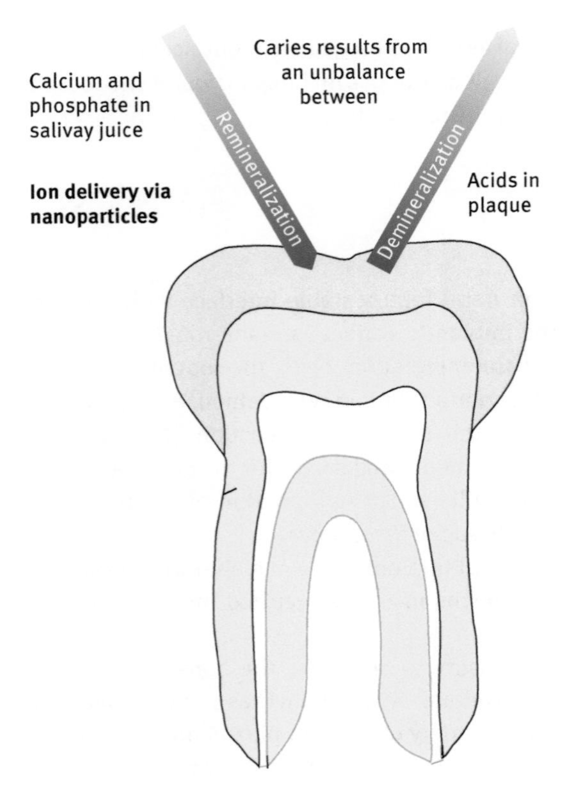

Figure 1.1: The enamel, given in gray, has direct contact to the oral cavity. Therefore, in acidic environment a demineralization takes place, which is converted by changes in the pH value.

invisible can be repaired. The penetration depth of these species is, however, limited, restricting their efficiency to surface incipient lesions. For slightly larger damages, termed white spots, which are visible because of their few hundred micrometers depth, products that are reported to improve the natural regeneration capacity of the crowns are on the market [19, 20]. Their mode of action, however, is not fully understood, since there is no driving force known that pushes the calcium ions into the layers far from the crown surface.

If the caries lesion reaches a critical size, remineralization is currently no longer possible. Here, the highly active inorganic bioactive nanoparticulate glasses are promising candidates to promote remineralization in deeper decayed layers. These materials have already been implemented to produce *bioactive* restorations with an antibacterial effect, reducing bacterial population in affected dentin [21]. Currently, however, the dentist mechanically removes the diseased region as well as some surrounding unaffected tissue, before an isotropic and, therefore, not biomimetic dental filling is implemented. Although the filling materials and their preparation procedures have constantly been improved, we do not know any procedure to build a filling with a micro- and nanostructure similar to that of enamel and dentin including their interface. Here, interdisciplinary teams of experts in dentistry and nanotechnology have to invent suitable compositions of materials and appropriate procedures to be compatible with a chair-side patient treatment in order to reach a crown repair in a biomimetic fashion.

For the dentin, recent studies have shown that even after severe demineralization due to caries the overall nanostructural framework of the crown tissue remains intact [22–24]. As a result, the remineralization of moderate carious lesions can become feasible soon.

1.3 Nanotechnology-Based Artificial Muscles for the Treatment of Severe Incontinence

1.3.1 Anatomy and Function of the Natural Continence Organ

From an engineering point of view, the continence organ acts as a simple switch. Usually the hollow organ is closed and just to pass water or for defecation the hollow organ is opened for a restricted period of time.

In the case of fecal continence, the closeness of the anus depends on the intact continence organ consisting of the internal anal sphincter and the external anal sphincter, the hemorrhoid cushion, and the puborectal muscle. The puborectal muscle surrounds the rectum and pulls it toward a ventral bone of the pelvis. When the puborectal muscle is activated, the rectum is closed, and feces cannot descent from the rectal ampulla to the anal canal. For defecation, the puborectal muscle is relaxed, the rectum straightens, and feces descent. Compared to the technical analogue, the anatomy and the function of the continence organs is complex. Therefore, it is not advised to build the medical device as an exact copy of the natural counterpart. Instead, the implant should be as simple as possible, but provide the full functionality of the natural continence organ.

1.3.2 Lack of Biomimetic Artificial Sphincters

Patients suffering from severe incontinence like to improve their quality of life even if the currently available devices are suboptimal. Most of these artificial muscles generate a constant pressure onto the hollow organ [25, 26]. If this pressure is low, the patients still loose urine and feces. If the pressure onto the hollow organ is high, rejection responses generally occur and atrophy as well as erosion give rise to further critical interventions often resulting in definitive removal of the medical device [25, 26].

The main reason for the disappointment is the missing feedback via the nervous system. Therefore, the artificial muscle should comprise not only the actuator but also a sensory feedback, which should be as fast as in the healthy situation. Millisecond response times are desired.

The device should be autonomous, that is, it should contain not only a battery with limited life time but also a component for energy harvesting from the human body.

As a consequence, the artificial muscle for incontinence treatment is a rather complex device that should reliably operate as actuator, sensor, and energy harvester, simultaneously.

1.3.3 Possible Physical Principle for the Biomimetic Artificial Sphincters

Currently, dielectric elastomer transducers (DETs) are promoted as powerful devices to simultaneously work as actuator and sensor. There are also applications as energy harvester. Thus, these physical principles can also be applied in the field of medicine (see Figure 1.2). The major drawback, however, is the operation voltages, which are usually in the kilovolt range and are reduced to several hundred volts in best cases [27]. As the power required to operate the artificial muscle is relatively large, the operation voltages have to be reduced to the well-known battery ranges of a few volts. Such a reduction is possible by a significant increase of the elastomer's permittivity [28, 29] and by reducing the elastomer film thickness to the nanometer range [30]. The preparation of such materials and films, however, is a critical challenge [31, 32].

1.3.4 Nanometer-Thin DETs for Artificial Muscles

Today, elastomer thin films or membranes with a thickness of several hundred nanometers and an appropriate elasticity can be fabricated [33]. The electrical contacts on both sides on the membranes, however, exhibit Young's moduli orders of magnitude larger than the elastomer membranes. These electrodes dominate the overall mechanical properties, although they are more than one order of magnitude thinner. As a consequence, research teams search for soft electrodes, which include conductive polymers and liquid metals. Nonconfluent metal films with a thickness above the percolation threshold, for example 7-nm-thin gold films, are a promising alternative [34].

Even more important is the generation of the necessary forces. Such thin DETs can only produce forces, which are about four orders of magnitude below the forces necessary for the actuation of the artificial muscle for continence [30]. Therefore,

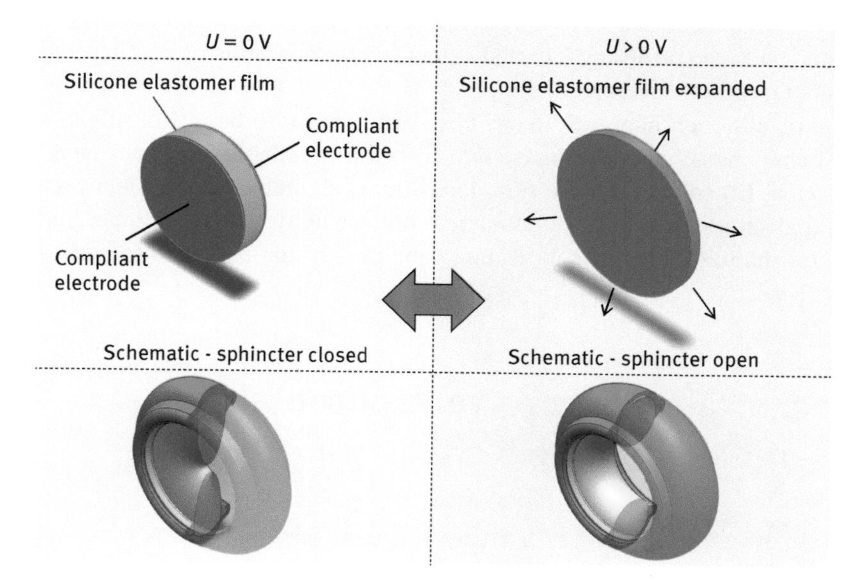

Figure 1.2: Operation principle of a low-voltage dielectric elastomer transducer: A several hundred nanometer-thin elastomer film, such as silicone, is sandwiched between two compliant electrodes. The elastomer is essentially incompressible but deformable. Therefore, the application of a voltage not only generates an electrostatic pressure, but reduces the thickness of the silicone film and is associated with an expansion parallel to the compliant electrodes. This actuation is fully reversible with response times of a few milliseconds. In the second row, a sphincter-like ring with a pre-stretched dielectric elastomer transducer (red color) and a liquid-filled cuff (yellow color) schematically shows the integration of the actuator into a prototyped implant.

multilayered nanostructures with a controlled thickness homogeneity have to be prepared. The related thin-film technology for both the elastomeric and the conductive films has to be developed.

Based on molecular beam deposition (MBD), the functional group density and chain length of polydimethylsiloxane (PDMS) can be tailored [35] and thus, enables to manipulate the elasticity and chemical integrity of the obtained nanometer-thin elastomer membranes [33, 36]. Elastomers with enhanced dielectric properties have been synthesized based on dedicated functional groups and subsequently thermally evaporated [37]. Furthermore, organic MBD enables to implement adhesion layers on the basis of thiol-functionalized PDMS – essential for the reliable binding of the gold electrode to the silicone layers [38]. Thiol-functionalized PDMS furthermore allows for photo-crosslinking [39]. Currently available deposition rates, however, are often below 1 µm per hour. As an alternative deposition technique, electrospray deposition (ESD) of PDMS with subsequent UV-based cross-linking has been introduced. The homogeneity of submicrometer-thin, ESD-prepared PDMS films, however, remains a challenge due to high defect level compared to MBD and the micrometer-rough film surface [40–42]. The homogeneous and soft polymer membrane with a thickness in the submicrometer range is the key for reliable low-voltage operation of multilayered DETs to qualify for artificial muscles.

In summary, although nanotechnology-based DETs seem to be technically feasible, several challenges have to be mastered: (i) fast and reliable elastomer membrane production, (ii) soft conductive film deposition, (iii) stable bonding between elastomer and electrode materials, (iv) thickness homogeneity of the elastomer and electrode better than 2%, and (v) self-healing capability of the device to tolerate a certain defect level.

1.4 Mechanically Responsive Nanocontainers for Targeted Drug Delivery

1.4.1 Emergency Treatment of Cardiovascular Diseases

It is well known that the formation of plaque in the arterial wall usually starts at the age of 20 years. The related cardiovascular disease, atherosclerosis, develops over years and decades. Abruptly, however, the plaque can rupture and can cause a myocardial infarction or stroke. Then, the time to treat the patient is critical. The shorter the time to medical treatment, the smaller the volumes of the heart muscle or brain tissue, which are lost. Currently available procedures are based on technologies such as endovascular devices for intra-arterial clot lysis, stent implantation, and arterial balloon dilatation – all of them are invasive and have to be performed in the catheterization room of a well-equipped hospital.

At pre-hospital level, the patient can systemically obtain a vasodilator drug often based on nitric oxides in the emergency vehicle. These drugs do not only widen the desired constricted artery but the entire blood vessel system. Because of the ensued blood pressure fall, which is a side effect, the maximal therapeutic doses of the administered medicine and the related success in an efficient treatment of the world-leading cause of death are restricted.

1.4.2 Constrictions in the Blood Vessel System

Constrictions within the vessel system, termed stenosis, diminish the blood flow to the end organ. As the cross section at the stenosis is significantly smaller than in the healthy situation, the blood velocity and the associated wall shear stress are substantially higher (see Figure 1.3) [43]. Recently, researchers proposed to take advantage of this behavior and to fabricate mechanoresponsive species, which release the vasodilator upon the disease-specific purely physical triggering [44, 45]. The preferential

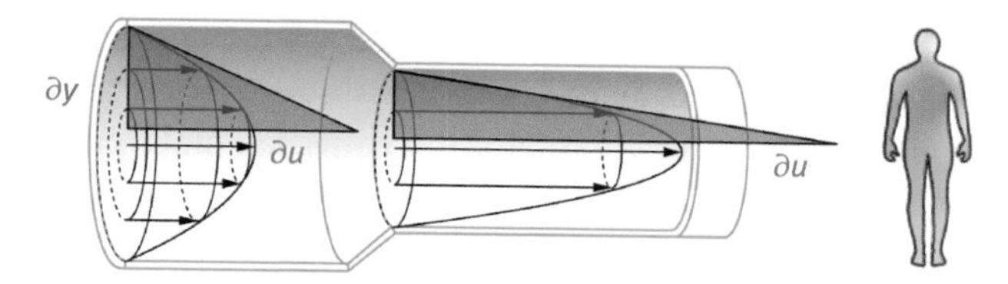

Figure 1.3: The shear stress in Newtonian fluid is proportional to the dynamic viscosity and the derivative of the velocity profile in direction of shear. If there is a disease-related constriction (stenosis) present, the flow of the velocity increases and the related wall shear stress may open the mechanoresponsive nanocontainer. The enclosed drug will be released near the stenosis and not in a systemic fashion.

release at the constrictions enhances the local concentration of the vasodilator and a considerably higher dose can be applied.

In order to determine the actual average wall shear stress at a critically constricted artery, the lumen has to be determined with necessary precision. In vivo examinations do not provide suitable data yet, as plaque-related artifacts and limited spatial resolution hinder the precise three-dimensional imaging. Therefore, the affected artery has to be identified and extracted for postmortem tomographic imaging, which also includes the decalcification. Once the three-dimensional data are available and the lumen segmented, flow simulations allow for the calculation of the local average wall shear stress values [46].

It is surprising that the most critical challenge relates to the localization of a critical stenosed artery, as the disease is the world-leading cause of death in the industrialized countries. Nonetheless, one can state that the threshold for the vasodilator release can be set one order of magnitude above the healthy situation, that is, to values above 20 Pa [47, 48].

1.4.3 Mechanoresponsive Vesicles for Targeted Vasodilator Release

Lipid bilayers can form various three-dimensional arrangements. Most interesting for medicine are liposomes or vesicles, where a bilayer membrane forms a sphere enclosing an aqueous inner cavity. As these vesicles are formed via self-assembly processes of phospholipids, synthetic molecules can substitute the natural phospholipids and this will lead to vesicles with unprecedented properties [49]. Using 1,3-diamidophospholipids, it is possible to formulate mechanoresponsive vesicles that are stable at rest but release their cargo upon mechanical stress. The threshold for release depends on the temperature and the applied stress. The vesicles need to be in the gel state, which is defined by the main phase transition of the bilayer membrane [44]. So far, mechanoresponsive liposomes, which do not release their cargo, for example the selected drug, at the healthy sites of an artificial blood vessel system, but yield the drug to the stenosed sites, have been developed. Their main phase transition temperature, however, lies slightly below body temperature. Therefore, their thermal stability is insufficient to successfully treat patients in an emergency vehicle [50].

These days, liposomal drugs are Food and Drug Administration approved and well established on the market, although adverse effects are noticed. The mechanoresponsive liposomes identified, however, show even less adverse effects than the established liposomal drugs [50, 51]. Therefore, a promising route for the acute treatment of myocardial infarction on the basis of nanotechnology has been discovered. It has to be further refined for the envisioned patient treatment.

1.5 Interdisciplinary Approaches for Nanoscience-Based Medicine

The three examples, that is, oral health, continence, and atherosclerosis therapy, demonstrate that nanomedicine is driven by the patient needs, managed by the medical doctors and dentists, but require close collaboration with a variety of engineers and natural scientists. This teamwork in the field of nanoscience and nanotechnology has given rise to a prominent improvement of the quality of life and has supported the great reduction of morbidity and mortality. Our society is in the era of the widespread use of nanotechnology for health and health care.

Acknowledgments: The author thanks the following scientists for their valuable contributions: Dr. Hans Deyhle (Southampton, UK), Dr. Simone E. Hieber (Basel, Switzerland), Dr. Bekim Osmani (Basel, Switzerland), Dr. med. Till Saxer (Geneva, Switzerland), Dr. Tino Töpper (Basel, Switzerland), and Prof. Dr. Andreas Zumbuehl (Fribourg, Switzerland).

References

1. Duncan R. Nanomedicine an ESF – European Medical Research Councils (EMRC) forward look report, Strasbourg, France, 2005, pp. 48.
2. Boisseau P, Loubaton B. Nanomedicine, nanotechnology in medicine. Comptes Rendus Physique 2011;12:620–36.
3. Müller B, Van de Voorde M. Nanoscience and nanotechnology for human health. Weinheim: Wiley, 2017.
4. Deyhle H, Hieber SE, Müller B. Nanodentistry. In Bhushan B, Winbigler HD, editors. Encyclopedia of nanotechnology. Berlin, Heidelberg: Springer, 2012, 1514–18.
5. Hieber SE, Müller B. Nanodentistry. In: Logothetidis S, editor. Nanomedicine and nanobiotechnology. Berlin, Heidelberg: Springer, 2012, 95–106.

6. Müller B, Zumbuehl A, Walter MA, Pfohl T, Cattin PC, Huwyler J, et al. Translational medicine: nanoscience and -technology to improve patient care. In: Van de Voorde M, Werner M, Fecht H-J editors. The nano-micro interface II – bridging the micro and nano worlds. Weinheim: Wiley-VCH, 2015, 291–308.
7. White SN, Luo W, Paine ML, Fong H, Sarikaya M, Snead, ML. Biological organization of hydroxyapatite crystallites into a fibrous continuum toughens and controls anisotropy in human enamel. J Dent Res 2001;80:321–6.
8. Lawn BR, Lee, J-W, Chai H. Teeth: among nature's most durable biocomposites. Ann Rev Mater Res 2010;40:55–75.
9. Wegst UG, Bai H, Saiz E, Tomsia AP, Ritchie RO. Bioinspired structural materials. Nat Mater 2014;14:23.
10. Marshall SJ, Balooch M, Habelitz S, Balooch G, Gallagher R, Marshall GW. The dentin-enamel junction – a natural, multilevel interface. J Eur Ceram Soc 2003;23:2897–904.
11. Imbeni V, Kruzic JJ, Marshall GW, Marshall SJ, Ritchie RO. The dentin–enamel junction and the fracture of human teeth. Nat Mater 2005;4:229.
12. Deyhle H, Bunk O, Buser S, Krastl G, Zitzmann N, Ilgenstein B, et al. Bio-inspired dental fillings. Proc SPIE 2009;7401:74010E.
13. Gordan VV, Garvan CW, Richman JS, Fellows JL, Rindal DB, Qvist V, et al. How dentists diagnose and treat defective restorations: evidence from the dental PBRN. Oper Dent 2009;34:664–73.
14. Sharif M, Catleugh M, Merry A, Tickle M, Dunne S, Brunton P, et al. Replacement versus repair of defective restorations in adults: resin composite. Cochrane Database Syst Rev 2010;pub2:CD005971. DOI:10.1002/14651858. CD005971
15. Müller B. Natural formation of nanostructures: from fundamentals in metal heteroepitaxy to applications in optics and biomaterials sciences. Surf Rev Lett 2001;8:169–228.
16. Müller B, Riedel M, Michel R, De Paul SM, Hofer R, Heger D, et al. Impact of nanometer-scale roughness on contact-angle hysteresis and globulin adsorption. J Vac Sci Technol B: Microelectron Nanometer Struct 2001;19:1715–20.
17. Riedel M, Müller B, Wintermantel E. Protein adsorption and monocyte activation on germanium nanopyramids. Biomaterials 2001;22:2307–16.
18. Stalder AK, Ilgenstein B, Chicherova N, Deyhle H, Beckmann F, Müller B, et al. Combined use of micro computed tomography and histology to evaluate the regenerative capacity of bone grafting materials. Int J Mater Res 2014;105:679–91.
19. Kind L, Stevanovic S, Wuttig S, Wimberger S, Hofer J, Müller B, et al. Biomimetic remineralization of carious lesions by self-assembling peptide. J Dent Res 2017;96:790–7.
20. Silvertown JD, Wong BP, Sivagurunathan KS, Abrams SH, Kirkham J, Amaechi, BT. Remineralization of natural early caries lesions in vitro by P11-4 monitored with photothermal radiometry and luminescence. J Invest Clin Dent 2017;8:e12257–n/a.
21. Imazato S. Bio-active restorative materials with antibacterial effects: new dimension of innovation in restorative dentistry. Dent Mater J 2009;28:11–19.
22. Deyhle H, Bunk O, Müller B. Nanostructure of healthy and caries-affected human teeth. Nanomed: Nanotechnol Biol Med 2011;7:694–701.
23. Deyhle H, White SN, Bunk O, Beckmann F, Müller B. Nanostructure of the carious tooth enamel lesion. Acta Biomater 2014;10:355–64.
24. Gaiser S, Deyhle H, Bunk O, White SN, Müller B. Understanding nano-anatomy of healthy and carious human teeth: a prerequisite for nanodentistry. Biointerphases 2012;7:4.
25. Al Adem KM, Bawazir SS, Hassen WA, Khandoker AH, Khalaf K, McGloughlin T, et al. Implantable systems for stress urinary incontinence. Ann Biomed Eng 2017:1–16. DOI: 10.1007/s10439-017-1939-9
26. Fattorini E, Brusa T, Gingert C, Hieber SE, Leung V, Osmani B, et al. Artificial muscle devices: innovations and prospects for fecal incontinence treatment. Ann Biomed Eng 2016;44:1355–69.

27. Carpi F, De Rossi D, Kornbluh R, Pelrine R, Sommer-Larsen P. Dielectric elastomers as electromechanical transducers. Amsterdam: Elsevier, 2011.
28. Madsen FB, Daugaard AE, Hvilsted S, Skov AL. The current state of silicone-based dielectric elastomer transducers. Macromol Rapid Commun 2016;37:378–413.
29. Madsen FB, Javakhishvili I, Jensen RE, Daugaard AE, Hvilsted S, Skov AL. Synthesis of telechelic vinyl/allyl functional siloxane copolymers with structural control. Polym Chem 2014;5:7054–61.
30. Töpper T, Weiss FM, Osmani B, Bippes C, Leung V, Müller B. Siloxane-based thin films for biomimetic low-voltage dielectric actuators. Sens Actuators A: Phys 2015;233:32–41.
31. Poulin A, Rosset S, Shea, HR. Fully printed 3 microns thick dielectric elastomer actuator. Proc SPIE 2016;9798:97980L
32. Rosset S, Shea H. Small, fast, and tough: Shrinking down integrated elastomer transducers. Appl Phys Rev 2016;3:031105.
33. Töpper T, Osmani B, Müller B. Polydimethylsiloxane films engineered for smart nanostructures. J Microelectron Eng 2017;submitted.
34. Osmani B, Töpper T, Deyhle H, Pfohl T, Müller B. Gold layers on elastomers near the critical stress regime. Adv Mater Technol 2017;2:1700105.
35. Töpper T, Lörcher S, Weiss FM, Müller B. Tailoring the mass distribution and functional group density of dimethylsiloxane-based films by thermal evaporation. APL Mater 2016;4:056101.
36. Töpper T, Osmani B, Lörcher S, Müller B. Leakage current and actuation efficiency of thermally evaporated low-voltage dielectric elastomer thin-film actuators. Proc SPIE 2017;10163:101631F.
37. Weiss FM, Madsen FB, Töpper T, Osmani B, Leung V, Müller B. Molecular beam deposition of high-permittivity polydimethylsiloxane for nanometer-thin elastomer films in dielectric actuators. Mater Des 2016;105:106–13.
38. Töpper T, Lörcher S, Deyhle H, Osmani B, Leung V, Müller B. Time-resolved plasmonics used to on-line monitor metal-elastomer deposition for low-voltage dielectric elastomer transducers. Adv Electron Mater 2017;3:1700073.
39. Goswami K, Daugaard AE, Skov AL. Dielectric properties of ultraviolet cured poly(dimethyl siloxane) sub-percolative composites containing percolative amounts of multi-walled carbon nanotubes. RSC Adv 2015;5:12792–9.
40. Osmani B, Töpper T, Siketanc M, Müller B. Electrospraying and ultraviolet light curing of nanometer-thin polydimethylsiloxane membranes for low-voltage dielectric elastomer transducers. Proc SPIE 2017;10163:101631E.
41. Weiss FM, Töpper T, Osmani B, Deyhle H, Kovacs G, Müller B. Thin film formation and morphology of electrosprayed polydimethylsiloxane. Langmuir 2016;32:3276–83.
42. Weiss FM, Töpper T, Osmani B, Peters S, Kovacs G, Müller B. Electrospraying nanometer-thin elastomer films for low-voltage dielectric actuators. Adv Electron Mater 2016;2:1500476.
43. Saxer T, Zumbuehl A, Müller B. The use of shear stress for targeted drug delivery. Cardiovasc Res 2013;99:328–33.
44. Holme MN, Fedotenko IA, Abegg D, Althaus J, Babel L, Favarger F, et al. Shear-stress sensitive lenticular vesicles for targeted drug delivery. Nat Nanotechnol 2012;7:536–43.
45. Korin N, Kanapathipillai M, Matthews BD, Crescente M, Brill A, Mammoto T, et al. Shear-activated nanotherapeutics for drug targeting to obstructed blood vessels. Science 2012;337:738–42.
46. Holme MN, Schulz G, Deyhle H, Weitkamp T, Beckmann F, Lobrinus JA, et al. Complementary X-ray tomography techniques for histology-validated 3D imaging of soft and hard tissues using plaque-containing blood vessels as examples. Nat Protoc 2014;9:1401–15.
47. Cheng C, Helderman F, Tempel D, Segers D, Hierck B, Poelmann R, et al. Large variations in absolute wall shear stress levels within one species and between species. Atherosclerosis 2007;195:225–35.

48. Knight J, Olgac U, Saur SC, Poulikakos D, Marshall Jr W, Cattin PC, et al. Choosing the optimal wall shear parameter for the prediction of plaque location, A patient-specific computational study in human right coronary arteries. Atherosclerosis 2010;211:445–50.
49. Mellal D, Zumbuehl A. Exit-strategies – smart ways to release phospholipid vesicle cargo. J Mater Chem B 2014;2:247–52.
50. Buscema M, Matviykiv S, Mészáros T, Gerganova G, Weinberger A, Mettal U, et al. Immunological response to nitroglycerin-loaded shear-responsive liposomes in vitro and in vivo. J Controlled Release 2017;264:14–23.
51. Bugna S, Buscema M, Matviykiv S, Urbanics R, Weinberger A, Meszaros T, et al. Surprising lack of liposome-induced complement activation by artificial 1,3-diamidophospholipids in vitro. Nanomedicine: NBM. 2016;12:845–9.

Ernst Wagner

2 Nanomedicines for targeted therapy

2.1 Introduction

"Why develop nanomedicines?" is a frequently raised question. In other biomedical fields such as genomics/genome medicine, new nanotechnologies have initially raised big enthusiasm and high expectations for fast translation for a better health care. This innovation fountain fueled public and private funding of the associated basic science, and triggered big investments by pharma companies and biotech companies. As often, real development proceeded continuously but not as fast as initial unrealistic hopes and science fiction. After an initial euphoric incubation phase, recognizing reality with all pros and cons of the subject, the field experienced negative feedback loops, critical debates on safety, failure of start-up companies, and withdrawals of big pharma. This did happen, for example, in the area of gene therapy, antisense oligonucleotides, or small interfering RNA (siRNA) therapeutics. These exploratory cycles are not special and evolved in many technology fields. What remained after critical self-evaluation has been a more solid ground as basis for continuous further developments [1–3]. History tells that for innovative biomedical agents, the translation from research to medical product often requires about three to four decades; we observed this for monoclonal antibodies, for therapeutic liposomes, and for gene therapy. This chapter addresses the current stage of nanomedicines, their previous history, and future prospects.

Nanomedicine is a very broad term, referring to the application of nanotechnology in health care. This area comprises the use of medical nanotechnology for ultrasensitive diagnosis, molecular bioimaging, surgical devices and nonsurgical treatment procedures, implants, and others. This chapter focuses on a subclass of nanomedicine called "nanodrugs", that is, pharmaceutical drug substances formulated into nanosized particles or captured into nanostructured material matrices. The following sections review classes of such nanodrugs and their unique material characteristics, which justify their development for medicine, taking also careful attention to their unique safety profile. Three main goals motivate the development of nanomedicines: (i) the targeted delivery of drugs in the body to the site of disease may provide a more specific action with less side effects; (ii) imitating the action of natural viruses, nanomedicines may enter target cells and subsequently release their therapeutic cargo into the cytosol, as required for RNA interference (RNAi) agents, or within the cell nucleus, as required for DNA vectors in gene therapy; and (iii) at the therapeutic site, microenvironmental or physical triggers can trigger a spatial and temporal controlled release of the active drug substance.

For medical translation, the predictable advantages request multidisciplinary efforts in material sciences, pharmacology and toxicology, and medical sciences.

https://doi.org/10.1515/9783110547221-002

Strategies of programmed drug delivery and chemical evolution of biomimetic nanomedicines stimulate optimization at the technology level. The human body handles nanoparticulate matter with special innate responses, which require careful attention by nanotoxicology and open unique pharmacological options. Clinical developments are flourishing, with the first gene and nucleic acid therapies having reached approval as medical drugs, and recombinant therapeutic proteins already dominating the pharmaceutical revenue by novel drugs.

2.2 Targeted Delivery

Paul Ehrlich, winner of Nobel Prize in Physiology or Medicine in 1908, stated: "We have to learn how to cast magic bullets, which behave like the magic bullets of a marksman and exclusively hit pathogens." His vision is persisting for more than a century and paved the ground for antibiotics and molecular therapeutics that very specifically interact with their molecular disease target. Over the years, the specific targeted recognition has been understood as a general natural mechanism, for example, in enzyme–substrate recognition and antibody–antigen recognition. Also T cells, as part of the cellular immune response, recognize and destroy antigen-presenting infected cells. In biotechnology, monoclonal antibody engineering [4, 5] brought Ehrlich's magic bullet concept of "active targeting" more close to realization than ever before. Basically, recombinant humanized antibodies can be generated against any human target protein; if such a target protein is accessible on the cell surface or in the extracellular space of the disease site, systemically administered antibodies can bind it, neutralize its action, or kill the affected target cell by immunological mechanisms [6, 7]. Alternatively, for anticancer therapy, the tumor-targeting antibody can be conjugated with a cytotoxic agent or radioisotope, which kills the tumor cell subsequently to cell binding [8]; antibody–drug conjugates are an enormously growing class of medicines in oncology [9–14]. Brentuximab vedotin and trastuzumab emtansine were approved in 2011 and 2013, respectively. History tells that antibody and other recombinant protein development for more than four decades resulted in medical protein products that exceed market revenues obtained with classical drugs [15]. Between 2011 and 2016, the Food and Drug Administration (FDA) and Center for Biologics Evolution and Research approved 62 recombinant therapeutic proteins [16]. The annual global revenue from protein drugs moves beyond 200 billion EUR.

Recombinant antibodies can be considered as rather small (150–170 kDa) nanomedicines acting via *active targeting*. Nevertheless, due to their 10–12 nm size, they face nanomedicine-typical barriers of limited extravasation out of the blood stream into the peripheral tissue as well as reduced migration within the target tissue, different from small drugs, which have a less restricted, broader distribution. The active targeting principle was extended from antibodies to numerous other molecular recognitions. In general, chemical conjugates or liposomes, polymer micelles [17–19], or related nanosized formulations were modified with targeting ligands that can recognize and bind cell surface receptors with sufficiently high affinity. Ligands can be small molecules such as the folic acid, which binds the folate receptor (FR) overexpressed in many tumor tissues [20], and small carbohydrates such as the GalNAc trimer recognized by the hepatocyte-specific asialoglycoprotein receptor [21]. Synthetic peptide-based ligands such as bombesin, RGD (arginine-glycine-aspartate, binding integrin), epidermal growth factor (EGF) or phage-library-derived GE11 (targeting the EGF receptor) [22–24], c-Met binding peptide (targeting the hepatocyte growth factor receptor) [25, 26], and angiopep (targeting low-density lipoprotein [LDL] receptor-related protein [LRP]) [27] were incorporated into nanoparticles. Moreover, whole proteins such as the iron transport transferrin (binding the Tf receptor overexpressed on many tumor cells, which require iron for DNA synthesis) or recombinant antibodies were integrated into nucleic acid and other delivery systems [28–30]. Aptamers (oligonucleotide sequences for specific protein recognition) were applied as nonpeptidic targeting ligands. Numerous publications (such as reviewed in [18, 31]) demonstrate the positive effect of targeting ligands. Active specific binding to receptor-positive cells, usually followed by receptor-mediated internalization of the targeted nanoparticle into endocytic vesicles (see also below) was observed in cell culture experiments. In several cases, in vivo targeting to the target organ (usually liver or tumor) was also demonstrated in experimental mouse models and some few clinical studies [30].

Apart from active targeting, selective drug delivery can be achieved by *passive targeting* or by various *physical targeting* technologies [32]. Passive targeting is based on the so-called EPR (enhanced permeability and retention) effect [33]. The vasculature of tumors or inflamed disease sites was found to be less tight than healthy vessel structures. Nanostructures, if they remain sufficiently long in the blood circulation and are not captured by macrophages of the reticuloendothelial system (RES), can occasionally extravasate from blood via tumor or inflammatory tissue via the leaky vasculature. These disease areas also lack sufficient lymphatic drainage; as a consequence, macromolecules and nanoparticles (but not small-molecule drugs) remain stuck in the disease site and exert favorably enhanced activity. The development and market approval of liposomal doxorubicin against Kaposi's sarcoma (Doxil, Caelyx) [34, 35] presents a pioneering success story in this field. Free anthracycline drugs such as doxorubicin induce treatment-limiting cardiotoxicity, whose liposomal formulation is preventing. Circulation of liposomes in the blood stream however is transient because of removal by the RES. Modification of the liposomal surface with the polymer polyethylene glycol (PEG), as found in Doxil, prevents the interaction with RES and results in long-term circulation "stealth" liposomes, demonstrating accumulation in well-perfused tumors such as Kaposi's sarcoma by the EPR effect. It should be noted that tumor types are very different and heterogeneous with regard to leakiness of vessels. Therefore, integration of additional measures (such as combination with active and/or physical targeting) is necessary for more general utility [36]. Passive targeting of anti-inflammatory liposomes to sites of inflammation evolves as an encouraging application [37, 38].

Physical targeting [32] comprises several technologies manipulating target cells and tissues with physical forces that can be applied in a spatially and temporally focused fashion. Short electric pulses (for electrophoresis or electroporation) [39–42], ultrasound (sonoporation, a combination with gas-filled microbubbles) [43–46], cell squeezing [47–49], or local hyperthermia [50] can be used for transiently opening cell and tissue barriers. Magnetic fields can attract superparamagnetic nanoparticles to the target tissue (magnetic drug targeting, magnetofection) [51–53]. Local application of light has been applied in photodynamic therapy or photochemical internalization of drug substances into the cytosol of target cells [54–57]. These technologies already made impact in medicine; electrochemotherapy of bleomycin is used for local treatment of chemoresistant tumors [39]; and electrotransfer of a human papilloma virus (HPV antigens E6 and E7) plasmid DNA (pDNA) vaccine displays encouraging efficacy in cervix cancer [41, 42] in clinical studies, and probably will soon reach the medical market.

In sum, a series of different targeting technologies has come into existence and demonstrated targeted delivery of drugs in cellular systems in vitro and to the site of disease in vivo, in both preclinical animal models and human patients. This targeted approach may enhance therapeutic efficacy and reduce side effects at nontarget sites. However, it must be kept in mind that none of the multiple target technologies is even close to perfect. A recent paper by the group of Warren Chan highlighted the real efficacy of targeting nanoparticles into tumors; reviewing published work they conclude that on average *only 0.7% of the dose* is accumulating at the target site [58]. For scientists the low value was no surprise, knowing that a large fraction can be immediately cleared from the blood system and filtered by RES, the liver, and the renal system; 1% of delivered dose can be therapeutic, and several systems are delivering 1–10%. More surprising was the public response, including dramatic responses such as "failure of nanomedicine," indicating that the public perception was apparently more based on fiction than current stage of real science. In any case, the study clearly showed the need for better communication with public and further optimization of targeted delivery.

2.3 Intracellular Transfer

Administration of therapeutic nucleic acids presents an exciting nanomedicine approach for the treatment of genetic and other life-threatening diseases [59, 60]. The first gene therapies [61], the first therapeutic oligonucleotides [62], and also RNAi therapeutics [63] have reached medical use or are very close to approval. Additional novel exciting opportunities include the use of chemically stabilized messenger RNA (mRNA) for gene expression without genes [64], and CRISPR-Cas9/sgRNA for site-specific genome modification [65]. These successful developments were not at all straightforward. Therapeutic nucleic acids such as antisense oligonucleotides, siRNA, microRNA (miRNA) or mRNA, and pDNA are medium-large to very large charged macromolecules that cannot pass cellular membranes. However, they only can act if they reach the nucleus (for gene expression by pDNA or genome editing by CRISPR Cas9) or cytosol (for mRNA expression, or RNAi by siRNA or miRNA). Moreover, oligonucleotide phosphodiester backbones are labile and enzymatically degradable by nucleases, and, as they resemble viral nucleic acids, they are attackable by the innate immune system. Transfer into target cells requires help by viral vectors or nonviral gene delivery systems.

Already 45 years ago, Ted Friedman and Richard Roblin discussed a concept of gene therapy [59]. The efficiency of *viral vectors*, which by far exceeded synthetic systems in terms of delivery, was paving the way for the first gene therapy performed in 1990, using retroviral vector for genetic modification of cells of severe combined immunodeficiency (SCID) patients *ex vivo* with the adenosine deaminase (ADA) gene (see Table 2.1). It took 26 years to obtain market authorization for an optimized ADA gene therapy. The path was not smooth; a most similar concept, retroviral cytokine subreceptor γ_c gene transfer for treatment of SCID-X1 babies, was successful in restoring the immune system, but resulted in a high percentage of patients in development of leukemia because of retroviral integration next to a host proto-oncogene [66, 67]. During the late 1990s, multiple adenoviral vector clinical trials were performed for genetic diseases and in cancer. The majority of adenoviral trials in genetic diseases halted due to a fatal event in gene transfer to the liver of an 18-year-old patient (Jesse Gelsinger) with ornithine transcarbamoylase deficiency in 1999. Cancer studies using adenoviral and other viral vectors continued, resulting in the approval of Gendicine (adenoviral vector for p53 tumor suppressor gene) in 2004 in China, and

Table 2.1: Gene therapy: From clinical trials to approved medical drug.

Clinical studies (gene, vector)		Approved gene therapy product	
1990	Severe combined immunodeficiency Adenosine deaminase gene Retroviral vector, ex vivo	2016	Strimvelis approval by EMA
late 1990s	Cancer p53 tumor suppressor gene Adenoviral vector, intra/peritumoral, plus radiation	2004	Gendicine approval in China
2007	Lipoprotein lipase gene deficiency Lipoprotein lipase gene, adeno-associated virus vector in vivo, delivery to muscle	2012	Glybera approval by EMA
2003	Melanoma Oncolytic herpes simplex virus 1 expressing immune-activating granulocyte macrophage colony stimulating factor	2016	T-VEC approval by FDA
1996	Cancer Chimeric antigen receptor T-cell therapy	2017	Kymriah approval for CD19 B-cell acute lymphoblastic leukemia
2007	LCA blindness, adeno-associated virus vector RPE65 gene, in vivo delivery sub-retinal	2017	Luxturna approval by FDA

recently for T-VEC, an oncolytic herpes simplex virus 1 expressing immune-stimulatory granulocyte macrophage colony stimulating factor in 2016. Another viral vector based on adeno-associated virus (AAV) triggers far less host reaction in humans. After production problems were solved, this vector has been clinically evaluated with great success in several diseases, including hemophilia B (factor IX delivery), several eye diseases including Leber congenital amaurosis (LCA), and lipoprotein lipase (LPL) deficiency. Glybera, delivering the LPL gene via an AAV vector, obtained approval by the European Medicines Agency (EMA) in 2012. Luxturna, delivering the RPE65 gene via AAV, obtained approval for treatment of LCA by the U.S. Food and Drug Administration (FDA) in 2017. Gene therapy using nonviral technology has not yet obtained market approval. However, electroporation-enhanced pDNA vaccines, such as against HPV antigens E6 and E7, for treatment of early-stage cervix cancer [41, 42] probably will soon reach the medical market. In addition, smaller synthetic therapeutic nucleic acids have already reached the medical market (see below).

The nanoscale character plays an important role for successful *nonviral nucleic acid transfer*. Compaction of negatively charged nucleic acids with polycationic carrier molecules into compact nanoparticles not only protects the cargo against biodegradation by nucleases, it also provides structures that can bind to target cells and be internalized by various endocytic pathways, similar as natural viruses or natural nanoparticles (such as lipoprotein nanoparticles) utilize. Incorporation of targeting ligands (see Section 2.2) for cell surface receptor binding not only enhances targeting specificity, but usually also triggers internalization via receptor-mediated endocytosis. There is a jungle of multiple different uptake pathways that can be utilized, depending on the cell type and tissue microenvironment, disease stage, the specific receptor–ligand pair, the size of the nanoparticle, and the multivalency/density of ligands presented on the nanoparticle [68, 69]. For example, endocytosis via clathrin-coated pits is the common uptake for transferrin; this uptake pathway has an upper size limit of below 200 nm. Clathrin-independent uptake via caveolae has a similar size range. Larger nanoparticles (>200 nm) however can be internalized via micropinocytosis in some cell types and stages. Intracellular routes (again depending on cell type and environment) include recycling to the cell surface, transcytosis, and retrograde transport to Golgi apparatus or endoplasmic reticulum. The most common pathway delivers the cargo to secondary endosomes and degradative lysosomes. For many drugs and macromolecules, delivery into the endolysosomal vesicle system cannot be the aim of a successful transfer. In case of nucleic acid transfer, lysosomal delivery results in cargo degradation and inactivation. It became clear already in early transfection studies [18, 70, 71] that nanoparticles need to escape from early endosomal stages for delivery to the cytosol or further on into the nucleus. Viruses use endosomal membrane destabilizing fusion peptides for their escape. Therefore, analogous functional domains [72–76] have to be incorporated into artificial virus-like systems [77].

The development of *oligonucleotide drugs* and a subclass of RNAi drugs proceeded different routes [78, 79]. Oligonucleotides can manipulate the expressed genome indirectly, commonly on the mRNA level (antisense or RNAi), sometimes also on the protein level (aptamers). They are completely chemical drugs, synthesized by solid-phase supported synthesis on automated oligonucleotide synthesizers applying phosphoramidite chemistry [80]. By chemical modification, nucleic acid analogs can be generated providing favorable characteristics over their natural counterparts (such as stability against nuclease degradation, or improved antisense mRNA target binding) [78, 79, 81]. Intracellular uptake of such stabilized analogs is still suboptimal (1% or less). Nevertheless, by application of larger doses, delivery has been demonstrated as sufficient for therapeutic action. The first therapeutic antisense drug Fomivirsen, with market approval in 1998 (see Table 2.2), is an antisense molecule targeting and interfering with cytomegalovirus (CMV) RNA by sequence complementarity. The molecule is stabilized against nuclease degradation by phosphorothioate chemistry, that is, including sulfur in the backbone. The drug was administrated via intraocular injection against CMV retinitis in human immunodeficiency virus (HIV) patients. Because of improved health care of HIV patients with other drugs, the use of Fomivirsen stopped with 2005. As can be seen from Table 2.2, developments of oligonucleotide therapeutics have been ongoing, with six drugs having obtained market approval. The molecular mechanisms include antisense action against disease-related target mRNAs, but also aptamers (oligonucleotides specifically recognizing target proteins), and exon skipping or including oligonucleotides, thus modifying the maturation of mRNA and subsequent protein sequence.

In sum, facilitated by formulation of their nanostructure into virus-like dimensions, nucleic acid nanomedicines may enter target cells and subsequently release their therapeutic cargo into the cytosol or within the cell nucleus, imitating the first steps of natural viral infection processes. Alternatively, the smaller synthetic oligonucleotide drugs can be generated in a chemically stabilized and more active form for direct entry into target cells without packaging. For the medium-sized siRNA or miRNA drugs, chemical modification has often been combined with conjugation and/or nanoparticle formulation.

Table 2.2: FDA-approved oligonucleotide drugs.

Oligonucleotide drug	Approval	Disease (mode of action, target gene)
Fomivirsen (Vitravene)	1998–2005	Cytomegalovirus retinitis in HIV patients (antisense, against cytomegalovirus)
Pegaptanib (Macugen)	2004	Wet macular degeneration of the retina (VEGF-165)
Mipomersen (Kynamro)	2013	Familial hypercholesterolemia (apoB)
Defibrotide (Defitelio)	2016	Hepatic veno-occlusive disease (aptamer, FGF2 protein binding)
Eteplirsen (Exondys 51)	2016	Duchenne's muscular dystrophy (exon skipping)
Nusinersen (Spinraza)	2016	Spinal muscular atrophy in infants (exon inclusion)

2.4 Controlled Release

Another unique option of nanomedicine is the design of polymer- or hydrogel-based materials as source for controlled local release of a drug from the carrier over an extended period. Such nanostructured materials can be applied as coating of stents, bone grafts, and other biomedical devices, or be injected as sponges into surgical cavities. For example, Gliadel [82] is an FDA-approved nanomedicine (approval in 1996) against glioblastoma (GBM), initially designed by Henry Brem and Bob Langer. Gliadel consists of wafers that contain the chemotherapeutic agent carmustine (Bis-Chlorethyl-Nitroso-Urea, BCNU) in a biodegradable polyanhydride microsphere matrix. These wafers are surgically implanted into the brain after glioma surgery. Upon in vivo erosion of the polyanhydride matrix, the implant slowly releases the drug over weeks [83, 84]. Recently, this type of localized chemotherapy was found to possess an additional advantage; in contrast to standard chemotherapy it did not suppress the immune system. Localized controlled release chemotherapy in GBM enhanced the anti-programmed cell death protein 1 (PD-1) antitumor immunotherapy, whereas systemic administration of chemotherapy abrogated the antitumor immune responses [85]. Subsequent work includes other polymer formulations, more potent drugs, and other tumors, such as brain metastasis of breast cancer [86].

Initially, researchers developed polymer matrices for controlled release of small-molecule drugs only. The release of macromolecules from polymer networks was considered as impossible. However, already in 1976, Bob Langer and Judah Folkman were able to demonstrate the sustained release of proteins and other macromolecules from polymer matrices [87], which they applied in their search for antiangiogenic proteins and peptides. These findings paved the way for many controlled release formulations of therapeutic proteins and other therapeutic macromolecules.

2.5 Programmed Delivery and Biomimetic Nanomedicines

Classical drugs distribute in the patient's body in a rather passive mode; pharmacokinetics and biodistribution are dictated by the pharmacological characteristics of the body and the disease site. The drug chemistry may (or may not) favorably influence the transfer to and persistence of drug at the disease site; but even chemical properties within the Lipinski's Rule of Five [88] remain a compromise for drugs that, due to their fixed static structure, can act sufficiently but not optimally at the multiple different pharmacological steps. The key aspect of programmed drug delivery [32] aims at a design that causes drugs to become dynamic in character. *Dynamic nanomedicines* are drug systems preprogrammed to alter their structure and properties during the drug delivery process. Ideally, these changes make them always best fitting for the

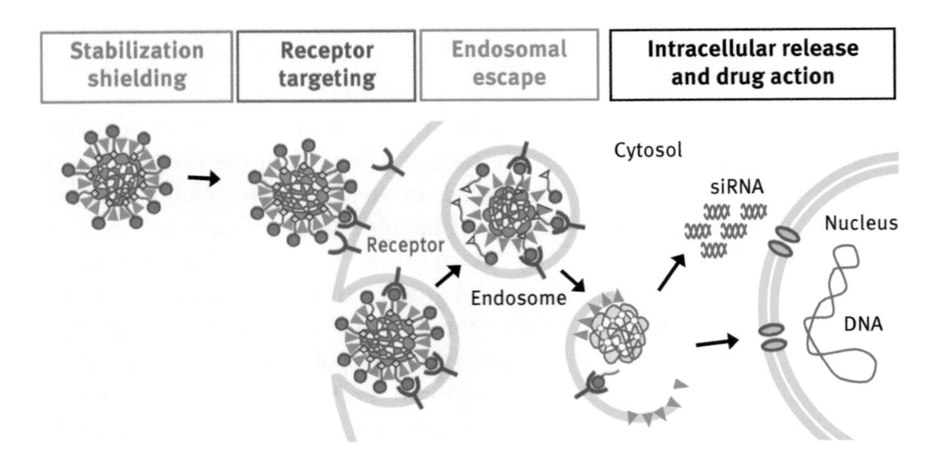

Figure 2.1: Dynamic nanosystems for overcoming cellular barriers. Nanoparticle stabilizing and shielding domains are in green, receptor targeting ligands in brown, activation of endosomolytic domains in red, and cleavable bonds in yellow (labile at endosomal pH, or bioreducible in the cytosol).

individual different delivery step (see Figure 2.1). For example, in the extracellular delivery step of blood circulation, the formulations need to protect the drug cargo against premature release and degradation and provide targeting to the disease site. After reaching this site, almost the contrary, the controlled release over an optimal time window, is required. Programming can be incorporated by the design of molecular sensors (e.g., cleavable bonds, dynamic conformations, and bioreversible noncovalent complexes) that are able to respond to natural pharmacological stimuli, including changes in pH, redox potential and enzymes, and artificial physical stimuli. Physical forces such as ultrasound and other mechanical triggers, electrical or magnetic fields, hyperthermia or light (see Section 2.2 on "physical targeting") may contribute to focusing triggered activation of dynamic nanosystems. Disease targeting principles may include systemic passive targeting and active receptor targeting.

A novel physical form of nanoprogrammed controlled release of drug presents the concept of *pharmacy on microchip*. Sima and colleagues developed such a microchip delivery by design of microelectromechanical systems (MEMS). The MEMS device consists of an array of reservoirs etched into a silicon substrate; drug release is achieved in a time-controlled manner (continuous or pulsed) by the electrochemical dissolution of gold membranes covering the reservoirs. Such MEMS devices were administered to locally deliver chemotherapeutic agents to experimental tumor models [89, 90]. An MEMS device containing the glioma chemotherapeutic temozolomide, when intracranially implanted next to a brain tumor, was found to be effective in reducing tumor development.

An alternative form of programmed nanosystems presents the design and synthesis of *artificial virus-like nanosystems* [72, 77, 91, 92]. Viruses present natural, dynamic nanoagents for potent extracellular and intracellular delivery of nucleic acids. Protein toxins are examples for most effective intracellular uptake of proteins as another class of macromolecules. Nature has optimized these nanocarriers that comprise multiple functions for overcoming delivery barriers, including blood–tumor barrier, blood–brain barrier (BBB), and tissue and cellular barriers such as cellular uptake and endosomal escape. In fact, natural viruses are fascinating masterpieces – but how can we learn from their sophisticated multifunctional precise structure and convert this knowledge into biomimetic nanomedicines? A significant effort was made over the last 25 years on synthesis of dynamic nanosystems (Figure 2.1) with encouraging results [72, 92–96]. These developments, however, were partly hampered because of two specific technical problems. First, macromolecular and supramolecular structures are far larger and more complex than classical chemical compounds. Polymer chemistry often generates products that are polydispersed in size and rather ill-defined with regard to modifications. Clean chemistry however is important to establish clear structure–activity relations (SARs). Apart from the challenge to synthesize precise macromolecules, an additional second formidable issue arose: how can we utilize simple SARs and drive the complicated macromolecular structures to an optimum?

One secret of life is that *information is stored in form of sequences*. The *natural evolution* process takes advantage of the definition of each protein as a precise amino acid sequence stored in form of a genetic sequence (of nucleotides). Refinement of sequences occurred by local variations, such as mutations, deletions, and additions, or larger rearrangements such as domain shuffling, followed by functional selection for a biological task in the set environment. This natural evolution has optimized viruses or toxins into macromolecular carriers that comprise multiple different functions for overcoming the delivery barriers. These basic design principles can also be applied for the generation of artificial dynamic nanosystems. A chemical evolution process may take advantage of combining empirical with rational design and utilizes a more diverse chemical design space than the natural biological variation of amino acids.

Chemical evolution (Figure 2.2) includes identification of chemical motifs for specific delivery steps and assembly of such microdomains into defined larger sequences. It may include rational design (based on previous knowledge or hypothetical models) or random variation and rearrangement into various sequences and (linear, branched) topologies, followed by screening for a predefined delivery task (e.g., targeted delivery into a tumor or across BBB). Chemical motifs may include but are not restricted to natural amino acid sequences. For example, polymer units like PEG, polysarcosine, or polyethylenimine (PEI), despite their simple chemical structure, can exert delivery functions such as shielding or endosomal escape, respectively, with similar efficacy as far more complex natural proteins. Building blocks are assembled into libraries of defined oligoaminoamide sequences by semimanual or automated solid-phase-assisted synthesis [97–101]. Selection of the most suitable sequences for intracellular delivery can be performed in cell culture. However, due to the additional

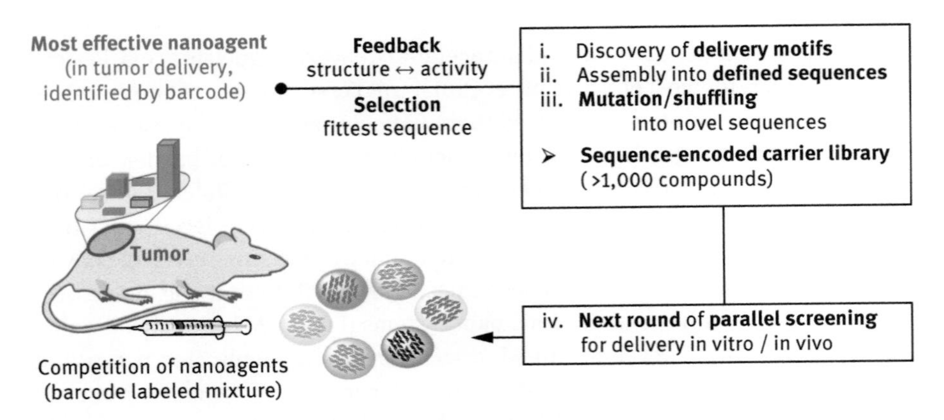

Figure 2.2: Bioinspired chemical evolution of nanomedicine carriers.

delivery barriers, in vivo efficacy often does not correlate with in vitro efficacy. Novel selection strategy such as nucleic acid barcode labeling of individual formulations enables parallel simultaneous screening of several types of nanoformulations within the same mouse [102].

In our own group, more than 1,200 precise, *sequence-defined oligomers* (Figure 2.3) were synthesized from combination of artificial amino acids such as Stp or Sph (containing the aminoethylene motif of PEI), natural amino acids, polymers such as PEG and lipidic residues such as oleic acid or cholanic acid [103, 104]. These oligomers were evaluated in various chemical evolution assays, screening for intracellular transfer or either pDNA, siRNA, protein, or natural products. Not surprising, the different cargos strongly influence the selection process. For example, for pDNA delivery, well-compacting four-arm oligomers mediated best gene transfer if endosomal-buffering histidines and terminal disulfide-forming cysteines were present [105]. For siRNA delivery, T-shaped lipo-oligomers provide convenient nanoparticle stabilization, which was optimized by integrated tyrosine trimers [104]. The efficacy/cytotoxicity ratio could be tuned by the various inserted fatty acids and precise redox-sensitive cleavage sites [101]. For the intracellular delivery of proteins, different carriers were most advantageous. For eGFP (green fluorescent protein) and RNase A, a tetra-oleic acid modified PEGylated Stp oligomer with cysteines for bioreversible coupling, resulting in proteomicelle or proteoliposomal nanostructures, was most effective in transduction of tumor cells [106, 107]. For the delivery of nanobodies (minimum size derived from camelid antibodies), another class of oligomers, which form a bioreducible oligomeric cage around the 2 nm small protein, was most favorable [108].

Combining core oligomers with shielding and targeting functions, delivery of nanomedicine into tumors, was demonstrated in mouse models. Functional activity was observed upon c-Met targeted pDNA delivery in distant hepatocellular carcinoma by marker gene expression [25]; using the theranostic sodium iodide symporter (NIS)

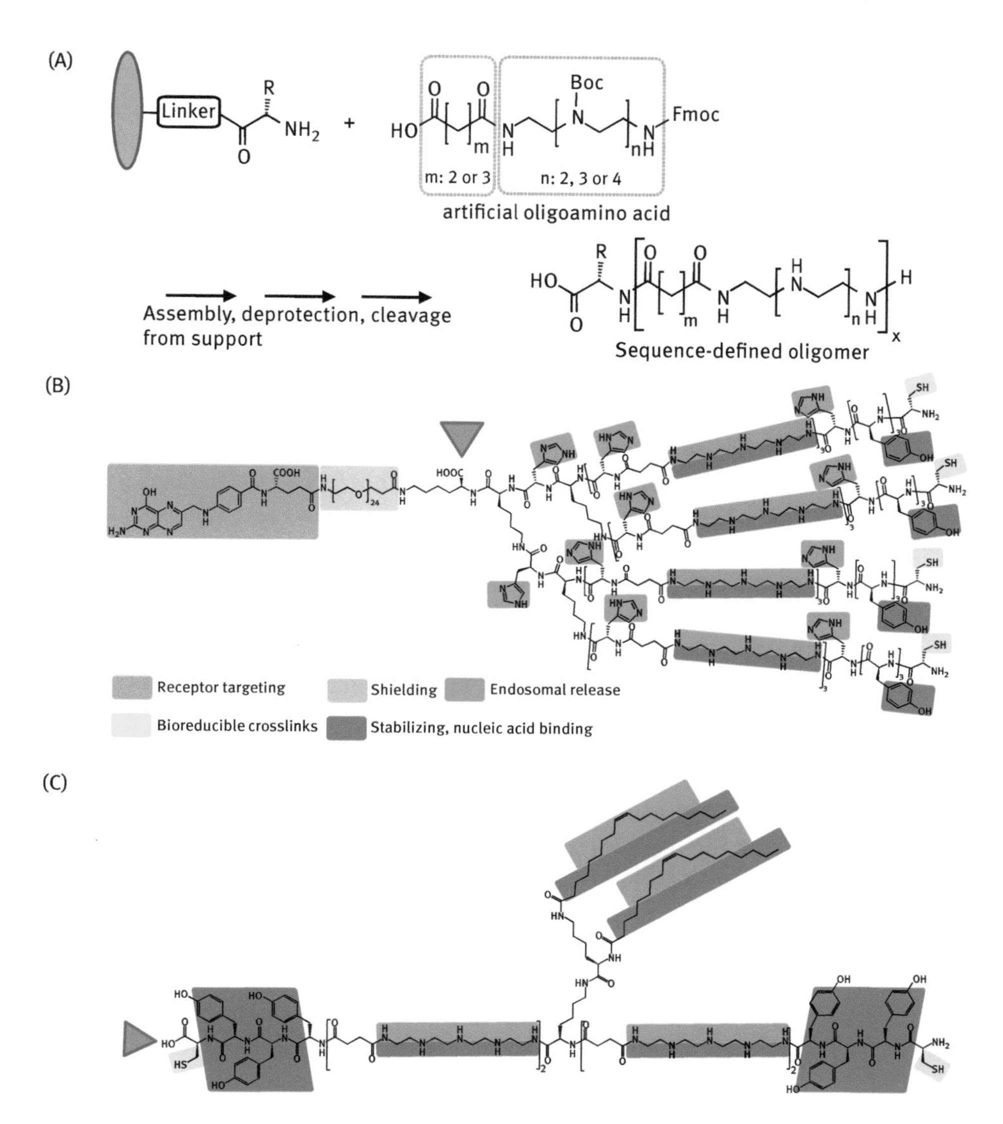

Figure 2.3: Examples of sequence-defined oligomers: (A) Solid-phase supported synthesis; (B) four-arm carrier for pDNA delivery; and (C) lipo-oligomer carrier for siRNA delivery.

gene, gene transfer into the tumor could be monitored by NIS-dependent accumulation of a diagnostic ^{123}I iodide radioisotope, and dosage of the therapeutic ^{131}I radioisotope reduced tumor growth and prolonged survival [26]. Using the same pDNA carrier concept, instead of c-Met targeting, an interleukin-6 (IL-6) derived peptide (I_6P_7) for targeting the IL-6 receptor, intravenous delivery of pDNA encoding for antitumoral ING4, and targeting of orthotopic brain tumors in mice were possible. The I_6P_7 peptide provides multiple functions, including the cascade-targeting potential represented by a combined BBB crossing and subsequent glioma-targeting ability, as well as a direct tumor-inhibiting effect by blockade of the IL-6 receptor. Together with the ING4 gene expression, significantly enhanced survival of mice was observed, better than with standard temozolomide chemotherapy [109]. For siRNA targeted delivery to glioma, a lipo-oligomer formulation containing angiopep as ligand for LRP was applied. Targeting of orthotopic brain tumors and silencing of the tumoral BAG3 gene were achieved [27]. The FR is overexpressed on a series of malignant tumors; FR-dependent gene silencing in tumors was successful using siRNA nanoplexes or polyplexes with folate or methotrexate as targeting ligands [110–112].

2.6 Risks and Safety of Nanomedicines

As for all novel drugs, development requests a thorough safety assessment at all stages before, during, and after clinical testing. This evaluation must take into account the previously known information, newly acquired data, and also predictions based on theoretical considerations [113]. Nanomedicines, in comparison to classical drugs, belong to a rather young class of drugs (only few decades old). Therefore special attention must be laid especially on properties that distinguish them from other medicines. First of all, the *nanoparticle dimension* must be taken under special focus case by case. This presents a significant challenge for all involved disciplines, including production, quality control, and drug regulatory affairs. The current perception of nanoparticle risk is illustrated as follows: classical and recombinant protein parenteral drugs are confronted with the request of a minimum (ideally none) nanoparticle burden. Though the request for the absence of drug-unrelated particles is well understood, the complete exclusion of drug-related nanoparticles (e.g., low fraction of protein aggregates) is not practical. The risk assessment is even more challenging for nanoparticle-based drugs.

Obviously, *nanotoxicology* has become an enormously important discipline to analyze nanoparticle-triggered health risks on an objective level. Although nanomedicines present a new class of drugs, our bodies have been interacting with nanoparticles for very long time. On the one hand, man-caused *environmental nanoparticles* (such as in air or smoke) were found to be associated with increased frequencies of severe illnesses such as cardiovascular and inflammatory lung diseases or cancer. On the other hand, the common topical exposure of skin to nanoparticle-containing protective sun lotions has not been found as risk. Mankind has been confronted with *viruses as natural nanoparticles* triggering serious infections and by evolution has acquired natural defense mechanisms such as the in-born (“innate”) immune response. Last but not least, our body contains *endogenous nanoparticles* such as LDL, high-density lipoprotein and related lipoprotein nanoparticles, or nanosized extracellular vesicles such as exosomes. Especially the natural nanoparticles illustrate very well the range of possible health risks. LDL presents an important endogenous nanoparticle required for cholesterol distribution, but in abundance and combined with local inflammatory responses it is a major risk factor in vascular diseases. In case of viruses and foreign artificial nanoparticles, it appears that our *innate immune system* (complement factors, innate defense cells) can recognize nanoparticles immediately, even before a specific (T-cell or antibody) immune defense has been built up. The recognition process is largely guided by the nanoparticle surface characteristics including surface charge and repetitive patterns (as found in microorganisms and viruses). Phagocytosis and destruction of the foreign nanoparticles follow, often associated with inflammatory and immunological host responses. While these responses may rapidly and effectively clear invading viruses from the body, over-reactions may unfortunately also be deleterious in viral diseases. Such an overwhelming host reaction against an adenoviral gene vector was also responsible for the fatal case of Jesse Gelsinger, who died after liver treatment with the highest vector dose in a dose-escalating phase I clinical gene therapy study (see Section 2.3). In sum, depending on their surface characteristics, nanoparticles may trigger innate and subsequently acquired immune responses; these might be life threatening (especially in systemic administrations), but can be beneficial in case of vaccines.

Virus-based nanomedicines can be associated with another type of risk. Ex vivo treatment of blood progenitor cells with retroviral vectors has resulted in leukemia in several SCID-X1 patients about three years after successful treatment (see Section 2.3). Retroviral vectors integrate their genome preferentially into the active part of the host genome; in the SCID-X1 cases, integration next to the LMO2 host proto-oncogene occurred, resulting in transformation into leukemic cells [66, 67]. Luckily, standard chemotherapy cured the majority of patients from this leukemia; but at least one patient died. Confronted with these results, vector scientists have developed the next generation of retroviral vectors that insulate their own transcriptional machinery from neighboring genomic areas. Insertional oncogenesis must also remain in mind as potential risk, when exciting novel technologies such as genome modification by CRISPR/Cas9 are considered as therapeutic strategy [65]. Although such sequence-specific nucleases cut with high fidelity, undesired genomic modifications cannot be excluded (note a comment by the genome-editing pioneer George Church: “most of what we call genome editing is really genome vandalism”).

Regrettably, we often face too simplistic presentations stating that “viral vectors are dangerous, but nonviral vectors are safe.” Protein-free *synthetic nanosystems* may lack immunogenicity of viral proteins or highly specific integration mechanisms, but there is no rationale guaranteeing their safety. In fact, like other drugs, dose escalation studies reveal significant toxicities at higher doses, and the therapeutic windows between efficacy and acceptable toxicity often are lower than desired. The main reason is the far lower efficacy of nonviral vectors at particle level. Optimization as outlined in Section 2.5 would result in far lower dosages to be administrated. Second, the nanomaterial properties of the *nanosystem core and shell* strongly influence short- and long-term toxicity and biosafety. For example, Otmane Boussif and Jean-Paul Behr had developed very effective polyplex formulations of DNA with the cationic polymer PEI, which subsequently developed into one of the most used transfection reagent. PEI has also been locally administrated in phase I and II clinical studies in cancer with promising results [114, 115]. In preclinical models, unmodified PEI polyplexes, depending on the dose, showed significant to lethal side effects; the positively charged nanoparticles activated the complement system, activated the lung endothelium, and increased adhesion of small cell-containing aggregates, which blocked fine vessels [116–118]. Coating such positively charged PEI polyplexes with PEG strongly reduced toxicity including several of the mentioned effects [118–120]. More sophisticated analyses of the toxicity of PEI polyplexes revealed several phases of cytotoxicity on the cellular level [19]; these include cell membrane destabilization, mitochondrial dysfunction, perturbations of glycolytic flux, and redox homeostasis. Learning about toxicity at refined level provides not only a better understanding for risk assessment but also a better handling for optimizing the carrier into a degradable, better tolerated polymer [19].

2.7 Conclusions and Prospects

In conclusion, nanotechnology comprises the option of more effective administration of innovative drugs with respect to localization and time. Accordingly, focusing nanomedicines to the disease site would reduce undesired side effects in healthy organs. In addition, biochemically or physically programming nanomedicines enable controlled continuous or pulsed release of the active drug substance from the carrier matrix. Currently, the available nanotechnology has demonstrated proof of concept for disease-targeted delivery and controlled release. Efficiencies, however, remained below the levels required for wide success and application in medicine. In comparison to normal small-molecule drugs, nanomedicines face size-restricted tissue access, as well as innate host responses directed against particulate matter. Application and combination of new technologies, such as combining biochemical with physical targeting and microelectronics, or generation of novel carriers by chemical evolution of sequence-defined macromolecules, mimicking artificial viruses, will be key measures for realizing further breakthrough in translation of nanomedicines.

The future prospects of nanomedicines can be projected from the previous and recent achievements. Retrospectively, from the view of around 1990, the darkest hypothetical forecasts, like nanomedicine-triggered oncogenesis, became real, and the biggest hopes, the permanent cures of life-threatening genetic diseases, were also fulfilled. After a more than 25 years period of preclinical and clinical trials with advances and setbacks, knowledge has continuously accumulated. Based on these multidisciplinary experiences, clinical developments directed against severe, otherwise incurable diseases are currently flourishing. The first six gene therapies have reached approval as medical drugs and some further gene therapy products (including anticancer DNA vaccines) are close to registration. In 2017, six oligonucleotide therapeutics are on the market; several RNAi-based nucleic acid therapeutics are in late-stage clinical trials and will most probably receive market authorization soon. Needless to tell that recombinant therapeutic protein nanomedicines are already dominating the pharmaceutical revenue. In parallel to refinement of technologies, the understanding on the biological interaction of the body with nanoparticles will further increase, providing additional tools and markers for nanomedicine safety and efficacy [121]. With these developments, apparently the critical mass of technology and experience has reached the status, where broader application for severe diseases with high incidence, such as cardiovascular diseases, lung diseases, and neurodegeneration, is foreseeable. Harmonization of standards for the required high-end technologies, sophisticated production, clinical trial procedures, safety assessment, as well as storage of the collected preclinical and clinical information on the international and European level will be of utmost importance.

References

1. Duncan R, Gaspar R. Nanomedicine(s) under the microscope. Mol Pharm 2011;8:2141.
2. Schroeder A, Heller DA, Winslow MM, Dahlman JE, Pratt GW, Langer R, et al. Treating metastatic cancer with nanotechnology. Nat Rev Cancer 2012;12:50.
3. Moghimi SM, Wagner E. Nanoparticle technology: having impact, but needing further optimization. Mol Ther 2017;25:1461–3.
4. Kohler G, Milstein C. Continuous cultures of fused cells secreting antibody of predefined specificity. Nature 1975;256:495–7.
5. Gilliland DG, Steplewski Z, Collier RJ, Mitchell KF, Chang TH, Koprowski H. Antibody-directed cytotoxic agents: use of monoclonal antibody to direct the action of toxin A chains to colorectal carcinoma cells. Proc Natl Acad Sci USA 1980;77:4539–43.
6. Yaziji H, Goldstein LC, Barry TS, Werling R, Hwang H, Ellis GK, et al. HER-2 testing in breast cancer using parallel tissue-based methods. JAMA 2004;291:1972–7.
7. Vincenzi B, Zoccoli A, Pantano F, Venditti O, Galluzzo S. CETUXIMAB: from bench to bedside. Curr Cancer Drug Tar 2010;10:80–95.
8. Kellogg BA, Garrett L, Kovtun Y, Lai KC, Leece B, Miller M, et al. Disulfide-linked antibody-maytansinoid conjugates: optimization of in vivo activity by varying the steric hindrance at carbon atoms adjacent to the disulfide linkage. Bioconjugate Chem 2011;22:717–27.
9. Elgundi Z, Reslan M, Cruz E, Sifniotis V, Kayser V. The state-of-play and future of antibody therapeutics. Adv Drug Deliv Rev 2017;122:2–19.
10. Kennedy PJ, Oliveira C, Granja PL, Sarmento B. Antibodies and associates: partners in targeted drug delivery. Pharmacol Ther 2017;177:129–145.
11. Beck A, Goetsch L, Dumontet C, Corvaia N. Strategies and challenges for the next generation of antibody-drug conjugates. Nat Rev Drug Discov 2017;16:315–37.
12. de Goeij BE, Lambert JM. New developments for antibody-drug conjugate-based therapeutic approaches. Curr Opin Immunol 2016;40:14–23.
13. Lambert JM, Morris CQ. Antibody-drug conjugates (ADCs) for personalized treatment of solid tumors: a review. Adv Ther 2017;34:1015–35.
14. Thomas A, Teicher BA, Hassan R. Antibody-drug conjugates for cancer therapy. Lancet Oncol 2016;17:e254–62.
15. Leader B, Baca QJ, Golan DE. Protein therapeutics: a summary and pharmacological classification. Nat Rev Drug Discov 2008;7:21–39.

16. Lagasse HA, Alexaki A, Simhadri VL, Katagiri NH, Jankowski W, Sauna ZE, et al. Recent advances in (therapeutic protein) drug development. F1000 Res 2017;6:113.
17. Miyata K, Nishiyama N, Kataoka K. Rational design of smart supramolecular assemblies for gene delivery: chemical challenges in the creation of artificial viruses. Chem Soc Rev 2012;41:2562–74.
18. Lachelt U, Wagner E. Nucleic Acid Therapeutics Using Polyplexes: A Journey of 50 Years (and Beyond). Chem Rev 2015;115:11043–78.
19. Hall A, Lachelt U, Bartek J, Wagner E, Moghimi SM. Polyplex evolution: understanding biology, optimizing performance. Mol Ther 2017;25:1476–90.
20. Xia W, Low PS. Folate-targeted therapies for cancer. J Med Chem 2010;53:6811–24.
21. Nair JK, Willoughby JL, Chan A, Charisse K, Alam MR, Wang Q, et al. Multivalent N-acetylgalactosamine-conjugated siRNA localizes in hepatocytes and elicits robust RNAi-mediated gene silencing. J Am Chem Soc 2014;136:16958–61.
22. Mickler FM, Mockl L, Ruthardt N, Ogris M, Wagner E, Brauchle C. Tuning nanoparticle uptake: live-cell imaging reveals two distinct endocytosis mechanisms mediated by natural and artificial EGFR targeting ligand. Nano Lett 2012;12:3417–23.
23. Muller K, Klein PM, Heissig P, Roidl A, Wagner E. EGF receptor targeted lipo-oligocation polyplexes for antitumoral siRNA and miRNA delivery. Nanotechnology 2016;27:464001.
24. Schmohl KA, Gupta A, Grunwald GK, Trajkovic-Arsic M, Klutz K, Braren R, et al. Imaging and targeted therapy of pancreatic ductal adenocarcinoma using the theranostic sodium iodide symporter (NIS) gene. Oncotarget 2017;8:33393–404.
25. Kos P, Lachelt U, Herrmann A, Mickler FM, Doblinger M, He D, et al. Histidine-rich stabilized polyplexes for cMet-directed tumor-targeted gene transfer. Nanoscale 2015;7:5350–62.
26. Urnauer S, Morys S, Krhac Levacic A, Muller AM, Schug C, Schmohl KA, et al. Sequence-defined cMET/HGFR-targeted polymers as gene delivery vehicles for the theranostic sodium iodide symporter (NIS) Gene Mol Ther 2016;24:1395–404.
27. An S, He D, Wagner E, Jiang C. Peptide-like Polymers Exerting Effective Glioma-Targeted siRNA delivery and release for therapeutic application. Small 2015;11:5142–50.
28. Wagner E, Curiel D, Cotten M. Delivery of drugs, proteins and genes into cells using transferrin as a ligand for receptor-mediated endocytosis. Adv Drug Del Rev 1994;14:113–36.
29. Wagner E, Zenke M, Cotten M, Beug H, Birnstiel ML. Transferrin-polycation conjugates as carriers for DNA uptake into cells. Proc Natl Acad Sci USA 1990;87:3410–14.
30. Davis ME, Zuckerman JE, Choi CH, Seligson D, Tolcher A, Alabi CA, et al. Evidence of RNAi in humans from systemically administered siRNA via targeted nanoparticles. Nature 2010;464 1067–70.
31. Ogris M, Wagner E. To be targeted: is the magic bullet concept a viable option for synthetic nucleic acid therapeutics? Hum Gene Ther 2011;22:799–807.
32. Wagner E. Programmed drug delivery: nanosystems for tumor targeting. Expert Opin Biol Ther 2007;7:587–93.
33. Maeda H. The enhanced permeability and retention (EPR) effect in tumor vasculature: the key role of tumor-selective macromolecular drug targeting. Adv Enzyme Regul 2001;41:189–207.
34. Northfelt DW, Dezube BJ, Thommes JA, Levine R, Von Roenn JH, Dosik GM, et al. Efficacy of pegylated-liposomal doxorubicin in the treatment of AIDS-related Kaposi's sarcoma after failure of standard chemotherapy. J Clin Oncol 1997;15:653–9.
35. Gabizon AA, Shmeeda H, Zalipsky S. Pros and cons of the liposome platform in cancer drug targeting. J Liposome Res 2006;16:175–83.
36. Ojha T, Pathak V, Shi Y, Hennink WE, Moonen, CTW, Storm G, et al. Pharmacological and physical vessel modulation strategies to improve EPR-mediated drug targeting to tumors. Adv Drug Deliv Rev 2017;119:44–60.
37. Watanabe A, Tanaka H, Sakurai Y, Tange K, Nakai Y, Ohkawara T, et al. Effect of particle size on their accumulation in an inflammatory lesion in a dextran sulfate sodium (DSS)-induced colitis model. Int J Pharm 2016;509:118–22.

38. Ozbakir B, Crielaard BJ, Metselaar JM, Storm G, Lammers T. Liposomal corticosteroids for the treatment of inflammatory disorders and cancer. J Controlled Release 2014;190:624–36.
39. Gothelf A, Mir LM, Gehl J. Electrochemotherapy: results of cancer treatment using enhanced delivery of bleomycin by electroporation. Cancer Treat Rev 2003;29:371–87.
40. Heller R, Heller LC. Gene electrotransfer clinical trials. Adv Genet 2015;89:235–62.
41. Trimble CL, Morrow MP, Kraynyak KA, Shen X, Dallas M, Yan J, et al. Safety, efficacy, and immunogenicity of VGX-3100, a therapeutic synthetic DNA vaccine targeting human papillomavirus 16 and 18 E6 and E7 proteins for cervical intraepithelial neoplasia 2/3: a randomised, double-blind, placebo-controlled phase 2b trial. Lancet (London, England) 2015;386:2078–88.
42. Morrow MP, Kraynyak KA, Sylvester AJ, Shen X, Amante D, Sakata L, et al. Augmentation of cellular and humoral immune responses to HPV16 and HPV18 E6 and E7 antigens by VGX-3100. Mol Ther Oncolytics 2016;3:16025.
43. Manome Y, Nakamura M, Ohno T, Furuhata H. Ultrasound facilitates transduction of naked plasmid DNA into colon carcinoma cells in vitro and in vivo 990. Hum Gene Ther 2000;11:1521–8.
44. Hauff P, Seemann S, Reszka R, Schultze-Mosgau M, Reinhardt M, Buzasi T, et al. Evaluation of gas-filled microparticles and sonoporation as gene delivery system: feasibility study in rodent tumor models. Radiology 2005;236:572–8.
45. Shen ZP, Brayman AA, Chen L, Miao CH Ultrasound with microbubbles enhances gene expression of plasmid DNA in the liver via intraportal delivery. Gene Ther 2008;15:1147–55.
46. Helfield B, Chen X, Watkins SC, Villanueva FS. Biophysical insight into mechanisms of sonoporation. Proc Natl Acad Sci USA 2016;113:9983–8.
47. Kollmannsperger A, Sharei A, Raulf A, Heilemann M, Langer R, Jensen KF, et al. Live-cell protein labelling with nanometre precision by cell squeezing. Nat Commun 2016;7:10372.
48. Sharei A, Trifonova R, Jhunjhunwala S, Hartoularos GC, Eyerman AT, Lytton-Jean A, et al. Ex vivo cytosolic delivery of functional macromolecules to immune cells. PLoS One 2015;10:e0118803.
49. Sharei A, Zoldan J, Adamo A, Sim WY, Cho N, Jackson E, et al. A vector-free microfluidic platform for intracellular delivery. Proc Natl Acad Sci USA 2013;110:2082–7.
50. Kong G, Braun RD, Dewhirst MW. Hyperthermia enables tumor-specific nanoparticle delivery: effect of particle size. Cancer Res 2000;60:4440–5.
51. Scherer F, Anton M, Schillinger U, Henke J, Bergemann C, Kruger A, et al. Magnetofection: enhancing and targeting gene delivery by magnetic force in vitro and in vivo. Gene Ther 2002;9:102–9.
52. Huttinger C, Hirschberger J, Jahnke A, Kostlin R, Brill T, Plank C, et al. Neoadjuvant gene delivery of feline granulocyte-macrophage colony-stimulating factor using magnetofection for the treatment of feline fibrosarcomas: a phase I trial. J Gene Med 2008;10:655–67.
53. Plank C, Zelphati O, Mykhaylyk O. Magnetically enhanced nucleic acid delivery. Ten years of magnetofection-progress and prospects. Adv Drug Deliv Rev 2011;63:1300–31.
54. Hogset A, Prasmickaite L, Selbo PK, Hellum M, Engesaeter BO, Bonsted A, et al. Photochemical internalisation in drug and gene delivery. Adv Drug Deliv Rev 2004;56:95–115.
55. Kloeckner J, Prasmickaite L, Hogset A, Berg K, Wagner E. Photochemically enhanced gene delivery of EGF receptor-targeted DNA polyplexes. J Drug Target 2004;12:205–13.
56. Berg K, Folini M, Prasmickaite L, Selbo PK, Bonsted A, Engesaeter BO, et al. Photochemical internalization: a new tool for drug delivery. Current Pharm Biotechnol 2007;8:362–72.
57. Berg K, Weyergang A, Prasmickaite L, Bonsted A, Hogset A, Strand MT, et al. Photochemical internalization (PCI): a technology for drug delivery. Methods Mol Biol 2010;635:133–45.
58. Wilhelm S, Tavares AJ, Dai Q, Ohta S, Audet J, Dvorak HF, Chan WC. Analysis of nanoparticle delivery to tumors. Nature Reviews Materials 2016;1:16014. DOI:10.1038/natrevmats.2016.14
59. Friedmann T, Roblin R. Gene therapy for human genetic disease? Science 1972;175:949–55.
60. Wirth T, Parker N, Yla-Herttuala S. History of gene therapy. Gene 2013;525:162–9.
61. Touchot N, Flume M. Early insights from commercialization of gene therapies in Europe. Genes 2017;8:78 (4 pages).

62. Stein CA, Castanotto D. FDA-approved oligonucleotide therapies in 2017. Mol Ther 2017;25:1069–75.
63. Titze-de-Almeida R, David C, Titze-de-Almeida SS. The race of 10 synthetic RNAi-based drugs to the pharmaceutical market. Pharm Res 2017;34:1339–63.
64. Kormann MS, Hasenpusch G, Aneja MK, Nica G, Flemmer AW, Herber-Jonat S, et al. Expression of therapeutic proteins after delivery of chemically modified mRNA in mice. Nat Biotechnol 2011;29:154–7.
65. Doudna JA, Charpentier E. Genome editing. The new frontier of genome engineering with CRISPR-Cas9. Science 2014;346:1258096.
66. Hacein-Bey-Abina S, Von Kalle C, Schmidt M, McCormack MP, Wulffraat N, Leboulch P, et al. LMO2-associated clonal T cell proliferation in two patients after gene therapy for SCID-X1. Science 2003;302:415–19.
67. Yi Y, Hahm SH, Lee KH. Retroviral gene therapy: safety issues and possible solutions. Curr Gene Ther 2005;5:25–35.
68. Duncan R, Richardson SC. Endocytosis and intracellular trafficking as gateways for nanomedicine delivery: opportunities and challenges. Mol Pharm 2012;9:2380–402.
69. Muro S. Challenges in design and characterization of ligand-targeted drug delivery systems. J Controlled Release 2012;164:125–37.
70. Cotten M, Langle-Rouault F, Kirlappos H, Wagner E, Mechtler K, Zenke M, et al. Transferrin-polycation-mediated introduction of DNA into human leukemic cells: stimulation by agents that affect the survival of transfected DNA or modulate transferrin receptor levels. Proc Natl Acad Sci USA 1990;87:4033–7.
71. Curiel DT, Agarwal S, Wagner E, Cotten M. Adenovirus enhancement of transferrin-polylysine-mediated gene delivery. Proc Natl Acad Sci USA 1991;88:8850–4.
72. Wagner E, Plank C, Zatloukal K, Cotten M, Birnstiel ML. Influenza virus hemagglutinin HA-2 N-terminal fusogenic peptides augment gene transfer by transferrin-polylysine-DNA complexes: toward a synthetic virus-like gene-transfer vehicle. Proc Natl Acad Sci USA 1992;89:7934–8.
73. Plank C, Oberhauser B, Mechtler K, Koch C, Wagner E. The influence of endosome-disruptive peptides on gene transfer using synthetic virus-like gene transfer systems. J Biol Chem 1994;269:12918–24.
74. Zauner W, Blaas D, Kuechler E, Wagner E. Rhinovirus-mediated endosomal release of transfection complexes. J Virol 1995;69:1085–92.
75. Boeckle S, Fahrmeir J, Roedl W, Ogris M, Wagner E. Melittin analogs with high lytic activity at endosomal pH enhance transfection with purified targeted PEI polyplexes. J Controlled Release 2006;112:240–8.
76. Dohmen C, Edinger D, Frohlich T, Schreiner L, Lachelt U, Troiber C, et al. Nanosized multifunctional polyplexes for receptor-mediated siRNA delivery. ACS Nano 2012;6:5198–208.
77. Wagner E. Strategies to improve DNA polyplexes for in vivo gene transfer: will "artificial viruses" be the answer? Pharm Res 2004;21:8–14.
78. Behlke MA. Chemical modification of siRNAs for in vivo use. Oligonucleotides 2008;18:305–19.
79. Wagner E. Biomaterials in RNAi therapeutics: quo vadis? Biomater Sci 2013;1:804–809.
80. Caruthers MH, Beaucage SL, Becker C, Efcavitch JW, Fisher EF, Galluppi G, et al. Deoxyoligonucleotide synthesis via the phosphoramidite method. Gene Amplification Anal 1983;3:1–26.
81. Parmar R, Willoughby JL, Liu J, Foster DJ, Brigham B, Theile CS, et al. 5'-(E)-Vinylphosphonate: a stable phosphate mimic can improve the RNAi activity of siRNA-GalNAc conjugates. Chembiochem 2016;17:985–9.
82. Perry J, Chambers A, Spithoff K, Laperriere N. Gliadel wafers in the treatment of malignant glioma: a systematic review. Curr Oncol (Toronto, Ont) 2007;14:189–94.

83. Wu MP, Tamada JA, Brem H, Langer R. In vivo versus in vitro degradation of controlled release polymers for intracranial surgical therapy. J Biomed Mater Res 1994;28:387–95.
84. Brem H, Ewend MG, Piantadosi S, Greenhoot J, Burger PC, Sisti M. The safety of interstitial chemotherapy with BCNU-loaded polymer followed by radiation therapy in the treatment of newly diagnosed malignant gliomas: phase I trial. J Neurooncol 1995;26:111–23.
85. Mathios D, Kim JE, Mangraviti A, Phallen J, Park CK, Jackson CM, et al. Anti-PD-1 antitumor immunity is enhanced by local and abrogated by systemic chemotherapy in GBM. Science Transl Med 2016;8:370ra180.
86. Upadhyay UM, Tyler B, Patta Y, Wicks R, Spencer K, Scott A, et al. Intracranial microcapsule chemotherapy delivery for the localized treatment of rodent metastatic breast adenocarcinoma in the brain. Proc Natl Acad Sci USA 2014;111:16071–6.
87. Langer R, Folkman J. Polymers for the sustained release of proteins and other macromolecules. Nature 1976;263:797–800.
88. Lipinski CA, Lombardo F, Dominy BW, Feeney PJ. Experimental and computational approaches to estimate solubility and permeability in drug discovery and development settings. Adv Drug Deliv Rev 2001;46:3–26.
89. Li Y, Ho Duc HL, Tyler B, Williams T, Tupper M, Langer R, et al. In vivo delivery of BCNU from a MEMS device to a tumor model. J Controlled Release 2005;106:138–45.
90. Masi BC, Tyler BM, Bow H, Wicks RT, Xue Y, Brem H, et al. Intracranial MEMS based temozolomide delivery in a 9L rat gliosarcoma model. Biomaterials 2012;33:5768–75.
91. Wagner E. Polymers for siRNA delivery: inspired by viruses to be targeted, dynamic, and precise. Acc Chem Res 2012;45:1005–13.
92. Zuber G, Dauty E, Nothisen M, Belguise P, Behr JP. Towards synthetic viruses. Adv Drug Deliv Rev 2001;52:245–53.
93. Plank C, Zatloukal K, Cotten M, Mechtler K, Wagner E. Gene transfer into hepatocytes using asialoglycoprotein receptor mediated endocytosis of DNA complexed with an artificial tetra-antennary galactose ligand. Bioconjugate Chem 1992;3:533–9.
94. Walker GF, Fella C, Pelisek J, Fahrmeir J, Boeckle S, Ogris M, et al. Toward synthetic viruses: endosomal pH-triggered deshielding of targeted polyplexes greatly enhances gene transfer in vitro and in vivo. Mol Ther 2005;11:418–25.
95. Mastrobattista E, van der Aa MA, Hennink WE, Crommelin DJ. Artificial viruses: a nanotechnological approach to gene delivery. Nat Rev Drug Discov 2006;5:115–21.
96. Cheng Y, Yumul RC, Pun SH. Virus-inspired polymer for efficient in vitro and in vivo gene delivery. Angew Chem Int Ed Engl 2016;55:12013–17.
97. Hartmann L, Hafele S, Peschka-Suss R, Antonietti M, Borner HG. Tailor-made poly (amidoamine)s for controlled complexation and condensation of DNA. Chemistry 2008;14:2025–33.
98. Schaffert D, Badgujar N, Wagner E. Novel Fmoc-polyamino acids for solid-phase synthesis of defined polyamidoamines. Org Lett 2011;13:1586–9.
99. Hartmann L. Polymers for control freaks: sequence-defined poly(amidoamine)s and their biomedical applications. Macromol Chem Phys 2011;212:8–13.
100. Schaffert D, Troiber C, Salcher EE, Frohlich T, Martin I, Badgujar N, et al. Solid-phase synthesis of sequence-defined T-, i-, and U-shape polymers for pDNA and siRNA delivery. Angew Chem Int Ed Engl 2011;50:8986–9.
101. Klein PM, Reinhard S, Lee DJ, Muller K, Ponader D, Hartmann L, et al. Precise redox-sensitive cleavage sites for improved bioactivity of siRNA lipopolyplexes. Nanoscale 2016;8:18098–104.
102. Dahlman JE, Kauffman KJ, Xing Y, Shaw TE, Mir FF, Dlott CC, et al. Barcoded nanoparticles for high throughput in vivo discovery of targeted therapeutics. Proc Natl Acad Sci USA 2017;114:2060–5.

103. He D, Muller K, Krhac Levacic A, Kos P, Lachelt U, Wagner E. Combinatorial optimization of sequence-defined oligo(ethanamino)amides for folate receptor-targeted pDNA and siRNA delivery. Bioconjug Chem 2016;27:647–59.
104. Troiber C, Edinger D, Kos P, Schreiner L, Klager R, Herrmann A, et al. Stabilizing effect of tyrosine trimers on pDNA and siRNA polyplexes. Biomaterials 2013;34:1624–33.
105. Lachelt U, Kos P, Mickler FM, Herrmann A, Salcher EE, Rodl W, et al. Fine-tuning of proton sponges by precise diaminoethanes and histidines in pDNA polyplexes. Nanomedicine 2014;10:35–44.
106. Zhang P, He D, Klein PM, Liu X, Röder R, Döblinger M, et al. Enhanced intracellular protein transduction by sequence defined tetra-oleoyl oligoaminoamides targeted for cancer therapy. Adv Funct Mater 2015;25:6627–36.
107. Zhang P, Steinborn B, Lachelt U, Zahler S, Wagner E. Lipo-oligomer nanoformulations for targeted intracellular protein delivery. Biomacromolecules 2017;18:2509–2520.
108. Roder R, Helma J, Preiss T, Radler JO, Leonhardt H, Wagner E. Intracellular delivery of nanobodies for imaging of target proteins in live cells. Pharm Res 2017;34:161–74.
109. Wang S, Reinhard S, Li C, Qian M, Jiang H, Du Y, et al. Antitumoral cascade-targeting ligand for IL-6 receptor-mediated gene delivery to glioma. Mol Ther 2017;25:1556–66.
110. Lee DJ, He D, Kessel E, Padari K, Kempter S, Lachelt U, et al. Tumoral gene silencing by receptor-targeted combinatorial siRNA polyplexes. J Controlled Release 2016;244:280–91.
111. Lee DJ, Kessel E, Edinger D, He D, Klein PM, Voith von Voithenberg L, et al. Dual antitumoral potency of EG5 siRNA nanoplexes armed with cytotoxic bifunctional glutamyl-methotrexate targeting ligand. Biomaterials 2016;77:98–110.
112. Lee DJ, Kessel E, Lehto T, Liu X, Yoshinaga N, Padari K, et al. Systemic delivery of folate-PEG siRNA lipopolyplexes with enhanced intracellular stability for in vivo gene silencing in leukemia. Bioconjug Chem 2017;28:2393–2409.
113. Mirshafiee V, Jiang W, Sun B, Wang X, Xia T. Facilitating translational nanomedicine via predictive safety assessment. Mol Ther 2017;25:1522–30.
114. Boussif O, Zanta MA, Behr JP. Optimized galenics improve in vitro gene transfer with cationic molecules up to 1000-fold. Gene Ther 1996;3:1074–80.
115. Neuberg P, Kichler A. Recent developments in nucleic acid delivery with polyethylenimines. Adv Genet 2014;88:263–88.
116. Chollet P, Favrot MC, Hurbin A, Coll JL. Side-effects of a systemic injection of linear polyethylenimine-DNA complexes. J Gene Med 2002;4:84–91.
117. Merkel OM, Urbanics R, Bedocs P, Rozsnyay Z, Rosivall L, Toth M, et al. In vitro and in vivo complement activation and related anaphylactic effects associated with polyethylenimine and polyethylenimine-graft-poly(ethylene glycol) block copolymers. Biomaterials 2011;32:4936–42.
118. Plank C, Mechtler K, Szoka FC, Jr., Wagner E. Activation of the complement system by synthetic DNA complexes: a potential barrier for intravenous gene delivery. Hum Gene Ther 1996;7:1437–46.
119. Kircheis R, Schuller S, Brunner S, Ogris M, Heider KH, Zauner W, et al. Polycation-based DNA complexes for tumor-targeted gene delivery in vivo. J Gene Med 1999;1:111–20.
120. Ogris M, Brunner S, Schuller S, Kircheis R, Wagner E. PEGylated DNA/transferrin-PEI complexes: reduced interaction with blood components, extended circulation in blood and potential for systemic gene delivery. Gene Ther 1999;6:595–605.
121. Anchordoquy TJ, Barenholz Y, Boraschi D, Chorny M, Decuzzi P, Dobrovolskaia MA, et al. Mechanisms and barriers in cancer nanomedicine: addressing challenges, looking for solutions. ACS Nano 2017;11:12–18.

Tiziana Di Francesco and Gerrit Borchard

3 Nanotechnology and pharmacy

3.1 Introduction

In hindsight, scientific progress appears to be pushed forward by visionary individuals having the time to ponder the extrapolation of the current state of the art beyond and above the event horizon of their time. One might argue that the application of specific and directed drug delivery to improve therapeutic effects while reducing side effects was first thought of by Paul Ehrlich and his co-worker Sahachiro Hata [1] during their development of the first effective chemotherapeutic agent Salvarsan. Incidentally, Ehrlich coined the now infamous term of "magic bullet" that would achieve such directed drug delivery, which one would imagine nanoparticles used in drug delivery would look like [2]. Taken together, Ehrlich unknowingly created the idea of nanopharmaceuticals, that is the application of nanotechnology to therapy.

The development of nanopharmaceuticals [3] in their various manifestations (Figure 3.1) from vision to actual implementation into therapy has been influenced and guided by a number of technologies over the last 50 years, as shown in Figure 3.2 [4].

According to a definition by Bawa [5], nanotechnology is defined as "The design, characterization, production and application of structures, devices and systems by controlled manipulation of size and shape at the nanometre scale (atomic, molecular and macromolecular scales) that produces structures, devices and systems with at least one novel/superior characteristic or property". This definition does include larger scale devices or systems (e.g. surfaces) having nanosized substructures. The definition of the "nanoscale", however, is still under discussion. While the NIH Common Fund Nanomedicine Initiative [6] defines nanotechnology to create "Products obtained through nanotechnology in the range of 1 to 100 nanometer", the commonly accepted scale of nanopharmaceuticals is often found to be in the range of 150–300 nm and may even exceed 1,000 nm. The relationship with and mutual impact of nanotechnology on other scientific disciplines has recently been mapped in accordance with publication as well as citation data from the Science Citation Index [7].

Nanomedicine, in turn, is defined as the "application of nanotechnology to medicine" [3], a "field uniquely focused on medically related, patient-centric nanotechnologies" [8] and as an "offshoot of nanotechnology", which is a "highly specific medical intervention at the molecular scale for curing diseases or repairing damaged tissue" [9]. Some authors group nanomedicine(s) into three categories [10], namely diagnostic systems and (microinvasive) surgical tools, imaging and monitoring systems, and nanoscale drug delivery systems and technologies (nanopharmaceuticals). Two or all three categories may be combined into a single system [11].

Although nanopharmaceuticals are not very common in public pharmacies, clinical pharmacists are increasingly involved in the (personalized) therapeutic

https://doi.org/10.1515/9783110547221-003

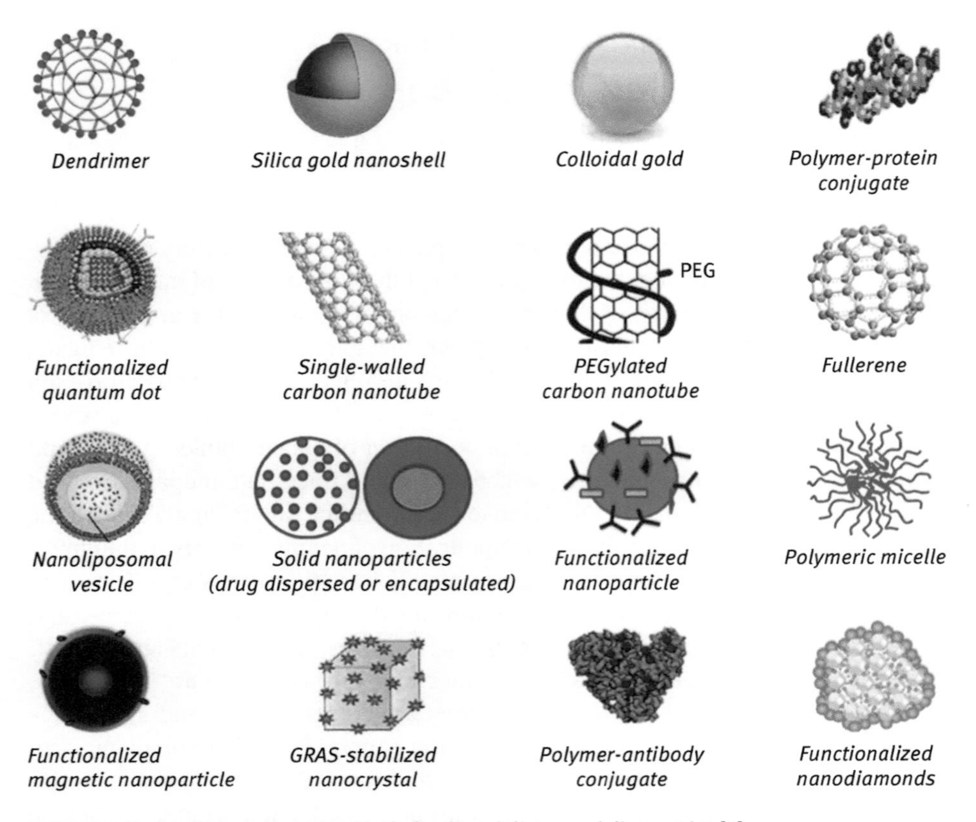

Figure 3.1: Types of nanopharmaceuticals for drug delivery and diagnostics [3].

application of such drugs in clinical care. Some of these drugs have been introduced into the market following regulatory strategies that, in the light of increasing scientific knowledge, may appear today to be insufficient to evaluate their efficacy and safety [12], as discussed in this chapter. In the following, an overview of nanopharmaceuticals used in clinical practice is given, and clinical, regulatory and practical aspects related to their approval and use are discussed.

3.2 Nanopharmacy and Nanopharmaceuticals on the Market

The term "nanopharmacy" has appeared during the last years [13]. In the absence of a straightforward official definition and in extension of the nanomedicine definitions mentioned above, it may be interpreted as the science(s) whose study objects would be nanopharmaceuticals. In addition, the complexity of such drugs requires

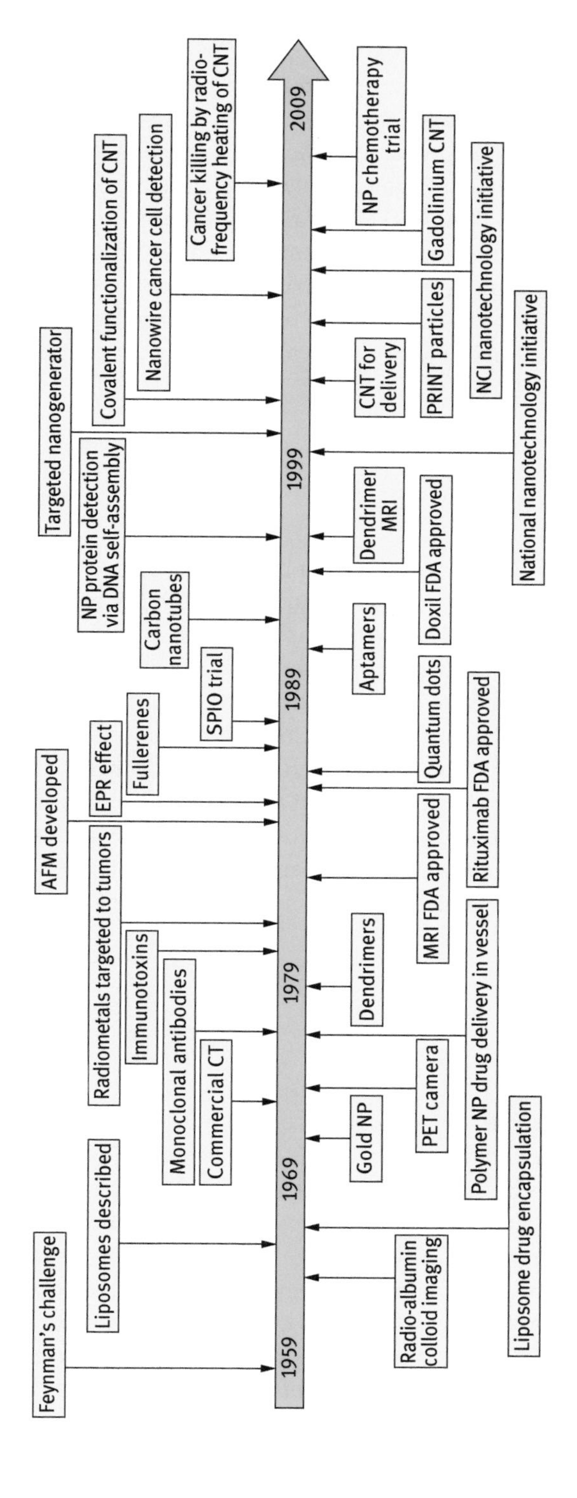

Figure 3.2: Timeline of technologies that have had an influence on the development of nanotechnology for (cancer) therapy (with permission from [4]).

standardized protocols for the manufacturing, handling and administration to ensure safety and efficacy. Such – in essence pharmaceutical – operations would naturally also be covered by the term "nanopharmacy".

Since 1975, with the approval of Gris-PEG® (griseofulvin nanocrystals), a number of nanopharmaceuticals have received EMA (European Medicine Agency) market authorization and FDA (US Food and Drug Administration) approval. Nanopharmaceutical formulations found in commercialized products today comprise drug nanocrystals, liposomes, polymeric drugs and nanoparticles. A list of nanopharmaceuticals on the market at the time of the revision of this chapter, together with their indication(s) and date of approval/marketing authorization, is given in Table 3.1.

3.2.1 Drug Nanocrystals

The occurrence of unfavourable properties of biopharmaceutical classification system [14] class IIa drug candidates such as low aqueous solubility and resulting low bioavailability is resolved by strategies to increase solubility [15]. Such strategies include the reduction in size or "nanonizing" [16] of the active pharmaceutical ingredient (API) principle, resulting in drug nanocrystals of increased overall surface area and surface energy [17]. Drug nanocrystals are manufactured by either precipitation techniques in a "bottom-up" process, or by a "top-down" approach by size reduction of larger drug particles or crystalline powders by application of attrition forces ("nanomilling"). In their final formulation, drug nanocrystal suspensions are stabilized against aggregation by the presence of stabilizers adsorbed to the nanocrystal surface [18], which may in the future also serve to adsorb moieties for cell-specific drug targeting functionalization [19, 20].

3.2.2 Liposomes

Liposomes are vesicles consisting of spontaneously forming phospholipid bilayers, which may be stabilized by the addition of other lipids (e.g. cholesterol) [21]. "Swollen phospholipid systems" were first described in 1965 [22], and soon the adaptation of such vesicles to drug delivery was suggested [23]. Drug encapsulation may occur in the hydrophilic liposomal core or intercalated within the phospholipid bilayers (hydrophobic drugs). Encapsulation into liposomes has been shown to alter drug pharmacokinetics [24], an effect even enhanced by the functionalization of the liposomal surface with the polymer polyethylene glycol (PEG). PEGylation [25] renders liposomes less recognizable by the immune system, probably by altering the extent and/or composition of the formation of a plasma protein corona at the liposomal surface through steric or electrostatic effects [26].

Table 3.1: A list of currently marketed nanopharmaceuticals, their indication(s) and year of approval.

Nanotechnology	Name	Drug	Indication	Approval year
Drug nanocrystals	Cesamet®	Nabilone	Anti-emetic	2006
	Cholib®	Fenofibrate/simvastatin	Dyslipidaemias	2013
	Emend®	Aprepitant	Anti-emetic	2003
	Gris-PEG®	Griseofulvin	Antifungal	1975
	Megace ES®	Megestrol acetate	Hypercholesterolaemia, hypertriglyceridaemia	2005
	Rapamune®	Rapamycin, formulated in tablets	Immunosuppression	2002
	Tricor®	Fenofibrate as nanocrystals	Hypercholesterolaemia, hypertriglyceridaemia	2004
	Triglide®	Fenofibrate as non-soluble drug delivery microparticles	Hypercholesterolaemia, hypertriglyceridemia	2004
	Xeplion®	Paliperidone	Schizophrenia	2011
	Zypadhera®	Olanzapine	Schizophrenia	2008
Liposomes	AmBisome®	Amphotericin B	Fungal infections	1990
	DepoCyt®	Cytarabine	Meningeal neoplasms	1999
	Exparel®	Bupivacaine	Anaesthetic	2011
	DaunoXome®	Daunorubicin	Cancer advanced HIV-associated Kaposi's sarcoma	1996
	Caelyx®/ Doxil® Lipodox®	Doxorubicin HCl (PEGylated)	Breast, ovarian neoplasms, multiple myeloma, Kaposi's	1995
	Myocet®	Doxorubicin HCl	Breast neoplasms	2000
	DepoDur®	Morphine	Pain relief	2004
	Mepact®	Mifamurtide	Osteosarcoma	2009
	Visudyne®	Verteporfin	Macular degeneration, myopia	2000
	Marqibo®	Vincristine	Lymphoblastic leukaemia	2013
Polymeric drugs	Copaxone® Glatopa®	Glatiramer acetate	Multiple sclerosis	1996 2016

(continued)

Table 3.1: (continued)

Nanotechnology	Name	Drug	Indication	Approval year
	VivaGel®	Dendrimer	Bacterial vaginosis	2015
Nanoparticles	Abraxane®	Nab-Paclitaxel	Metastatic breast cancer Adv. non-small cell lung Metastatic pancreatic Gastric cancer	2005 2012 2013 2013
	Maltofer®	Iron polymaltose	Iron deficiency	1964
	Ferinject®/ Injectafer®	Ferric carboxymaltose	Iron deficiency	2007
	Rienso®/ FeraHeme®	Ferumoxytol	Iron deficiency	2009
	Dexferrum®	High-molecular-weight iron dextran	Iron deficiency	1996
	Cosmofer®	Low-molecular-weight iron dextran	Iron deficiency	2001
	Ferrlecit®	Sodium ferric gluconate	Iron deficiency	2009
	Monofer®	Iron isomaltoside	Iron deficiency	2009
	Venofer®	Iron sucrose	Iron deficiency	1992
Virus-like particles	Cervarix®	HPV-type 16L1 and 18L1 antigens	Prevention of HPV-induced cancers	2007
	Gardasil®	Major capsid protein L1 of HPV types 6, 11, 16 and 18	Prevention of HPV-induced cancers	2006
	Engerix B®	Recombinant hepatitis B surface antigen (HBsAg)	Prevention against hepatitis B infection	1986
Virosomes	Inflexal® V	Haemagglutinin, neuramidase antigens	Influenza	1997
	Epaxal®	Formalin inactivated HAV	Prevention of hepatitis A infection	1993

HIV, human immunodeficiency virus; HPV, human papillomavirus; HAV, hepatitis A virus.

This alteration of the surface properties is potentially impacting on the liposomal uptake by the reticuloendothelial system (RES) [27], leading to longer systemic circulation times. Liposomal technology has led to the development of first Ambisome® (liposomal amphotericin B) and later liposomal doxorubicin (Doxil®/Caelyx®) [25]. Drug encapsulation into

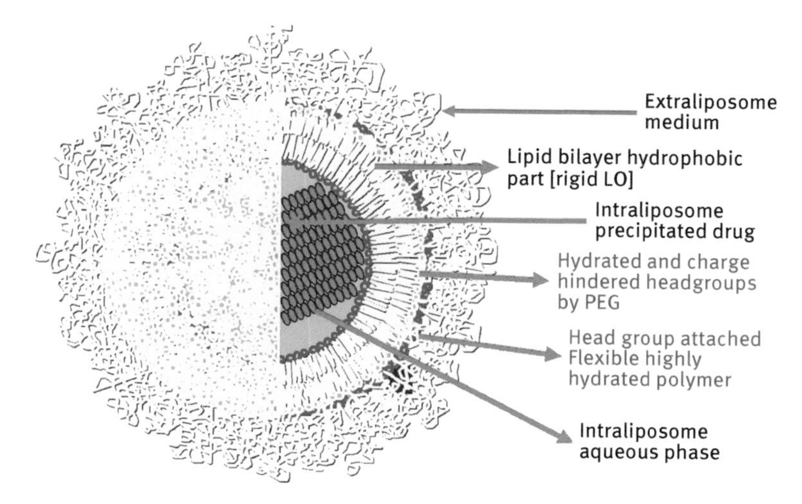

Figure 3.3: Schematic representation of liposomal doxorubicin (Doxil®) [25].

liposomes has significantly reduced unwanted side effects attributed to the plasma levels of free drug such as cardiotoxicity in the case of Doxil® [28] (Figure 3.3).

3.2.3 Polymeric Drugs

Nanopharmaceuticals defined as polymeric drugs currently available on the market are the glatiramer acetate (GA) [29] and one dendrimer product (VivaGel®) [30]. GA is an immunomodulator used to treat patients with relapsing–remitting form of multiple sclerosis (MS), reducing the frequency of relapses [31]. GA products (Copaxone® and Glatopa®) may include up to 10^{29} different peptide structures, synthesized from four amino acids (alanine, lysine, glutamic acid and tyrosine) [32]. Upon subcutaneous injection, these polypeptide chains are partially hydrolysed, the resulting fragments are taken up by local antigen-presenting cells and specific epitopes are presented to T lymphocytes [33], thus attenuating autoimmune activity in MS leading to demyelination and inflammation in the brain. The highly complex structure of GA makes pharmacokinetic modelling in the absence of an exactly defined API (which polypeptide(s)? at what ratio/concentration?) impossible and requires a tightly controlled manufacturing process to ensure both safety and efficacy.

Dendrimers are synthetic macromolecules of narrow size distribution on the nanoscale, characterized by their branched three-dimensional structure allowing for drug incorporation and surface functionalization for targeting purposes [34]. While drug–dendrimer conjugates have been developed as pro-drugs [35], the VivaGel® product appears to exert its activity through binding of the highly negatively charged surface to pathogens, thus reducing their mucosal attachment and invasion [36]. Approved for therapeutic use in bacterial vaginosis in 2015, a clinical phase I/II trial revealed that the dendrimer formulation appears to also have a measurable effect against vaginal infections with human immunodeficiency virus 1 and herpes simplex virus 2 [36].

3.2.4 Nanoparticles

Currently commercialized nanopharmaceuticals in the submicron range are albumin nanoparticles loaded with paclitaxel (Abraxane®), superparamagnetic iron oxide nanoparticles coated with a low-molecular-weight semisynthetic carbohydrate (Ferumoxytol®) and the group of nanoparticles composed by iron cores coated with various saccharides (isomaltoside, dextran, gluconate, polymaltose and sucrose). The development of Abraxane® (Figure 3.4) as paclitaxel-loaded albumin nanoparticles for therapy of several solid tumours (Table 3.1) is essentially based on the observation of the generally high plasma protein binding of cytotoxic drugs [37].

Abraxane® exploits these properties of albumin to reversibly bind paclitaxel, actively and specifically transport it across the endothelium by interaction with gp60/ caveolin-1 [39] and accumulate it in tumour cells due to the interaction with the albumin-binding protein SPARC (secreted protein, acidic and rich in cysteine, osteonectin or BM-40) [40]. Formulation of paclitaxel in 130 nm albumin nanoparticles avoids the use of cremophor EL (CrEL), which contributes to serious toxicity (e.g. hypersensitivity and axonal degeneration) [41] and requires special infusion tubing, premedication and prolonged infusion. While CrEL forms micelles that entrap paclitaxel, leading to decreased unbound drug fraction, decreased drug clearance and lack of dose-dependent anti-tumour activity, Abraxane® has been shown to have improved bioavailability and linear pharmacokinetics. Thus, drug exposure becomes predictable even after modification of the dosing regimen [42].

Iron carbohydrate nanoparticles are applied in iron replenishment therapy of anaemia as a consequence of chronic kidney disease (CKD), pregnancy, autoimmune

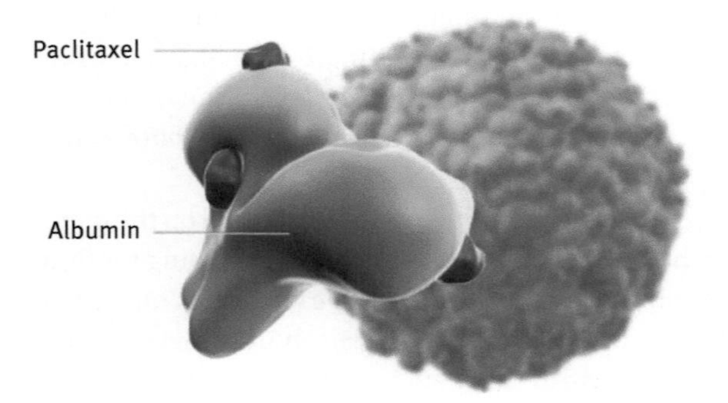

Figure 3.4: Nab-Paclitaxel (Abraxane®), 130 nm albumin nanoparticles loaded with cytostatic paclitaxel. Modified from Ref. [38].

disorders, chronic infections or cancer [43]. With approximately 1.6 billion of new cases per year, anaemia is considered the most important nutritional disease in the world [44]. Treatment of iron deficiency includes oral preparations of both ferrous and ferric salts. However, when the administration of salts is ineffective or not tolerated, the use of iron carbohydrate complexes is preferred [45]. These complex nanocolloidal suspensions are composed of an iron-(III)-oxyhydroxide core, which is covered by a carbohydrate shell and are administered either orally or intravenously [46].

Maltofer® is a highly complex colloidal suspension composed of iron-(III)-oxyhydroxide cores surrounded by a polymaltose shell. Its molecular weight, determined by gel permeation chromatography is described to be about 52 kDa. Recent size determination by dynamic light scattering and transmission electron microscopy performed in our laboratory revealed a hydrodynamic diameter of about 9 nm (unpublished results). Maltofer® is a second generation oral iron supplement administered to boost iron body levels in the presence of iron deficiency anaemia [47], and taken up into systemic circulation through the intestinal epithelium by passive diffusion [48]. Through coating with polymaltose unwanted side effects of orally applied iron preparations are significantly reduced, although haemoglobin levels achieved were comparable to those after treatment with iron sulphate [49].

The other iron carbohydrate products mentioned in Table 3.1 are administered by i.v. injection or infusion. Once administered, a corona of plasma proteins is formed on their surface [50], which leads to phagocytosis by the RES. The cores are processed intracellularly and the iron utilized to replenish iron storage and erythrocyte synthesis (Figure 3.5) [51]. They differ in the polysaccharide used for coating the iron cores, which does have an influence on their particle size, stability and degradation upon

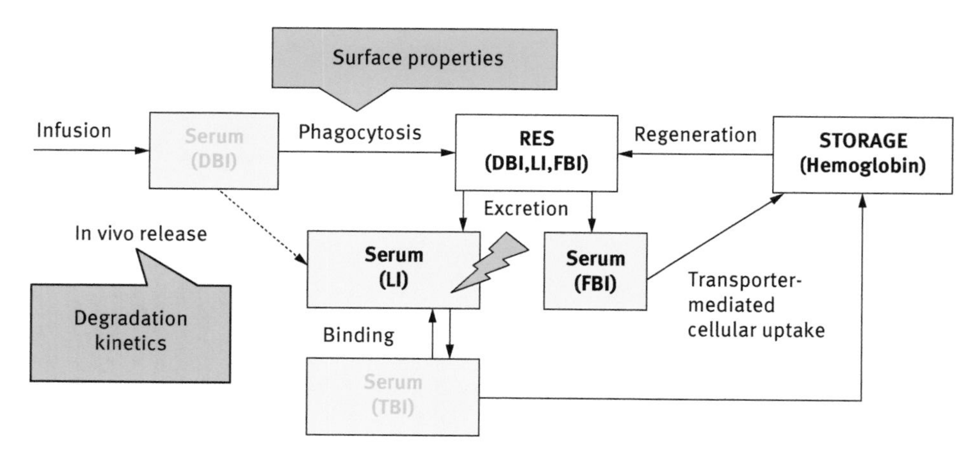

Figure 3.5: Fate of iron carbohydrate drugs upon injection. Surface properties are suggested to influence kinetics of phagocytosis, while degradation kinetics would impact on release of labile iron, which is cause of toxicity. DBI: drug bound iron, LI: labile iron, FBI: ferritin-bound iron, TBI: transferrin-bound iron. Adapted from Zhen et al. [51].

injection. Iron sucrose (IS, Venofer®), as an example, needs to be stabilized by keeping the nanosuspension at a pH of 11, while dextran-coated particles (Dexferrum®, Cosmofer®) are stable at physiological pH. As shown in Figure 3.5, surface parameters of the particles may influence extent and rate of phagocytosis. Degradation kinetics may determine release of labile iron, which is supposed to cause toxicity through the formation of reactive oxygen species (ROS) [52]. However, physicochemical properties of these products have as yet not been directly linked (in terms of critical quality attributes) to differences observed in clinical outcome, which will be discussed below.

The field of nanopharmaceuticals is constantly evolving, with novel candidates and formats being put through the development pipelines. A review by Etheridge et al. [53] reported 213 clinical trials at various stages (including 14 terminated/discontinued studies), and 150 commercialized nanomedicine products at the date of publication. A report by the US Center for Drug Evaluation and Research (CDER) [54] on the number of applications of nanomaterial containing drug products showed a steady increase over the last 40 years (Figure 3.6). The term "nanomaterials" included liposomes, nanocrystals, (nano)emulsions, iron–polymer complexes and micelles as the five largest groups, making up about 80% of applications submitted.

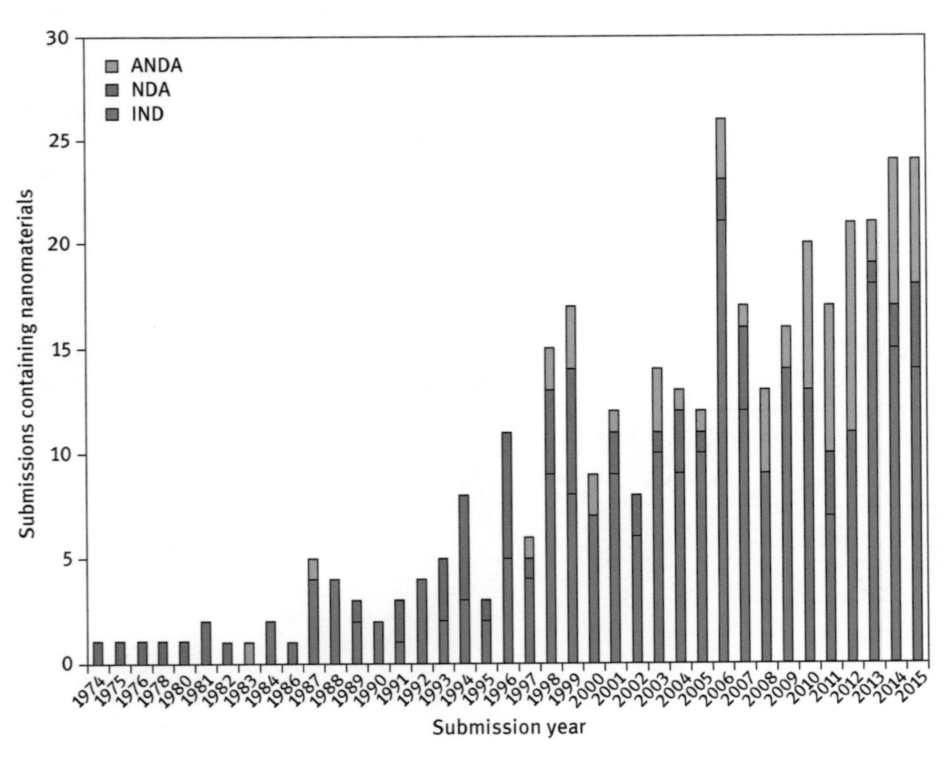

Figure 3.6: Number of nanomaterial product applications submitted to CDER by year [54].

Aspects of the regulatory and compendial framework for approval of these nanopharmaceuticals are discussed below.

3.2.5 Virus-Like Particles and Virosomes

Nanotechnology is also contributing to the development of novel “nanovaccines”, mimicking the properties of viruses in terms of shape and size, repetitive surface structures and antigen presentation [55]. A plethora of different polymeric materials, including chitosan, polylactide glycolic acid (PLGA), polyethyleneimine, dextran and alginate, have been utilized to produce nanovaccines [56]. These efforts so far have remained on the pre-clinical and early clinical study level, with the most advanced systems being PLGA-based DNA vaccines [57]. By contrast, nanovaccines based on virus-like particles (VLPs) and virosomes have successfully been developed and entered the market in the 1980s [58, 59], as shown in Table 3.1.

VLPs are highly ordered nanostructures resembling genuine viruses in terms of size (22–150 nm) and protein envelope composition, but lacking viral genetic material [60]. Due to their virus-like appearance and repetitive structural surface features, VLPs are highly immunogenic and antigenic [60], they are interacting with the immune system through similar pathways as the original pathogens.

VLPs are prepared by self-assembly of the viral structural proteins, when expressed as recombinant proteins, into structures similar to the parental viruses. They are manufactured by expression of the viral building block from viral or plasmid vectors [61], which may be supplemented with specific epitopes, and include one or multiple proteins. While non-enveloped single- or multiple-capsid VLPs result from self-assembly of virus-derived capsid proteins, enveloped VLPs result from particle budding from the host cell (*E. coli*, yeast, insect cells or plant based) [62]. VLPs have been derived from a wide variety of viruses, including hepatitis B [63], human papillomavirus [64], and Norwalk virus [65].

Virosomes are VLPs that are generated from inactivated virus produced in embryonic chicken eggs or cell culture [66], the same material used for the production of influenza vaccines. The virus is dissolved in an appropriate detergent and the solubilized viral envelope components purified by removal of the insoluble matter. Unilamellar virosomes are subsequently reassembled by detergent removal. The addition of selected lipids prior to assembly is key to a robust, industrial-scale manufacturing process, which generates virosomes that are called immunopotentiating reconstituted influenza virosomes [66].

Along with payload antigens, the reconstituted viral envelopes present the same surface proteins haemagglutinin (HA) and neuraminidase (NA) at the same ratios as the parental virus [66]. The suitability of the virosomal platform as an efficient adjuvant and carrier system in parenteral vaccine formulations is illustrated by the commercialization of two products, Epaxal® and Inflexal® [67].

3.3 Regulatory Aspects of Nanopharmaceuticals and Nanosimilars

The regulatory landscape for the approval of nanopharmaceuticals is characterized by the number of products having been introduced into the market as early as the 1950s (Figure 3.7), the development of increasingly sophisticated nanopharmaceuticals as well as current progress in analytical methods used to characterize such drugs [68]. In addition, progress in nanopharmaceutical development has led to the coining of the term "non-biological complex drugs" (NBCDs) [69]. The term acknowledges the complexity of nanopharmaceuticals of mostly synthetic origin. NBCDs share aspects of complex structure, manufacturing procedures, potential immunogenicity and impossibility of full characterization by physicochemical methods alone with their biological counterparts (therapeutic antibodies, proteins, peptides, etc. [70]), as shown in Table 3.2.

Table 3.2: Properties of NBCDs in comparison to biologics and low molecular weight drugs, based on [71].

	Small-molecule drugs	Biologics	NBCDs
Molecular weight	Low (<500 Da)	High (5–900 kDa)	
Structure	Well defined	Complex, heterogeneous, defined by manufacturing process	
Manufacturing	Chemical synthesis	Produced in living cells or organisms	Synthetic technologies (including nanotech)
Characterization	Complete characterization	Not fully characterized	
Copy characteristics	Identical copies can be made	Impossible to ensure identical copy versions	

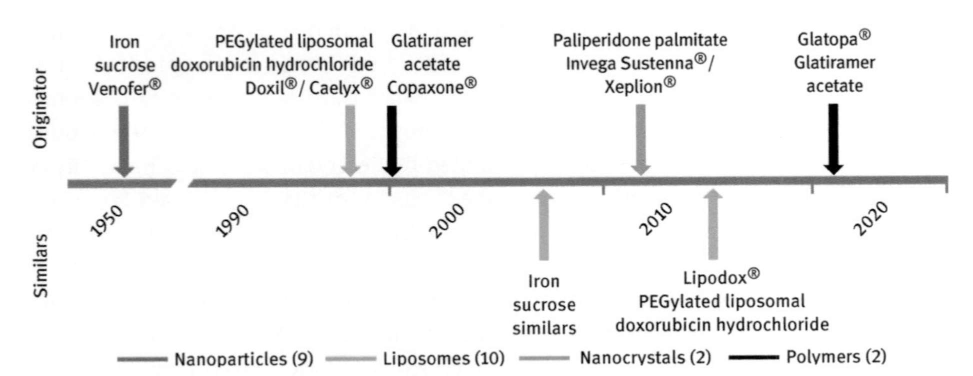

Figure 3.7: Time of market introduction of original nanopharmaceuticals and their "similars" (courtesy of Prof. Stefan Mühlebach, Vifor Pharma Ltd.).

Moreover, recent years have seen the advent of follow-on versions of NBCDs, which, in accordance with the "biosimilar" approach for biological complex drugs, are suggested to be termed "nanosimilars".

Unlike the biosimilar pathway for therapeutic protein follow-on products, no such regulatory approach has been put in place yet [72]. While "nanosimilars" have received marketing authorization through the generic pathway in the past, recent retrospective and prospective studies have reported differences in clinical outcomes when switching from the original to the follow-on nanopharmaceutical. One example is the replacement of Venofer® (IS) with an IS similar (ISS) in the therapy of 75 chronic kidney disease (CKD) patients undergoing haemodialysis in France [73]. The exchange led to a significant decrease in haemoglobin levels that were stable under treatment with IS. In addition, significantly reduced transferrin saturation and serum ferritin levels required to increase dosing of ISS as well as erythropoietin stimulating agents to keep patients in correct/acceptable blood parameters. These observations were recently confirmed in a prospective study involving 66 patients [74].

Moreover, several different intended copies of Maltofer® were introduced on various markets under the name of iron polymaltose similars. Clinical [75] and non-clinical evidences [76] suggested that the treatment with these nanosimilars may not lead to the same clinical outcome as to be expected using the originator drug.

Another example is the non-equivalence of "generic" liposomal doxorubicin, introduced into the US market due to drug shortage of the originator drug, Doxil® [77]. The same product is commercialized in Europe under the Caelyx®. According to the Committee for Medicinal Products for Human Use assessment report by EMA [78], the applicant (SUN Pharmaceutical Industries Europe B.V.) could show that the amount of unencapsulated doxorubicin in their product is comparable to Doxil® (the US reference product), but not to its European equivalent, Caelyx®. The report came to the conclusion that " it cannot be concluded that the test formulation of liposomal doxorubicin is essentially similar to the reference product".

These few examples bear evidence that NBCDs and their follow-on products may neither be simply regarded to be of therapeutic equivalence, nor as interchangeable. This notion is currently carried forward by several institutions, including EMA, FDA and ICH (International Conference on Harmonisation of Technical Requirements for Registration of Pharmaceuticals for Human Use), building on existing specific guidance on selected NBCDs (e.g. micelles, liposomal products, and nanoparticle iron medicinal products) [79, 80]. Ehmann and Pita [81], in an editorial response on the regulatory strategies for NBCDs and their follow-on versions [82] expressed their opinions that the current regulatory framework in Europe is "robust and efficient for the evaluation of NBCDs, including nanomedicines...". This publication, though, was critically discussed by a successive letter to the editor of the publishing journal [83] as a sign of the vivid and deeply scientific discussion on NBCD regulation [84].

The current thinking in Europe with regard to follow-on NBCD products appears to follow the suggestion of the "nanosimilar" approach along the biosimilar strategy for therapeutic proteins, as expressed in a recent reflection paper published by EMA [85]. However, this strategy is not entirely followed by the FDA, where approval of generic parenteral drug products containing nanomaterials is based on the 505(j) (ANDA) pathway and existing data of already approved products [51]. FDA is relying more heavily on the physicochemical characterization of nanomedicines and their similars, requiring such products in general to have qualitative (Q1) and quantitative sameness (Q2). FDA therefore, in contrast to EMA, does currently not require non-clinical and clinical studies for such products. In the meantime, FDA has supported a post-approval study in healthy human subjects to look into the comparison of several clinical parameters achieved with therapy using generic and reference sodium ferric gluconate [86]. In addition, in December 2015 members of the US House of Representatives have asked Government Accountability Office (GAO) to conduct a study on how FDA is reviewing generic versions of NBCDs [87]. The GAO report is currently pending.

3.4 Nanopharmaceuticals in Clinical Practice

Therapeutic equivalence of nanopharmaceuticals and thus exchangeability is a field of broad scientific and regulatory discussion, as it may have profound repercussions in clinical practice. Nanopharmaceuticals are complex drugs, whose entire composition affects their quality, safety and efficacy. Alteration in the composition, for example by variation of manufacturing methods or mishandling, may have severe repercussions in patients. After successfully having introduced Copaxone® into the market, the manufacturer attempted to develop an improved version of their glatiramer (TV-5010). Although no toxicity was observed in short-term clinical phase I studies, severe long-term toxicity [88] was reported in pre-clinical studies in non-human primates and rats. It was therefore surprising that in 2015 FDA approved a "generic" version of Copaxone®, Glatopa®, in the absence of clinical trial data [89].

Changes in clinical outcome for IS products due to their supposedly different physicochemical properties were already reported earlier [73, 74]. However, also handling of these complex drugs in the hospital may lead to complications, which appear to be differing between originator and similar complex drugs. A report on postpartum and gynaecologic post-operative patients treated with IS or ISS by Lee et al. [90] revealed differences in the adverse event patterns. Injection site reactions (ISRs) were significantly ($p < 0.02$) enhanced for the ISS (6.2%) in comparison to IS (1.8%) when the drug was diluted in 100 mL saline. Occurrence of ISRs was even increased once the ISS dose was diluted in 200 mL saline for injection to 8.2%. In clinical practice, the addition of the IS dose to an infusion bag of physiological NaCl solution may result in the same complication. Higher dilution of the ISS dose lead to the destabilization of the nanoparticles, and the resulting premature release of labile iron then caused generation of ROS at the injection site.

A first question is: are such nanosimilars adequately evaluated and substituted correctly in clinical practice? A recent publication [91] reported on the practice of dispensing IS and ISSs in 70 French and 70 Spanish hospitals. It was revealed that there was little awareness toward the potential differences in therapeutic efficacy and safety profiles of this class of nanopharmaceuticals and their characteristics of being "similar" rather than "generic". Interchange and substitution are decided rather on financial aspects, and rarely so on scientific and clinical data. In Spain, 34% of all hospital pharmacists confirmed to have the authority to switch from the originator drug to a similar (France: 43%). As a consequence, in 26% (France) and 52% (Spain) of the cases where a medication change to a similar took place (France: 38%, Spain: 47% of all prescriptions), the prescribing physician was not informed.

How can nanopharmaceuticals and their similars be selected in the clinical practice, maintaining therapeutic efficacy and safety for the patient, when interchangeability and substitutability of such complex drugs cannot be taken for granted? While selection criteria for biosimilars have been published by the American Society of Health-System Pharmacists [92], no such guidelines exist for the selection of nanosimilars. Therefore, a consensus roundtable with hospital pharmacists from six countries (Belgium, France, Germany, Spain, Switzerland and USA) [93] suggested formulary selection criteria for nanosimilars based on the biosimilar document, which is presented in Table 3.3. This table takes into account the specific criteria related to pharmaceutical quality, efficacy

Table 3.3: Selection criteria for similars of nanopharmaceuticals, based on the formulary selection criteria for biosimilars [92].

Pharmaceutical quality	Efficacy/safety	Manufacturer considerations	Product considerations	Hospital and patient factors

- Chemical composition
- Identity
- Quantity
- Pharmacopoeial specifications
- Particle size and size distribution
- Particle surface characteristics
- Uncaptured Pharmacological active moiety fraction
- Storage stability
	- Pharmacokinetics
	  - Uptake
	  - Distribution
	- Clinical data
	- Range of indications
	- Immunogenicity
	- Potential for therapeutic interchange
	- Number of similar agents on formulary
	- Pharmacovigilance requirements
		- Supply reliability
		- History of drug shortages
		- Supply chain security
		- Anti-counterfeit measures
		- Patient assistance programs
		- Reimbursement support
		- Manufacturer services, expertise
			- Product packaging and labeling
			- Bar coding
			- Compatibility with CSTDs*, robotics
			- Ready-to-use preparation and administration
			  - Stability for ready-to-use administration
			- Storage requirements
				- Economic considerations
				  - Hospital
				  - Payer
				  - Patient
				- Transition of care
				- IT and medication system changes
				- Educational requirements
				- Pharmacovigilance requirements

*CSTDs: Closed System Transfer Devices

and safety, and product stability. This table and mentioned criteria will be updated with progress in fundamental and clinical research in nanopharmacy.

3.5 Future Vision

Mentioning the terms "vision" and "regulatory strategies" in one sentence may appear inappropriate. However, the rocky road from vision to medical reality must be paved by adding bricks of fundamental and clinical data, as well as education and training of regulatory persons, prescribers and dispensers. Nanopharmacy and nanopharmaceuticals are here to stay, with their importance especially in individualized medicine steadily increasing. Lest we only want to continue reading about "very promising nanocarrier systems", we should bring all stakeholders to the table and tackle the scientific, clinical, regulatory and practical challenges of nanopharmacy and their products together. Given their qualifications and mindset, pharmaceutical scientists are ideally posed to do just that.

References

1. Bosch F, Rosich L. The contributions of Paul Ehrlich to pharmacology: A tribute on the occasion of the centenary of his Nobel prize. Pharmacol 2008;82:171–9.
2. Rahman M, Ahmad MZ, Kazmi I, Akhter S, Afzal M, Gupta G, Sinha VR. Emergence of nanomedicine as cancer targeted magic bullets: recent development and need to address the toxicity apprehension. Curr Drug Discov Technol 2012;9:319–29.
3. Tinkle S, McNeil SE, Mühlebach S, Bawa R, Borchard G, Barenholz Y, Tamarkin L, Desai N. Nanomedicines: addressing the scientific and regulatory gap. Ann NY Acad Sci 2014;1313:35–56.
4. Scheinberg DA, Villa CH, Escorcia FE, McDevitt MR. Conscripts of the infinite armada: systemic cancer therapy using nanomaterials. Nat Rev Clin Oncol 2010;7:266–76.
5. Bawa R. Patents and nanomedicine. Nanomedicine (London). 2007;2:351–74.
6. National Nanotechnology Initiative, Washington, DC, USA. Available at https://www.nano.gov/nanotech-101/what/definition. Accessed: 23 July 2017.
7. Leydesdorff L, Rafols I. A global map of science based on the ISI subject categories. J Am Soc Inf Sci Technol 2009;60:348–62.
8. Nanomedicine, An ESF-European Medical Research Councils (EMRC) forward look report, European Science Foundation 2005, ISBN: 2-912049-52-0.
9. Webster TJ. Nanomedicine: what's in a definition? Int J Nanomed 2006;1:115–16.
10. Duncan R, Gaspar R. Nanomedicine(s) under the microscope. Mol Pharmaceutics 2011;8:2101–41.
11. Rutka JT, Kim B, Etame A, Diaz RJ. Nanosurgical resection of malignant brain tumors: Beyond the cutting edge. ACS Nano 2014;8:9716–22.
12. Hussaarts L, Mühlebach S, Shah VP, McNeil S, Borchard G, Flühmann B, Weinstein V, Neervannan S, Griffiths E, Jiang W, Wolff-Holz E, Crommelin DJA, De Vlieger JSB. Equivalence of complex drug products: advances in and challenges for current regulatory frameworks. Ann NY Acad Sci 2017. DOI: 10.1111/nyas.13347

13. Cornier J, Owen A, Kwade A, Van de Voorde M. Pharmaceutical nanotechnology: Innovation and production. John Wiley & Sons, New York, 2017.
14. Butler JM, Dressman JB. The developability classification system: application of biopharmaceutics concepts to formulation development J Pharm Sci 2010;99:4940–4954.
15. Williams HD, Trevaskis NL, Charman SA, Shanker RM, Charman WN, Pouton CW, Porter CJ. Strategies to address low drug solubility in discovery and development. Pharmacol Rev 2013;65:315–499.
16. Junghanns JUAH, Müller RH. Nanocrystal technology, drug delivery and clinical applications. Int J Nanomed 2008;3:295–309.
17. Gao L, Liu G, Ma J, Wang X, Zhou L, Li X, Wang F. Application of drug nanocrystal technologies on oral drug delivery for poorly soluble drugs. Pharm Res 2013;30:307–24.
18. Borchard G. Drug nanocrystals. In: Crommelin DJA, De Vlieger JSB, eds. Non-biologic complex drugs. 1st ed. Cham: Springer International Publishing Switzerland, 2015:171–91.
19. Åkerman ME, Chan WCW, Laakkonen P, Bhatia SN, Ruoslahti E. Nanocrystal targeting in vivo. Proc Nat Acad Sci USA 2002;99:12617–21.
20. Fuhrmann K, Gauthier MA, Leroux JC. Targeting of injectable nanocrystals. Mol Pharm 2014;11:1762–71.
21. Allen TM, Cullis PR. Liposomal drug delivery systems: from concept to clinical applications. Adv Drug Deliv Rev 2013;65:36–48.
22. Bangham AD, Standish MM, Watkins JC. Diffusion of univalent ions across the lamellae of swollen phospholipids. J Mol Biol 1965;13:238–52.
23. Gregoriadis G. Drug entrapment in liposomes. FEBS Lett. 1973;36:292–96.
24. Drummond DC, Noble CO, Hayes ME, Park JW, Kirpotin DB. Pharmacokinetics and in vivo drug release rates in liposomal nanocarrier development. J Pharm Sci 2008;97:4969–740.
25. Barenholz Y. Doxil® – The first FDA-approved nano-drug: lessons learned. J Controlled Release 2012;160:117–34.
26. Palchetti S, Colapicchioni V, Digiacomo L, Caracciolo G, Pozzi D, Capriotti AL, La Barbera G, Laganà A. The protein corona of circulating PEGylated liposomes. Biochim Biophys Acta 2016;1858:189–96.
27. Immordino ML, Dosio F, Cattel L. Stealth liposomes: review of the basic science, rationale, and clinical applications, existing and potential. Int J Nanomed 2006;1:297–315.
28. Ansari L, Shiehzadeh F, Taherzadeh Z, Nikoofal-Sahlabadi S, Momtazi-borojeni AA, Sahebkar A, Eslami S. The most prevalent side effects of pegylated liposomal doxorubicin monotherapy in women with metastatic breast cancer: a systematic review of clinical trials. Cancer Gene Ther 2017;24:189–93.
29. Weinstock-Guttman B, Nair KV, Glajch JL, Ganguly TC, Kantor D. Two decades of glatiramer acetate: from initial discovery to the current development of generics. J Neurol Sci 2017;376:255–9.
30. Rupp R, Rosenthal SL, Stanberry LR. VivaGel™ (SPL7013 Gel): A candidate dendrimer – microbicide for the prevention of HIV and HSV infection. Int J Nanomed. 2007;2:561–566.
31. Johnson KP, Brooks BR, Cohen JA, Ford CC, Goldstein J, Lisak RP, Myers LW, Panitch HS, Rose JW, Schiffer RB, Vollmer T, Weiner LP, Wolinsky JS. Extended use of glatiramer acetate (Copaxone) is well tolerated and maintains its clinical effect on multiple sclerosis relapse rate and degree of disability. Neurology 1998;50:701–8.
32. Weinstein V, Schwartz R, Grossman I, Zeskind B, Nicholas JM. Glatiramoids. In: Crommelin DJA, De Vlieger JSB, eds. Non-biologic complex drugs. 1st ed. Cham: Springer International Publishing Switzerland, 2015:107–48.
33. Lalive PH, Neuhaus O, Benkhoucha M. Glatiramer acetate in the treatment of multiple sclerosis: emerging concepts regarding its mechanism of action. CNS Drugs 2011;25:401–14.

34. Nanjwade BK, Bechra HM, Derkar GK, Manvi FV, Nanjwade VK. Dendrimers: emerging polymers for drug-delivery systems. Eur J Pharm Sci 2009;38:185–96.
35. Najlah M, D'Emanuele A. Synthesis of dendrimers and drug-dendrimer conjugates for drug delivery. Curr Opin Drug Discov Dev 2007;10:756–67.
36. Price CF, Tyssen D, Sonza S, Davie A, Evans S, Lewis GR, Xia S, Spelman T, Hodsman P, Moench TR, Humberstone A, Pauli JRA, Tachedijan G. SPL7013 gel (VivaGel®) retains potent HIV-1 and HSV-2 inhibitory activity following vaginal administration in humans. PLoS ONE 2011;6:e24095.
37. Sonnichsen DS, Relling MV. Clinical pharmacokinetics of paclitaxel. Clin Pharmacokin 1994;27:256–69.
38. Abraxane information, Celgene, Summit, NJ, USA. Available at: https://www.abraxanepro.com/about-abraxane/. Accessed: 18 October 2017.
39. Desai N, Trieu V, Yao Z, Louie L, Ci S, Yang A, Tao C, De T, Beals B, Dykes D, Noker P, Yao R, Labao E, Hawkins M, Soon-Shiong P. Increased antitumor activity, intratumor paclitaxel concentrations, and endothelial cell transport of cremophor-free, albumin-bound paclitaxel, ABI-007, compared with cremophor-based paclitaxel. Clin Cancer Res 2006;12:1317–24.
40. Desai N, Trieu V, Damascelli B, Soon-Shiong P. SPARC expression correlates with tumor response to albumin-bound paclitaxel in head and neck cancer patients. Transl Oncol 2009;2:59–64.
41. Gelderblom H, Verweij J, Nooter K, Sparreboom A. Cremophor EL: the drawbacks and advantages of vehicle selection for drug formulation. Eur J Cancer 2001;37:1590–98.
42. Ibrahim NK, Desai N, Legha S, Soon-Shiong P, Theriault RL, Rivera E, Esmaeli B, Ring SE, Bedikian A, Hortobagyi GN, Ellerhorst JA. Phase I and pharmacokinetic study of ABI-007, a Cremophor-free, protein-stabilized, nanoparticle formulation of paclitaxel. Clin Cancer Res 2002;8:1038–44.
43. Camaschella C. Iron-deficiency anemia. N Engl J Med 2015;372:1832–43.
44. De Benoist B, McLean E, Egli I, Cogswell M. Worldwide prevalence of anaemia 1993–2005. WHO Global Database on Anaemia. 2008: WHO. Available at: http://apps.who.int/iris/bitstream/10665/43894/1/9789241596657/ eng.pdf. Accessed: 25 July 2017,
45. Cancado RD, Munoz M. Intravenous iron therapy: how far have we come? Rev Bras Hematol Hemoter 2011;33:461–9.
46. Danielson BG. Structure, chemistry, and pharmacokinetics of intravenous iron agents. J Am Soc Nephrol. 2004;15(2):S93–8.
47. Geisser P. Safety and efficacy of iron(III)-hydroxide polymaltose complex. Drug Res 2007;57:439–52.
48. Geisser P, Muller A. Pharmacokinetics of iron salts and ferrichydroxide-carbohydrate complexes. Arzneimittelforschung 1987;37:100–4.
49. Toblli JE, Brignoli R. Iron(III)-hydroxide polymaltose complex in iron deficiency anemia: review and meta-analysis. Arzneimittelforschung 2007;57:431–8.
50. Sakulkhu U, Mahmoudi M, Maurizi L, Salaklang J, Hofmann H. Protein corona composition of superparamagnetic iron oxide nanoparticles with various physico-chemical properties and coatings. Scientific Reports. 2014, 4, Article number: 5020.
51. Zheng N, Sun DD, Zou P, Jiang W. Scientific and regulatory considerations for generic complex drug products containing nanomaterials. AAPS J 2017:619–31.
52. Van Wyck DB. Labile Iron: Manifestations and Clinical Implications. J Am Soc Nephrol 2004;15:S107–11.
53. Etheridge ML, Campbell SA, Erdman AG, Haynes CL, Wolf SM, McCullough J. The big picture on nanomedicine: the state of investigational and approved nanomedicine products. Nanomedicine Nanotechnol Biol Med 2013;9:1–14.
54. D'Mello SR, Cruz CN, Chen ML, Kapoor M, Lee SL, Tyner KM. The evolving landscape of drug products containing nanomaterials in the United States. Nature Nanotechnol 2017;12:523–9.

55. Karch CP, Burkhard P. Vaccine technologies: from whole organism to rationally designed protein assemblies. Biochem Pharmacol 2016;120:1–14.
56. Poecheim J, Borchard G. Immunotherapy & vaccines. In: Forcada J, van Herk A, Pastorin G, editors. Controlled release systems: advances in nanobottles and active nanoparticles. Pan Stanford Publishing, Singapore, 2016:425–47.
57. Bolhassani A, Javanzad S, Saleh T, Hashemi M, Aghasadeghi MR, Sadat SM. Polymeric nanoparticles. Hum Vaccin Immunother 2014;10:321–32.
58. Zhao Q, Li S, Yu H, Xia N, Modis Y. Virus-like particle based human vaccines: quality assessment based on structural and functional properties. Trends Biotechnol 2013;31:654–63.
59. Vacher G, Kaeser M, Moser C, Gurny R, Borchard G. Potential of virus-like particles and virosome-based vaccines in mucosal immunization. Mol Pharm 2013;10:1596–609.
60. Grgacic EV, Anderson DA. Virus-like particles: passport to immune recognition. Methods 2006;40:60–5.
61. Almanza H, Cubillos C, Angulo I, Mateos F, Caston JR, van der Poel WH, Vinje J, Barcena J, Mena I. Self-assembly of the recombinant capsid protein of a swine norovirus into virus-like particles and evaluation of monoclonal antibodies cross-reactive with a human strain from genogroup II. J Clin Microbiol 2008;46:3971–9.
62. Roldao A, Mellado MC, Castilho LR, Carrondo MJ, Alves PM. Virus-like particles in vaccine development. Expert Rev Vaccines 2010;9:1149–76.
63. Ellis RW. The new generation of recombinant viral subunit vaccines. Curr Opin Biotechnol 1996;7:646–52.
64. Hines JF, Ghim S, Schlegel R, Jenson AB. Prospects for a vaccine against human papillomavirus. Obstet Gynecol 1995;86:860–6.
65. Bertolotti-Ciarlet A, White LJ, Chen R, Prasad BV, Estes MK. Structural requirements for the assembly of Norwalk virus-like particles. J Virol 2002;76:4044–55.
66. Moser C, Amacker M, Zurbriggen R. Influenza virosomes as a vaccine adjuvant and carrier system. Expert Rev Vaccines 2011;10:437–46.
67. Glück R, Metcalfe IC. New technology platforms in the development of vaccines for the future. Vaccine 2002;20(5):B10–6.
68. Borgos SE. Characterization methods: Physical and chemical characterization techniques. In: Cornier J, Owen A, Kwade A, Van de Voorde M, editors. Pharmaceutical nanotechnology: innovation and production. Weinheim: Wiley-VCH Verlag GmbH & Co. KGaA, 2017:135–56.
69. Schellekens H, Stegemann S, Weinstein V, De Vlieger JSB, Flühmann B, Mühlebach S, Gaspar R, Shah VP, Crommelin DJA. How to regulate nonbiological complex drugs (NBCD) and their follow-on versions: Points to consider. AAPS J 2014;16:15–21.
70. Morrow T, Hull Felcone L. Defining the difference: What makes biologics unique. Biotechnol Healthcare 2004;1:24–26, 28–29.
71. Schellekens H, Klinger E, Mühlebach S, Brin JF, Storm G, Crommelin DJA. The therapeutic equivalence of complex drugs. Regul Toxicol Pharmacol 2011;59:176–83.
72. Borchard G, Flühmann B, Mühlebach S. Nanoparticle iron medicinal products – Requirements for approval of intended copies of non-biological complex drugs (NBCD) and the importance of clinical comparative studies. Regul Pharmacol Toxicol 2012;64:324–8.
73. Rottembourg J, Kadri A, Leonard E, Dansaert A, Lafuma A. Do two intravenous iron sucrose preparations have the same efficacy? Nephrol Dial Transplant 2011;26:3262–7.
74. Rottembourg J, Guerin A, Diaconita M, Kadri A. The complete study of the switch from iron-sucrose originator to iron-sucrose similar and vice versa in hemodialysis patients. J Kidney 2016;2:1. doi.org/10.4172/jok.1000110
75. Schümann K, Solomons NW, Orozco M, Romero-Abal ME, Weiss G. Differences in circulating non-transferrin-bound iron after oral administration of ferrous sulfate, sodium iron EDTA, or iron polymaltose in women with marginal iron stores. Food Nutr Bull 2013;34:85–93.

76. Toblli JE, Cao G, Oliveri L, Angerosa M. Effects of iron deficiency anemia and its treatment with iron polymaltose complex in pregnant rats, their fetuses and placentas: oxidative stress markers and pregnancy outcome. Placenta 2012;33:81–7.
77. FDA approves generic version of Doxil to address shortage. Atlanta, GA, USA: American Cancer Society, 2013. Available at: https://www.cancer.org/latest-news/fda-approves-generic-version-of-doxil-to-address-shortage.html. Accessed: 26 July 2017.
78. Withdrawal of the marketing authorisation application for Doxorubicin SUN (doxorubicin). London, UK: European Medicines Agency, Committee for Medicinal Products for Human Use (CHMP), 2011. Available at: http://www.ema.europa.eu/docs/en_GB/document_library/Medicine_ QA/2011/09/WC500112024.pdf. Accessed: 27 July 2017.
79. Pita R, Ehmann F, Papaluca M. Nanomedicines in the EU – regulatory overview. AAPS J 2016;18:1576–82.
80. Draft guidance on iron sucrose. Bethesda, MA, USA: Food and Drug Administration, 2012. Available at: https://www.fda.gov/downloads/Drugs/GuidanceComplianceRegulatoryInformation/Guidances/UCM297630.pdf. Accessed: 27 July 27, 2017.
81. Ehmann F, Pita R. The EU is ready for non-biological complex medicinal products. GaBi J 2016;5:30–5.
82. Flühmann B, Crommelin DJA, De Vlieger JSB, Borchard G, Mühlebach S, McNeil SE, Weinstein V, Shah VP. Non-biological complex drugs (NBCDs) and their follow-on versions: time for an editorial section. GaBI J 2015;4:167–70.
83. De Vlieger JSB, Borchard G, Shah VP, Flühmann B, Neervannan S, Mühlebach S. Is the EU ready for non-biological complex drug products? GaBI J 2016;3:101–2.
84. Garattini L, Padula A. Why EMA should provide clearer guidance on the authorization of NBCDs in generic and hybrid applications. Exp Rev Clin Pharmacol 2017:10;243–5.
85. Reflection paper on the data requirements for intravenous iron-based nano-colloidal products developed with reference to an innovator medicinal product. London, UK: European Medicines Agency, 2015. Available at: http://www.ema.europa.eu/docs/en_GB/document_library/Scientific_guideline/2015/03/WC500184922.pdf. Accessed: 27 July 2017
86. Evaluation of iron species in healthy subjects treated with generic and reference sodium ferric gluconate. Bethesda, MD, USA: Food and Drug Administration, 2014. Available at: https://grants.nih.gov/grants/guide/rfa-files/RFA-FD-14-019.html. Accessed: 27 July 2017,
87. Request letter to GAO. Washington, DC, USA: House of representatives, Committee on Energy and Commerce, 2015. Available at: https://energycommerce.house.gov/sites/republicans.energycommerce.house.gov/files/114/Letters/20151211GAO. Accessed: 27 July 27, 2017.
88. Ramot Y, Rosenstock M, Klinger E, Bursztyn D, Nyska A, Shinar DM. Comparative long-term preclinical safety evaluation of two glatiramoid compounds (glatiramer Acetate, Copaxone(R), and TV-5010, protiramer) in rats and monkeys. Toxicol Pathol 2012;40:40–54.
89. Anderson J, Bella C, Bishop J, Capila I, Ganguly T, Glajch J, Iyer M, Kaundinya G, Lansing J, Pradines J, Prescott J, Cohen BA, Kantor D, Sachleben R. Demonstration of equivalence of a generic glatiramer acetate (Glatopa™). J Neurol Sci 2015;359:24–34.
90. Lee EL, Park BR, Kim JS, Choi GY, Lee JJ, Lee IS. Comparison of adverse event profile of intravenous iron sucrose and iron sucrose similar in postpartum and gynecologic operative patients. Curr Med Res Opin 2013:141–7.
91. Knoeff J, Flühmann B, Mühlebach S. Medication practice in hospitals: are nanosimilars evaluated and substituted correctly? Eur J Hosp Pharm 2017. DOI: 10.1136/ejhpharm-2016-001059
92. Griffith N, McBride A, Stevenson JG, Green L. Formulary selection criteria for biosimilars: Considerations for US health-system pharmacists. Hosp Pharm 2014;49:813–25.
93. Astier A, Barton Pai A, Bissig M, Crommelin DJ, Flühmann B, Hecq JD, Knoeff J, Lipp HP, Morell-Baladrón A, Mühlebach S. How to select a nanosimilar. Ann NY Acad Sci 2017. DOI: 10.1111/nyas.13382

Karolina Jurczyk, Urs Braegger and Mieczyslaw Jurczyk

4 Nanotechnology in dental implants

4.1 Introduction

Prostheses on dental osseointegrated implants have become a standard of care in the management of edentulous patients. Titanium and titanium alloys have been the preferred materials in the production of fixed substitutes for roots of lost teeth over the past 30 years [1–3]. A satisfactory clinical outcome depends on the capability of the implant to bear loads, which is obtained by primary stability immediately following implantation. However long-term outcomes are the result of solid osseointegration of the implant into the host bone. The characteristics of the implant surface itself are of critical importance for the progression toward osseointegration [4]. Cell biological as well as biomechanical properties of the biomaterial play a crucial role in the initial stability of the bone–implant system. An open-porous structure and adequate pore sizes are the determinants for bone ingrowth [5, 6].

The mechanical properties of the implants acting as a scaffold for bone ingrowth should be adjusted to the mechanical properties of the surrounding tissue. Several factors have been demonstrated to have an influence on bone ingrowth into porous implant surfaces, such as the porous structure (pore size, pore shape, porosity, and interconnecting pore size) of the implant, duration of implantation, biocompatibility, implant stiffness, and micromotion between the implant and adjacent bone [7–11].

Developments in material engineering resulted in the fabrication of porous scaffolds that mimic the architecture and mechanical properties of natural bone. The porous structure functions as a framework for bone cell ingrowth into the pores and therefore leading to the integration with host tissue, known as osseointegration [12]. Appropriate mechanical properties, in particular, a low elastic modulus similar to the bone may minimize the stress-shielding problem.

Preparation of porous metals can be obtained by several methods [10, 13–24]. Aiming at bone ingrowth, long-term stability, and load-bearing capacity, porous titanium should represent the following features [10]: (i) high porosity and interconnected pore structure for sufficient space enabling attachment and proliferation of new bone tissue and the transport of body fluids; (ii) a critical pore size range usually within 150–500 μm; and (iii) appropriate mechanical properties adjusted to the surrounding bone tissue for load-bearing and transforming.

In the past, efforts have been made to enhance the topography of implant surfaces in order to accelerate the healing process. Titanium surfaces with microscopic scale roughness have been proposed as an alternative to more conventional implant surfaces produced by machining or titanium plasma spraying. Various techniques,

https://doi.org/10.1515/9783110547221-004

Table 4.1: Present dental Ti-type implants and new Ti-type materials with nanostructure under investigation for dental implant applications.

Present α-Ti-type implants	Bulk nanostructured α-Ti-type implants	Bulk α-Ti-type nanocomposites	Bulk β-Ti-type alloys with nanostructure	Implants with nanosurface
Machined	SPD	Ti-HA	Ti23Mo	Ti-1Ag
TPS	ECAP	Ti-45S5 Bioglass	Ti23Mo-45S5 Bioglass	Ti-HA
SLA®	ECAP with TMP	Ti-SiO2	–	–
SLActive®	Nanoimplants®	Ti-Bioglass-Ag	–	–
Roxolid®	–	–	–	–
TiUnite®	–	–	–	–

TPS, titanium plasma spray layers; SLA®, sandblasted, large-grit, acid-etched; SLActive®, SLA® chemically modified; Roxolid®, metal alloy composed of 15% zirconium and 85% titanium; TiUnite®, Nobel Biocare's proprietary titanium oxide dental implant surface; SPD, severe plastic deformation; ECAP, equal-channel angular pressing process; TMP, thermomechanical processing; Nanoimplants® Timplant Ltd; HA, hydroxyapatite.

such as sandblasting, acid etching, and combinations of both, have been applied to obtain microrough surfaces used nowadays (Table 4.1).

Current research focuses on improving the mechanical performance and biocompatibility of metal-based systems through changes in alloy composition, microstructure, and surface treatments [13, 23–31]. In the case of titanium, a lot of attention is paid to enhance the biocompatibility of commercial purity grades in order to avoid potential biotoxicity of alloying elements, especially in dental implants [25, 32]. In the past few years, application of nanomaterials has become very popular in medicine [33]. Nanostructured materials can exhibit enhanced mechanical, biological, and chemical properties compared to their conventional counterparts [31, 34].

Improvement of the physicochemical and mechanical performance of Ti-based implant materials can be achieved through microstructure control, the top-down approaches known as severe plastic deformation (SPD), and mechanical alloying (MA) techniques [29, 30, 34–36]. Recent studies showed clearly that nanostructuring of titanium can considerably improve not only the mechanical properties but also the biocompatibility [24, 31, 37]. Moreover, this approach also has the benefit of enhancing the biological response to the cp titanium surface [31, 37].

An alternative method for changing the properties of Ti and Ti-based alloys is the production of a composite, which will combine the favorable mechanical properties of titanium and the excellent biocompatibility and bioactivity of ceramics [15, 16, 23, 30, 36–40]. The main ceramics used in medicine are hydroxyapatite (HA, $Ca_{10}(PO_4)_6(OH)_2$) and 45S5 Bioglass (44.8% SiO_2, 24.9% Na_2O, 24.5% CaO, 5.8% P_2O_5) [41]. The ceramic coating on titanium improves surface bioactivity, but often flakes off as a result of poor ceramic/metal interface bonding, which may cause an early or late failure. However, the nanocomposite materials containing metal and bioceramic as a reinforced phase are promising alternatives compared to conventional materials, because they can potentially be designed to match the properties of bone tissue in order to enhance bone healing.

Table 4.2: Application of nanomaterials in dentistry [44].

Specialty of dentistry	Materials
Restorative dentistry	Ketac™, Ketac N100; Nano-ionomers, Filtek Supreme XT, Fuji IX GP, Nano-primer, Premise™ , Adper™ Single bond plus Adhesive, Ceram X™
Regenerative dentistry and tissue engineering	Ostim®, VITOSSO™, Nano-Bone®
Periodontics	Arestin®, Nanogen®
Preventive dentistry	NanoCare® Gold
Orthodontics	Ketac™ N100 Light Curing Nano-Ionomers, Filtek Supreme Plus Universal
Prosthodontics	Nanotech elite H-D plus, GC OPTIGLAZE color®
Oral implantology	Nanotite™ Nano-coated implant, SLA®, SLActive®, Roxolid®, TiUnite®
Endodontic	AH plus™, Epiphany, Guttaflow®

The branches in dentistry where nanotechnology has become popular, apart from implantology, are tooth regeneration, periodontal therapy, soft and hard tissue reconstruction, bone repair, plaque control, caries diagnostics, and caries treatments. There are two key approaches (top down and bottom up) in nanotechnology for creating smaller or better materials and use of smaller constituents into more complex assembling [42–46]. Top-down approach is based on solid-state processing of materials. Regardless of the defects produced by this approach, they will continue to play an important role in the synthesis of nanostructures. The top-down approaches such as chemical vapor deposition (CVD), monolithic processing, and wet and plasma etching are used to fabricate functional structures at micro and nanoscales. These methods are successfully used for coatings of medical implants and stent using CVD technology to enhance blood flow and biocompatibility. Currently, there is a wide range of nanomaterial's applications in different subspecialties of dentistry (Table 4.2).

This chapter reviews recently published and ongoing research on Ti-based nanomaterials for tooth replacement. The following aspects are in focus: the development of a new generation of titanium-based implants with a strictly specified chemical and phase compositions, porosity, and surface morphology. Possible future ideas for studies are discussed, which could lead to tailoring the properties of Ti-based implants for specific indications.

4.2 Nanotechnology in Dental Implants

Nanotechnology is the development and application of techniques enabling the production of structures in the physical size ranging from 1 to 100 nm, as well as the incorporation of these specimens into dental applications [42, 43]. The manufacturing of Ti-based nanomaterials and their upscaling for industrial use is still challenging [31, 44–47].

New types of bulk three-dimensional (3D) porous Ti-based nanocomposite biomaterials with preferred size of porous and 3D capillary-porous coatings were developed. The aim of ongoing research is to produce new generations of titanium-bioceramic nanocomposites by constructing porous structures with a strictly specified chemical and phase composition, porosity, and surface morphology, which will promote good adhesion of the substrate and show increased hardness, high resistance to biological corrosion, and good biocompatibility [13–16, 23, 24, 37]. Materials with nanoscaled grains would offer new structural and functional properties for innovative products in dental applications.

4.2.1 Nanostructured Implants

4.2.1.1 Bulk Nanostructured Implants

Elias et al. introduced a so-called equal-channel angular pressing (ECAP) process, which is a viable processing route to grain refinement and property improvement [2]. In their study they obtained long-sized rods of nanostructured titanium (n-Ti), with superior mechanical and biomedical properties applicable for dental implants. It can be noted that the extreme grain refinement of the metal bulk down to nanoscale creates surface morphology that turns out to be conducive for enhanced adhesion and growth of living cells.

Commercially pure titanium (Grade 4) of the following composition was used: 0.052% C, 0.34% O_2, 0.3% Fe, 0.015% N, base material Ti (wt%). In the as-received condition, billets produced by hot rolling had an average grain size of 25 μm. Nanostructuring was performed using SPD by ECAP with subsequent thermomechanical processing, which made it possible to manufacture rod semiproducts with a length of 3 m and a diameter of 7 mm [48–50].

Combination of SPD and TMP processes resulted in a large reduction in grain size, from the 25 μm equiaxed grain structure of the initial titanium rods to 150 nm. The ultimate strength of the nanostructured cp Grade 4 titanium is nearly twice compared to conventional cp titanium [49]. Additionally, it has been shown that the fatigue strength of nanostructured cp titanium at 10^6 cycles is almost two times higher than for conventional cp titanium and exceeds that of the Ti–6Al–4V alloy.

Cytocompatibility tests utilizing fibroblast mice cells L929 were performed to verify the previously reported benefits of nanostructured cp titanium compared to conventionally processed coarse-grained cp Ti [49]. The results proved that the nanostructuring process causes dramatic increase in the fibroblast colonization. The surface cell occupation for conventional cp Ti was 53% after 72 h, whereas nanostructured CP (Grade 4) reached 87.2%. Therefore, a high osseointegration rate is predicted for nanostructured cp Grade 4 titanium in comparison to conventional titanium. Nanostructured (Nanoimplants® Timplant Ltd, Sjednoceni 77, Ostrava, Czech Republic) implants have been successfully designed and fabricated [51].

The certified system of Timplant®, manufactured according to the standard EN ISO 13485:2003, was used during the development of the Nanoimplant®. A 2.4 mm diameter intraosseous nanoimplant has the same strength as a 3.5 mm diameter conventional one [51]. Superior mechanical performance of the nanoimplants in comparison to conventional cp Grade 4 titanium and Ti-6Al-4V alloy implants is achieved, thanks to the nanostructuring of titanium by SPD processing.

4.2.1.2 Titanium Composites

Many researchers, focusing on the synthesis of nanoscale Ti-based biocomposites, have achieved better mechanical and corrosion properties of nanomaterials compared to microcrystalline titanium [13–16, 23, 30, 37, 39, 40]. For the Ti-HA nanocomposites, the Vickers hardness increased (Ti-20 wt% HA nanocomposites [1030 $HV_{0.2}$]) and was four times higher compared to pure microcrystalline Ti metal (250 $HV_{0.2}$) [37]. No significant difference in corrosion resistance among Ti-3 wt% HA ($i_c = 9.06 \times 10^{-8}$ A/cm^2, $E_c = -0.34$ V) and Ti-20 wt% HA ($i_c = 8.5 \times 10^{-8}$ A/cm^2, $E_c = -0.55$ V) was noted although there was a significant difference in porosity.

A Ti-45S5 Bioglass nanocomposite with a unique microstructure, higher hardness, and better corrosion resistance was produced by means of MA and a powder metallurgy process [39, 40]. Microhardness tests showed that the obtained material exhibited Vickers microhardness as high as 770 $HV_{0.2}$ for Ti-20 wt% 45S5 Bioglass, which was more than three times higher than that of a conventional microcrystalline titanium. Young's modulus of the obtained Ti-10 wt% 45S5 Bioglass composite was measured to be 110 GPa, whereas for the microcrystalline titanium the value was 150 GPa. The corrosion resistance tests proved that Ti-10 wt% of 45S5 Bioglass nanocomposites ($i_c = 1.20 \times 10^{-7}$ A/cm^2, $E_c = -0.42$ V vs SCE) was more corrosion resistant than microcrystalline titanium.

Nanograined materials are characterized by large surface energy due to very high number of atoms on the surface. This property can explain their entirely different behavior compared to the micron-sized grains. Osteoblasts adhere to the surface with a roughness in the nanometer range. However, not only the roughness but also the composition and the surface energy affect the initial contact and spreading of cells [47, 52, 53].

In vitro biocompatibility tests were performed with Ti-10 wt% 45S5 Bioglass nanocomposites [16, 39. 40] (Figure 4.1). The morphology of the cell cultures obtained on the tested nanocomposite was similar to those obtained on microcrystalline titanium. On the other hand, on porous scaffolds, the cells adhered with their entire surface

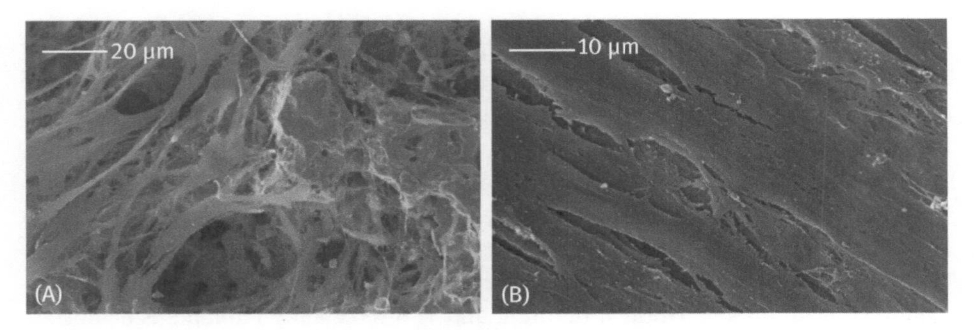

Figure 4.1: Scanning electron micrographs of osteoblasts cultured on Ti-10 wt% 45S5 Bioglass scaffolds with 67% porosity (A) and bulk Ti-10 wt% 45S5 Bioglass sample (B) after 5 days.

to the insert penetrating the porous structure, while on the polished surface, more spherical cells with a smaller surface of adhesion were noticed. The study has demonstrated that Ti-10 wt% 45S5 Bioglass scaffold nanocomposite is a promising biomaterial for dental tissue engineering.

The surface roughness R_a of commercial Ti dental implants produced by mechanical cutting is in the range of 0.08–1.3 mm [16]. Recent studies show that surface roughness influences cell spreading and proliferation but not cell attachment of human osteoblast-like cells [8, 9, 52, 54]. Therefore, it seems reasonable to search for modifications of surface roughness, which would provide a more suitable microenvironment for early osteoblast response to implant materials.

Silica (SiO_2) represents a bioactive material with a high corrosion resistance. Due to the formation of apatite at their surfaces when immersed in simulated body fluid (SBF), silica bioceramics found their application as bone substitute and dental implants [55, 56]. Carlilse et al. in their study explained the role of silicon in bone formation [56]. Addition of SiO_2 to the Ti-based material enhances its bioactivity by the formation of Si–OH groups on the material surface.

Ti-10 wt% SiO_2 nanocomposites and their scaffolds were synthesized (Figure 4.2) [15]. The Vickers hardness of the Ti-10 wt% SiO_2 nanocomposites reached 670 $HV_{0.2}$. The in vitro cytocompatibility of these materials was evaluated. The intensity of cell growth on the surface depends on the surface structure of the sample. The osteoblasts that grew on the inserts exhibited adhesion to the material surface after 1 day and covered the majority of the surface after 5 days. The collected data reveal a significant difference in the morphological characteristics of the cells on the porous and polished materials even after 1 day of cell culturing. On the porous surface, the cells adhered with their entire surface to the insert that penetrates the porous structure, whereas on the polished surface, more spherical cells were observed with a smaller surface of adhesion but with more filopodia. The ability to adhere and grow on a porous material is a specific characteristic of osteoblasts.

The bioactivity of silica is attributed to the formation of a hydroxycarbonated apatite (HCA) layer on its surface [56]. The rate of tissue bonding appears to depend

Figure 4.2: Optical micrographs of the Ti-10 wt% SiO_2 nanocomposite scaffold with a 48% porosity after sintering under a vacuum of 10^{-4} Torr in two steps: at 175°C for 2 h and at 1,150°C for 10 h (MA 20 h, heat treatment 1,150°C/2 h).

on the rate of HCA formation, which follows a sequence of reactions between the implanted material and the surrounding tissues and physiologic fluids. Precipitation of the calcium and phosphate ions released from the glass together with those from the solution form a calcium phosphate (CaP)-rich layer on the surface.

The surface roughness of the microcrystalline titanium, polished bulk Ti-10 wt% SiO_2 nanocomposite, and the Ti-10 wt% SiO_2 scaffold with a 48% porosity was studied [15]. The pores in the polished surface resulted in the formation of valleys and peaks. The microcrystalline titanium presented a smooth surface with very low values. The bulk Ti-10 wt% SiO_2 nanocomposite had an arithmetic mean roughness (R_a), maximum height of the profile (R_t), and 10-point mean roughness (R_z) of around 0.80, 8.03, and 7.03 µm, respectively. The Ti-10 wt% SiO_2 scaffold with 48% porosity had average surface roughnesses, with R_a, R_t, and R_z in the 58–445 µm range (Table 4.3, Figure 4.3). It has been documented that the optimal pore size for the cell attachment, differentiation, and ingrowth of osteoblasts and for vascularization is approximately 200–500 µm [57].

The in vitro cytocompatibility tests for Ti-45S5 Bioglass nanocomposites were also carried out on the established cell line of human fibroblasts CCD-39Lu to check their survival rate and proliferation activity [37]. No cytotoxic effect was noted of the tested materials on the CCD-39Lu cell line. The quantity of suspended material influenced the survival rate of CCD-39Lu fibroblasts cultured in the presence of both nano- and microcrystalline materials. The decrease of the viability of cells in the culture was observed with the increase of the tested material in the culture. Fibroblasts survival rate at 24 and 48 h of culture was more intensively inhibited in the presence of a microcrystalline Ti compared to the nanocrystalline Ti-10 wt% 45S5 Bioglass powder.

The study showed that the survival rate of CCD-39Lu fibroblasts was potentiated by the nanocrystalline Ti-10 wt% 45S5 Bioglass scaffold with the porous structure. Nanocrystalline scaffold material in the form of a disc acts in a similar way as the powder form on the survival rate of treated cells, intensifying survival rate of CCD-39Lu fibroblasts.

The design of Ti-10 wt% 45S5 Bioglass-1.5 wt% Ag composite foams with 70% porosity and pore diameter of about 0.3–1.1 mm is one of the most promising

Table 3: Roughness and topography surface parameters for the polished microcrystalline titanium, polished bulk Ti-10 wt% SiO_2 nanocomposite, and the Ti-10 wt% SiO_2 scaffold with a 48% porosity on different processing routes; parameters taken from surface area of 0.069 mm^2.

Sample	R_a (µm)	R_t (µm)	R_z (µm)	S_{sc} ($µm^{-1}$)	S_{dq} (deg)	S_{dr} (%)
Microcrystalline Ti*	0.21	6.35	4.43	2.71	37.12	23.86
Bulk Ti-10 wt% SiO_2 nanocomposite*	0.80	8.03	7.03	0.89	27.70	12.63
Ti-10 wt% SiO_2 scaffold with 48% porosity	57.74	444.97	431.60	17.63	87.45	14,201.13

*Polished.

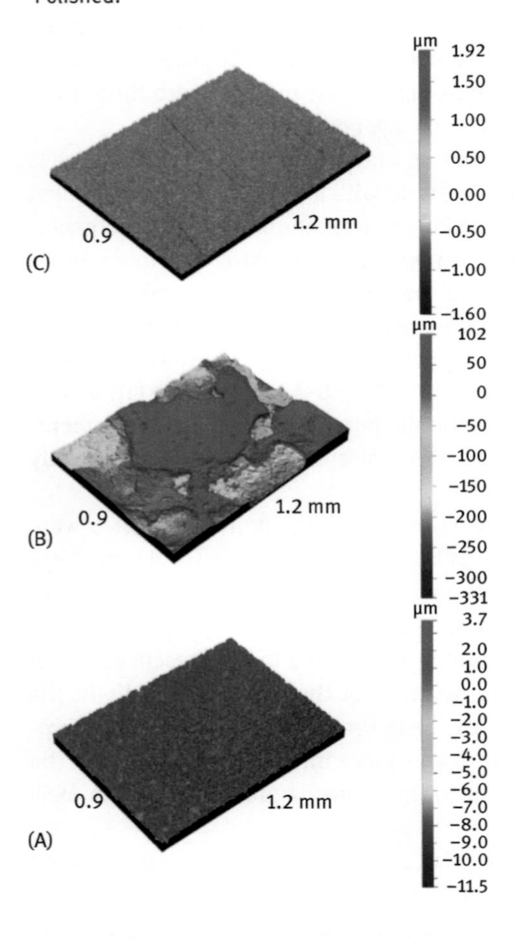

Figure 4.3: Optical profiler 3D topography of polished bulk Ti-10 wt% SiO_2 nanocomposite (A), Ti-10 wt% SiO_2 scaffold with a 48% porosity (B), and polished microcrystalline titanium (C); 0.9 × 1.2 mm scan size.

approaches to achieve optimal antibacterial activities against *Staphylococcus aureus* and *Staphylococcus mutans* [58, 59]. The Ti-Bioglass-Ag composite showed the highest antibacterial activity against *S. aureus* and *S. mutans*. In both cases, the biofilm formation was reduced by more than 90% in comparison to microcrystalline titanium.

According to the previously published research, when Ti-45S5 Bioglass-Ag material can be immersed in body fluid, the metallic Ag particles on the surface of the Ti-45S5 Bioglass-Ag could react with the body fluid and release ionized Ag into the surrounding fluid [60]. A steady and prolonged release of the silver biocide in a concentration level 0.1×10^{-9} is capable of rendering antibacterial efficacy [60]. Silver particles had the highest antibacterial effect with minimal inhibition concentration of 4.86 ± 2.71 µg/mL and minimal bactericidal concentration of 6.25 µg/mL, respectively [61].

The mechanism for bacterial toxicity of tested Ti-based nanocomposites may include free metal ion toxicity arising from the dissolution of metals from the surface of the silver particles (e.g., Ag^+ from Ag) [62]. A different hypothesis is oxidative stress via the generation of reactive oxygen species on crystal surfaces of some nanoparticles (e.g., silica [SiO_2], 45S5 Bioglass) [63]. *S. mutans* in this study was killed not only by Ag particles but also by SiO_2 particles. Silica has been found to inhibit bacterial adherence to oral biofilms [64]. The silica induced an unfavorable change in the biofilm, causing reduction of the adhesion, and therefore less proliferation of bacteria.

4.2.1.3 β-Type Nanostructured Alloys and Composites

Titanium β-type alloys attract attention as biomaterials for dental applications. Nanostructured β-type Ti23Mo – x wt% 45S5 27 Bioglass ($x = 0$, 3, and 10) composites were synthesized [65]. The crystallization of the amorphous material upon annealing led to the formation of a nanostructured β-type Ti23Mo alloy with a grain size of approximately 40 nm. With the increase of the 45S5 Bioglass content in Ti23Mo nanocomposite, an increase of the α-phase was noticeable. The electrochemical treatment in phosphoric acid electrolyte resulted in a porous surface, followed by a bioactive ceramic CaP deposition. Implants, due to the corrosive environment of the tissue and body fluids, may undergo unexpected local corrosion attacks, leading to a release of the corrosion products into the tissue with a toxic effect.

Corrosion resistance potentiodynamic tests in Ringer solution at 37°C demonstrated a positive effect of porosity and CaP deposition on nanostructured Ti23Mo 3 wt% 45S5 Bioglass nanocomposite. This nanocomposite shows best corrosion resistance after electrochemical etching and CaP deposition ($i_{cor} = 1.68 \times 10^{-8}$ A/cm^2, $E_{cor} = -0.44$ V) [65]. Contact angles of glycerol on the nanostructured Ti23Mo alloy were determined and revealed visible decrease for bulk Ti23Mo 3 wt% 45S5 Bioglass and etched Ti23Mo 3 wt% 45S5 Bioglass nanocomposites. In vitro test cultures of normal human osteoblast (NHOst) cells showed very good cell proliferation, colonization, and multilayering. The key factors for the success of implant integration with the surrounding hard tissues are the surface topography and the chemical composition. Therefore, the biofunctionalized nanocrystalline Ti23Mo 3BG composite may represent an important step forward in the development of such a structure, which will support the primary retention and initial healing of the implant.

4.2.2 Implants with Nanosurface

Many studies focused on how to improve the bone/implant interface. Two different approaches were discussed to obtain the enhanced interface: (i) chemically by incorporating inorganic phases, such as calcium phosphate, on or into the TiO_2 layer and (ii) physically by modifying the architecture of the surface topography [8, 12, 14, 23, 27, 54, 66, 67]. Both methods aimed at stimulating bone formation and therefore shortening the healing time and in consequence patient's rehabilitation time.

At the micrometer level, the concept of changing the surface topography is that a rough surface having a higher developed area than a smooth surface increases bone anchorage and reinforces the biomechanical interlocking of the bone with the implant, at least up to a certain level of roughness [68]. At the nanometer level, however, the roughness increases the surface energy, and thus improves matrix protein adsorption, bone cell migration and proliferation, and finally osseointegration [31, 69].

There are numerous reports demonstrating that the surface roughness of titanium implants affects the rate of osseointegration and the biomechanical fixation [70]. The biological properties of titanium depend on its surface oxide film [8]. The modification of the surface morphology and properties of titanium dental implants can be obtained by several mechanical and chemical treatments. One possible method of improving dental implant biocompatibility is to increase surface roughness and decrease the contact angle. Surface profiles in the nanometer range play an important role in the adsorption of proteins, adhesion of osteoblastic cells, and thus the rate of osseointegration [71].

Improvement of the interfacial properties between the surrounding tissues and the existing implants, for example, Ti and Ti-based alloy, is still in interest of many researchers. The electrochemical technique, a more simple and fast method, can be used as a potential alternative for producing porous Ti-based metals for medical implants. Good corrosion resistance of the titanium is provided by the passive titanium oxide film on the surface, which has the thickness of a few nanometers. This layer is important for the good biocompatibility. Its thickness can be increased up to the micrometer range by anodic oxidation and is dependent on the electrochemical etching conditions, for example, current density, voltage, and electrolyte composition [14, 23]. In the electrochemical etching of titanium, electrolytes containing H_3PO_4, CH_3COOH, and H_2SO_4 are used. In the Ti anodization, the dissolution is enhanced by HF- or NH_4F-containing electrolytes, which results in pore or nanotubes formation. The current density in this case is much higher than in electrolyte without HF or NH_4F [72]. Fluoride ions form soluble $[TiF_6]^{2-}$ complexes resulting in dissolution of the titanium oxides, limiting the thickness of the porous layer. Porous implant layers have a lower density than respective bulk structure; therefore, good mechanical strength is ensured by the characteristics of the latter.

It has been shown that a surface with an appropriate chemical composition and topography after combined electrochemical anodic and cathodic surface treatment supports osteoblast adhesion and proliferation on the Ti-6Zr-4Nb sintered nanocrystalline alloy (Figure 4.4) [23]. The porous surface of Ti-6Zr-4Nb sintered nanocrystalline material was produced by anodic oxidation in 1 M H_3PO_4 + 2% HF electrolyte at 10 V for 30 min. Then onto the formed porous surface, the CaP layer was deposited, using cathodic potential of −5 V kept for 60 min in 0.042 M $Ca(NO_3)_2$ + 0.025 M $(NH_4)_2HPO_4$ + 0.1 M HCl electrolyte. The deposited CaP layer anchored in the pores. In vitro test cultures of NHOst cells resulted in very good cell proliferation, colonization, and multilayering.

Recently, much attention has been paid to the TiO_2 nanotube preparation on dense titanium implants using anodization [73]. Porous titanium scaffolds with a porosity of 70% and pore sizes in the range of 200–300 µm were prepared by means of the space holder method using titanium and ammonium hydrogen carbonate particles. Finally, the bioactive anatase nanotubes with the size of approximately 100 nm were successfully fabricated on the titanium scaffold by anodization and heat treatment, which improved the biocompatibility obviously, as assessed by apatite-formation ability. Furthermore, the compressive strength of 36.8 MPa was similar to the mechanical requirement of cancellous bone.

4.2.3 Biomimetic Porous Scaffolds

Surface biomodification has been applied to titanium to improve its poor surface activity [23, 71–75]. Moreover, new techniques have been invented for the production of biomimetic porous titanium scaffolds for bone substitution [23, 71–74]. As it is well known, for osteoconductive function, the pore size distribution within the range of 200–500 µm is essential and can be provided by the space holder sintering method, which adjusts the pore shape and the porosity.

Deposition of even coatings on porous implants can be also achieved by the biomimetic process, due to its non-line-of-sight characteristics. The process was based on the heterogeneous nucleation of CaP from SBF, in which titanium implants were soaked directly in the SBF solution and a CaP layer was coated on the surface.

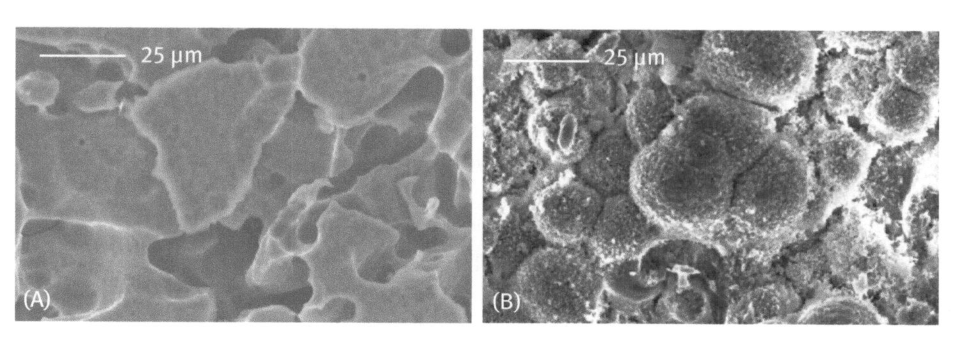

Figure 4.4: Surface of the nano-Ti–6Zr–4Nb after anodic oxidation (A) and additional CaP deposition (B).

Another effective method of improving bioactive properties is HA coating on the surface of Ti implants [76]. Synthetic HA is known as a bioactive and biocompatible material and bonds directly to the bone without the formation of connective tissue on the interface. Apatite is the main component of bone crystal, and also preferentially adsorbs proteins that serve as growth factors.

Zhang and Zou used fiber sintering process for the production of porous titanium with a complete 3D interconnected structure of pore size of 150–600 µm, porosity of 67%, and a high-yield strength of 100 MPa [77]. In order to enhance the surface bioactivity, Si-substituted HA (Si-HA) was coated on the surface by a biomimetic technique. New bone formation occurred in the uncoated porous titanium after 2 weeks of implantation and after 4 weeks, significant increase ($p < 0.05$) in the bone ingrowth rate (BIR) was noted, revealing the good osteoconductivity of the structure. Both HA-coated and Si-HA-coated porous titanium exhibited a significantly higher BIR than the uncoated titanium at all intervals.

Adjusting the mechanical properties of porous titanium with a 3D interconnected pore structure can be obtained by selecting the proper porosity enabling bone ingrowth. Additionally, the 3D pore structure can function as a transport channel for body fluids within the pore network.

The influence of nanoscale implant surface features on osteoblast differentiation was studied by Mendonça et al. [9, 78]. Human mesenchymal stem cells (hMSCs) were cultured on the discs for 3–28 days. The levels of ALP, BSP, Runx2, OCN, OPG, and OSX mRNA and a panel of 76 genes related to osteogenesis were evaluated. Topographical and chemical evaluation confirmed nanoscale features present on the coated surfaces only. Bone-specific mRNAs were increased on surfaces with superimposed nanoscale features compared to machined (M) and acid etched (Ac). At day 14, OSX mRNA levels were increased by 2-, 3.5-, 4-, and 3-fold for anatase (An), rutile (Ru), alumina (Al), and zirconia (Zr), respectively. OSX expression levels for machined and acid-etched surfaces approximated baseline levels. At 14 and 28 days, the BSP relative mRNA expression was significantly upregulated for all surfaces with nanoscale-coated features (up to 45-fold increase for Al). The polymerase chain reaction array showed an upregulation on Al-coated implants when compared to machined surface. An improved response of cells adhering to nanostructured-coated implant surfaces was represented by increased OSX and BSP expressions. It was demonstrated that the aluminum oxide nanoscale feature surface significantly changed the hMSCs gene expression pattern toward an upregulation in osteoblast differentiation. These surfaces may be able to improve the osseointegration response, providing a faster and more reliable bone to implant contact.

The surface of the TiUnite® implant (Nobel Biocare, Sweden) is a highly crystalline, phosphate-enriched titanium oxide characterized by open pores in the low micrometer range [79]. In comparison to machined implant surfaces, this surface has repeatedly proven to elicit a more enhanced bone response. Furthermore, the TiUnite® surface maintains primary stability better than the machined surfaces and shortens the healing time needed to accomplish secondary stability. Shibuya et al. determined the success of Brånemark System's TiUnite® implants placed in partially or completely edentulous jaws restored with fixed or removable prostheses [80]. A total of 131 jaws from 110 patients (64 maxillae and 67 mandibles) received 472 implants from July 2003 to March 2008. The TiUnite® implant employed in Shibuya's study had an overall success rate of 96.56% up to 6 years after implantation.

The SLActive surface is a development of the large grit-blasted and acid-etched SLA surface and is further processed to a high degree of hydrophilicity [81, 82]. SLActive® has been developed to optimize early implant stability and to reduce the risk during the critical early treatment. The in vitro and in vivo studies of the SLActive surface demonstrate a stronger cell and bone tissue response than for the predecessor, the SLA surface, produced by the same company. Immediate and early loading with Straumann SLActive® implants yields excellent survival rates (98% and 97% after 1 year).

Wennerberg et al.'s research summarizes the present documentation for the SLActive surface, a hydrophilic and nanostructured surface produced by Straumann Company in Switzerland, and covers the results from 15 in vitro, 17 in vivo, and 16 clinical studies [83]. In the clinical studies, a stronger bone response was reported for the SLActive surface during the early healing phase when compared with the SLA surface. However, the later biological response was quite similar for the two surfaces and both demonstrated very good clinical results.

4.3 Future Trends

Many new nanostructured Ti-based alloys have been developed and successfully been tested in vivo and in clinical trials. Available products in the market however are sparse.

Improvement of both short- and long-term tissue integration of nano-titanium implants can be achieved by modification of the surface roughness at the nanoscale level for increasing protein adsorption and cell adhesion, by biomimetic calcium phosphate coatings for enhancing osteoconduction, and by the addition of biological drugs for accelerating the bone healing process in the peri-implant area.

Better understanding of the interactions between proteins, cells, tissues, and implant material is the key factor for developing new strategies. The future implant should possess a surface with a controlled and custom-made topography and chemistry. The local release of bone stimulating drugs may help in advanced and complex clinical situations with poor bone quality and quantity. Acceleration of osseointegration time for immediate loading cases will offer patients shorter and safer total rehabilitation time.

More research on the environmental impacts of nanoparticles and effects are essential [84, 85]. As well it is time to look at the toxicity, epidemiology, persistence, and bioaccumulation of manufactured nanoparticles, including their exposure pathways in treated patients and in the environment

The ideal surface of an implant should provide adequate adhesion and initiation of the cell subdivision process. However, these processes should be controlled to avoid excessive growth rates which could lead to escape of cells from the surveillant agents, thus exceeding barriers of carcinogenesis. It still seems very important to develop such titanium-type alloys for the production of dental implants, which will demonstrate a reduced susceptibility to bacterial colonization and will not elicit pathogenic effects. It seems that the nanocrystalline Ti-type implants may be tailored to the production of structures, which support the process of continuous adaptation to the implant by the host organism.

References

1. Long M, Rack HJ. Titanium alloys in total joint replacement – a materials science perspective. Biomaterials 1998;19:1621–38.
2. Elias CN, Lima JH, Valiev R, Meyers MA. Biomedical applications of titanium and its alloys. JOM 2008;60:46–9.
3. Wang K. The use of titanium for medical applications in the USA. Mater Sci Eng A 1996;213:134–7.
4. Albrektsson T, Brånemark P-I, Hansson HA, Lindstrom J. Osseointegrated titanium implants. Requirements for ensuring a long lasting, direct bone-to-implant anchorage in man. Acta Orthop Scand 1981;52:155–70.
5. Li JP, Wijn JR, van Blitterswijk CA, de Groot K. Comparison of porous Ti6Al4V made by sponge replication and directly 3D fiber deposition and cancellous bone. Key Eng Mater 2007;330–332:999–1002.
6. Spoerke ED, Murray NG, Li H, Brinson LC, Dunand DC, Stupp SI. A bioactive titanium foam scaffold for bone repair. Acta Biomater 2005;1:523–33.
7. Ehrenfest DMD, Coelho PG, Kang BS, Sul YT, Albrektsson T. Classification of osseointegrated implant surfaces: materials, chemistry and topography. Rev Trends Biotechnol 2010;28:198–206.
8. Le Guéhennec L, Soueidan A, Layrolle P, Amouriq Y. Surface treatments of titanium dental implants for rapid osseointegration. Rev Dental Mater 2007;23:844–54.
9. Mendonça G, Mendonça DB, Aragaõ, FJ, Cooper LF. Advancing dental implant surface technology – from micron to nanotopography. Rev Biomaterials 2008;29:3822–35.
10. Singh R, Lee P, Dashwood R, Lindley T. Titanium foams for biomedical applications. Rev Mater Technol 2010;25:127–36.
11. Cachinho SC, Correia RN. Titanium scaffolds for osteointegration: mechanical, in vitro and corrosion behaviour. J Mater Sci Mater Med 2008;19:451–7.
12. Brånemark PI. Osseointegration and its experimental background. J Prosthetic Dent 1983;50:399–410.
13. Adamek G, Jakubowicz J. Mechanoelectrochemical synthesis and properties of porous nano Ti-6Al-4V alloy with hydroxyapatite layer for biomedical applications. Electrochem Commun 2010;12:653–6.

14. Jakubowicz J, Jurczyk K, Niespodziana K, Jurczyk M. Mechanoelectrochemical synthesis of porous Ti-based nanocomposite biomaterial. Electrochem Commun 2009;11:461–5.
15. Jurczyk MU, Jurczyk K, Niespodziana K, Miklaszewski A, Jurczyk M. Titanium-SiO_2 nanocomposites and their scaffolds for dental applications. Mater Charact 2013; 77:99–108.
16. Jurczyk MU, Jurczyk K, Miklaszewski A, Jurczyk M. Nanostructured titanium-45S5 Bioglass scaffold composites for medical applications. Mater Design 2011;32:4882–9.
17. Bram M, Laptev A, Buchkremer HP, Stover D. Near-net-shape manufacturing of highly porous titanium parts for biomedical applications. Materialwiss Werkstofftech 2004;35:213–8.
18. Dunand DC. Processing of titanium foams. Adv Eng Mater 2004;6:369–76.
19. Chino Y, Dunand DC. Directionally freeze-cast titanium foam with aligned, elongated pores. Acta Mater 2008;56:105–13.
20. Xue W, Krishna BV, Bandyopadhyay A, Bose S. Processing and biocompatibility evaluation of laser processed porous titanium. Acta Biomater 2007;3:1007–18.
21. Heinl P, Rottmair A, Körner C, Singer RF. Cellular titanium by selective electron beam melting. Adv Eng Mater 2007;9:360–4.
22. Sobral JM, Caridade SG, Sousa RA, Mano JF, Reis RL. Three-dimensional plotted scaffolds with controlled pore size gradients: Effect of scaffold geometry on mechanical performance and cell seeding efficiency, Acta Biomater 2011;7:1009–18.
23. Jakubowicz J, Adamek G, Jurczyk MU, Jurczyk M. 3D topography study of the biofunctionalized-nanocrystalline Ti-6Zr-4Nb/Ca-P. Mater Charact 1012;70:55–62.
24. Miklaszewski A, Jurczyk MU, Kaczmarek M, Paszel-Jaworska A, Romaniuk A, Lipinska N, Zurawski J, Urbaniak P, Jurczyk M. Nanoscale size effect in in situ titanium based composites with cell viability and cytocompatibility studies. Mater Sci Eng C 2017;73:525–36.
25. Pye AD, Lockhart DE, Dawson MP, Murray AA, Smith AJ. A review of dental implants and infection. J Hospital Infect 2009;72:104–10.
26. Hu K, Yang XJ, Cai YL, Cui ZD, Wei Q. Preparation of bone-like composite coating using a modified simulated body fluid with high Ca and P concentrations. Surf Coat Technol 2006;201:1902–6.
27. Choi J, Bogdanski D, Koller M, Esenwein SA, Muller D, Muhr G, Epple M. Calcium phosphate coating of nickel–titanium shape-memory alloys. Coating procedure and adherence of leukocytes and platelets. Biomaterials 2003;24:3689–96.
28. Estrin Y, Kasper C, Diederichs S, Lapovok R. Accelerated growth of preosteoblastic cells on ultrafine grained titanium. J Biomed Mater Res A 2008;90:1239–42.
29. Jakubowicz J, Adamek G. Preparation and properties of mechanically alloyed and electrochemically etched porous Ti-6Al-4V. Electrochem Commun 2009;1:1772–5.
30. Niespodziana K, Jurczyk K, Jakubowicz J, Jurczyk M. Fabrication and properties of titanium – hydroxyapatite nanocomposites. Mater Chem Phys 2010;123:160–5.
31. Webster TJ, Ejiofor JU. Increased osteoblast adhesion on nanophase metals: Ti, Ti6Al4V, and CoCrMo. Biomaterials 2004;25:4731–9.
32. Matusiewicz H. Potential release of in vivo trace metals from metallic medical implants in the human body: From ions to nanoparticles – A systematic analytical review. Acta Biomater 2014;10:2379–403.
33. Pathan DS, Doshi SB, Muglikar SD. Nanotechnology in implants: the future is mall. Univ Res J Dent 2015;5:8–13.
34. Jurczyk M. (ed) BIONANOMATERIALS FOR DENTAL APPLICATIONS. Singapore: Pan Stanford Publishing, 2012.
35. Park JW, Kim YJ, Park CH, Lee DH, Ko YG, Jang JH, Lee CS. Enhanced osteoblast response to an equal channel angular pressing-processed pure titanium substrate with microrough surface topography. Acta Biomater 2009;5:3272–80.

36. Tulinski M, Jurczyk M. Nanostructured nickel-free austenitic stainless steel composites with different content of hydroxyapatite. Appl Surf Sci 2012;260:80–3.
37. Kaczmarek M, Jurczyk MU, Rubis B, Banaszak A, Kolecka A, Paszel A, Jurczyk K, Murias M, Sikora J, Jurczyk M. *In vitro* biocompatibility of Ti-45S5 Bioglass nanocomposites and their scaffolds, J Biomed Mater Res A 2014;102:1316–24.
38. Papargyri SA, Tsipas D, Stergioudis G, Chlopek J. Production of titanium and hydroxyapatite composite biomaterial for use as biomedical implant by mechanical alloying process. Eng Biomater 2005;46:27–34.
39. Jurczyk K, Niespodziana K, Jurczyk MU, Jurczyk M. Synthesis and characterization of titanium-45S5 Bioglass nanocomposites, Mater Design 2011;32:2554–60.
40. Jurczyk K, Niespodziana K, Jurczyk MU, Jakubowicz J, Jurczyk M. Titanium-10 wt% 45S5 Bioglass nanocomposite for biomedical applications. Mater Chem Phys 2011;131:540–6.
41. Cao WP, Hench L. Bioactive materials. Ceramics Int 1966;22:493–507.
42. Van de Voorde M, Tulinski M, Jurczyk M. Engineered nanomaterials: a discussion of the major categories of nanomaterials. In: Mansfield E, Kaiser D, Fujita D, Van de Voorde M, editors. Metrology and standardization of nanomaterials: protocols and industrial innovations. Weinheim: Wiley-VCH, 2017:49–73.
43. Tulinski M, Jurczyk M. Nanomaterials synthesis methods. In: Mansfield E, Kaiser D, Fujita D, Van de Voorde M, editors. Metrology and standardization of nanomaterials: protocols and industrial innovations. Wiley-VCH, 2017:75–98.
44. Khurshid Z, Zafar M, Qasim S, Shahab S, Naseem M, Abu Reqaiba A. Advances in nanotechnology for restorative dentistry, review. Materials 2015;8:717–31
45. Zhang L, Webster TJ. Nanotechnology and nanomaterials: promises for improved tissue regeneration. Nano Today 2009;4:66–80.
46. Ozak ST, Ozkan P. Nanotechnology and dentistry. Eur J Dent 2013;7:145–151.
47. Ward BC, Webster TJ. Increased functions of osteoblasts on nanophase metals. Mat Sci Eng C 2007, 27, 575–8.
48. Valiev RZ, Islamgaliev RK, Alexandrov IV. Bulk nanostructured materials from severe plastic deformation. Prog Mater Sci 2000;45:103–89.
49. Valiev RZ, Semenova IP, Jakushina E, Latysh VV, Rack H, Lowe RC, J. Petruželka J, Dluhoš L, Hrušák D, Sochová J. Nanostructured SPD processed titanium for medical implants. Mater Sci Forum 2008;584–586:49–54.
50. Valiev RZ, SemenovaIP, Latysh VV, Shcherbakov AV, Yakushina EB. Nanostructured titanium for biomedical applications: new developments and challenges for commercialization. Nanotechnol Russ 2008, 3, 593–601.
51. Petruzelka J, Dluhos L, Hrusak D, Sochova J. Nanostructured titanium. Application in dental implants, Sbornikvedeckychpraci Vysokeskolybanske (Technicke University Ostrava) 2006;1:177–85.
52. Mueller U, Imwinkelried T, Horst M, Sievers M, Graf-Hausner U. Do human osteoblasts grow into open-porous titanium? Eur Cells Mater 2006;11:8–15.
53. Webster TJ, Siegel RW, Bizios R. Nanoceramic surface roughness enhances osteoblast and osteoclast functions for improved orthopaedic/dental implant efficacy. Scr Mater 2001;44:1639–42
54. Albrektsson T, Wennerberg A. Oral implant surfaces: Part 2 – review focusing on clinical knowledge of different surfaces. Int J Prosthodont 2004;17:544–64.
55. Hench LL. Bioceramics: from concept to clinic. J Am Ceram Soc 1991;74:1487–510.
56. Carlisle EM. Silicon: a possible factor in bone calcification. Science 1970;167:279–80.

57. Ellingsen JE, Johansson CB, Wennerberg A, Holmen A. Improved retention and bone implant contact with fluoride modified titanium implant. Int J Oral Maxillofac Implants 2004;19:659–66.
58. Jurczyk K, Miklaszewski A, Niespodziana K, Kubicka M, Jurczyk MU, Jurczyk M. Synthesis and properties of Ag-doped titanium-10 wt.% 45S5 Bioglass nanostructured scaffolds. Acta Metall Sin (Engl Lett) 2015;28:467–76
59. Jurczyk K, Kubicka MM, Ratajczak M, Jurczyk MU, Niespodziana K, Nowak DM, Gajecka M, Jurczyk M. Antibacterial activity of nanostructured Ti-45S5 Bioglass-Ag composite against *Streptococcus mutans* and *Staphylococcus aureus*. Trans Nonferrous Met Soc China 2016;26:118–25.
60. Kumar R, Münstedt H. Silver ion release from antimicrobial polyamide/silver composites. Biomaterials 2005;26:2081–8.
61. Hernández-Sierra JF, Ruiz F, Pena DC, Martinezgutiérrez F, Martinez AE, Guillén Ade J, Tapia-Pérez H, Castañón GM. The antimicrobial sensitivity of streptococcus mutans to nanoparticles of silver, zinc oxide, and gold. Nanomedicine 2008;4:237–40.
62. Kim JS, Kuk E, Yu KN, Kim JH, Park SJ, Lee HJ, Kim SH, Park YK, Park YH, Hwang CY, Kim YK, Lee YS, Jeong DH, Cho MH. Antimicrobial effects of silver nanoparticles. Nanomedicine 2007;3:95–101.
63. Wong MS, Chu WC, Sun DS, Huang HS, Chen JH, Tsai PJ, Lin NT, Yu MS, Hsu SF, Wang SL, Chang HH. Visible-light-induced bactericidal activity of a nitrogen-doped titanium photocatalyst against human pathogens. Appl Environ Microbiol 2006;72:6111–6.
64. Besinis A, De Peralta T, Handy R D. The antibacterial effects of silver, titanium dioxide and silica dioxide nanoparticles compared to the dental disinfectant chlorhexidine on Streptococcus mutans using a suite of bioassays. Nanotoxicology 2014;8:1–16.
65. Jurczyk K, Miklaszewski A, Jurczyk MU, Jurczyk M. Development of βtype Ti23Mo-45S5 Bioglass nanocomposites for dental applications. Materials 2015;8:8032–46.
66. Bonfante EA, Witek L, Tovar N, Suzuki M, Marin C, Granato R, Coelho PG. Physicochemical Characterization and *in vivo* evaluation of amorphous and partially crystalline calcium phosphate coatings fabricated on Ti-6Al-4V implants by the plasma spray method. Int J Biomaterials 2012;Article ID 603826:8 p. DOI:10.1155/2012/603826.
67. Albrektsson T, Wennerberg A. Oral implant surfaces: part 1 – review focusing on topographic and chemical properties of different surfaces and *in vivo* responses to them. Int J Prosthodont 2004;17:536–43.
68. Wennerberg A, Albrektsson T. On implant surfaces: a review of current knowledge and opinions. Int J Oral Maxillofac Implants 2010;25:63–74.
69. Elias CN, Rocha FA, Nascimento AL, Coelho PG. Influence of implant shape, surface morphology, surgical technique and bone quality on the primary stability of dental implants. J Mech Behav Biomed Mater 2012;16:169–80.
70. Wennerberg A, Hallgren C, Johansson C, Danelli S. A histomorphometric evaluation of screw-shaped implants each prepared with two surface roughnesses. Clin Oral Implants Res 1998;9:11–9.
71. Brett PM, Harle J, Salih V, Mihoc R, Olsen I, Jones FH, Tonetti M. Roughness response genes in osteoblasts. Bone 2004;35:124–33.
72. Jakubowicz J. Formation of porous TiO_x biomaterials in H_3PO_4 electrolytes. Electrochem Commun 2008;10:735–9
73. Fan XP, Feng B, Weng J, Wang JX, Lu X. Processing and properties of porous titanium with high porosity coated by bioactive titania nanotubes. Mater Lett 2011;65:2899–901.

74. Li F, Feng QL, Cui FZ, Li HD, Schubert H. A simple biomimetic method for calcium phosphate coating. Surf Coatings Technol 2002;154:88–93.
75. Zhang EL, Yang K. Biomimetic coating of calcium phosphate on biometallic materials. T Nonferr Metal Soc 2005;15, 1199–205.
76. Nouri A, Hodgson PD, Wen C. Biomimetic porous titanium scaffolds for orthopedic and dental applications. In: Mukherjee A, editor. Biomimetics learning from nature. ISBN: 978-953-307-025-4, InTech, 2010.
77. Zhang E, Zou C. Porous titanium and silicon-substituted hydroxyapatite biomodification prepared by a biomimetic process: Characterization and *in vivo* evaluation. Acta Biomater 2009;5:1732–41.
78. Mendonça G, Mendonça DBS, Simões LGP, Araújo AL, Leite ER, Duarte WR, Aragaõ FJL, Cooper LF. The effects of implant surface nanoscale features on osteoblast-specific gene expression. Biomaterials 2009;30:4053–062.
79. Rocci A, Martignoni M, Gottlow J. Immediate loading of Brånemark System TiUnite and machined-surface implants in the posterior mandible: A randomized open-ended clinical trial. Clin Implant Dent Res 2003;5(1):57–63.
80. Shibuya Y, Kobayashi M, Takeuchi J, Asai T, Murata M, Umeda M, Komori T. Analysis of 472 Brånemark system TiUnite implants: a retrospective study. Kobe J Med Sci 2009;55:E73–E81.
81. Buser D, Broggini N, Wieland M, Schenk RK, Denzer AJ, Cochran DL, Hoffmann B, Lussi A, Steinemann SG. Enhanced bone apposition to a chemically modified SLA titanium surface. J Dent Res 2004;83:529–33.
82. Ganeles J, Zöllner A, Jackowski J, Bruggenkate C, Beagle J, Guerra F. Immediate and early non-occlusal loading of Straumann® implants with SLActive® surface: 1 year results from a prospective multicenter randomized-controlled study. Clin Oral Impl Res 2008;19:1119–28.
83. Wennerberg A, Galli S, Albrektsson T. Current knowledge about the hydrophilic and nanostructured SLActive surface. Rev Clin Cosm Invest Dent 2011;3:59–67.
84. Hoet PHM, Brüske-Hohlfeld I, Salata OV. Nanoparticles – known and unknown health risks, review. *J Nanobiotechnol* 2004;2:12–27.
85 Tsuda A, Gehr P. Nanoparticles in the lung-environmental exposure and drug delivery. Boca Raton, FL: CRC Press, 2015.

Cecilia Bartolucci

5 Nanotechnology and food

5.1 Introduction

The development of new technologies should not be a goal in itself since technologies ultimately are only tools and as such should serve a purpose and provide innovation apt to address new challenges. So, which are the challenges regarding food production, food distribution, consumption, and nutrition we are facing already, and which will we be facing in the future? Why do we need nanotechnologies, and in particular, why do we need them applied to food?

Currently 3 billion people from 193 countries have low-quality diets and one in three people is malnourished [1]. In particular, globally, more than 2.5 billion people are overweight or obese, conditions that are linked to an increase in chronic diseases such as diabetes, cardiovascular disease, and cancer. Furthermore, an estimated 155 million children under 5 years of age (23% prevalence) experience stunted growth, reflecting chronic undernutrition during the early stages of life, while 41 million children (6%) are overweight [2]. The World Health Organizations estimated that currently, we are able to provide enough calories to feed the entire world population, and while enough calories are no guarantee for health, with the expectancy of a growing population, even this goal will be unreachable unless we change the ways we produce and distribute our food, as well as how we manage waste. In fact, roughly one-third of the food produced in the world for human consumption every year – approximately 1.3 billion tons – gets lost or wasted [3].

It is evident that the world is facing a nutritional crisis and the current food systems are inadequate to provide a nutritionally high-quality diet. Population growth and climate change will further stress our food systems and natural resources. According to a recent report [4], 33% of the world's soil is moderately to highly degraded, at least 20% of the world's aquifers are overexploited, 60% of global terrestrial biodiversity loss is related to food production, and over 80% of the input of minerals does not reach consumers' plates, implying very large nutrient losses to the environment. We need to find new approaches and new solutions, aimed at supplying nutritionally high-quality food, while using resources in a sustainable way.

The application of nanotechnologies to food production at all steps of the food chain as well as the use of nanoparticles in food can greatly contribute to addressing all these challenges. During the past years, it was possible to acquire knowledge about how physicochemical characteristics of nanosized substances can be used to modify structure, texture, and quality of foodstuffs [5]. This knowledge can be now applied to produce better quality food and to reduce the negative impacts on environment and natural resources that the current food production has.

https://doi.org/10.1515/9783110547221-005

According to the EC recommendation [6], nanomaterial is "a natural, incidental, or manufactured material containing particles, in an unbound state or as an aggregate or as an agglomerate and where, for 50% or more of the particles in the number size distribution, one or more external dimensions are in the size range 1–100 nm". Such materials, natural or engineered, show characteristics that can greatly differ from those of the same material on the macroscale. In food, there are many naturally occurring organic nanoparticles and nanostructures, such as proteins, polysaccharides, and lipids, which come directly from raw products (plant and animal), or may be generated during processing (e.g., cooking or milling), and humans have been consuming them for decades. In the past 10 years however, artificial organic and inorganic nanostructures, hence engineered nanoparticles, structures, and materials, have been created in order to provide food with greater functionality and value. One should talk about engineered nanomaterials obtained through nanotechnological applications, only if the correlation between the nanostructure of the novel materials and the resulting highly unique properties is recognized and deliberately applied. According to the EU regulation on the provision of food information to consumers [7], "all ingredients present in the form of engineered nanomaterials must be indicated in the list of ingredients with the word 'nano' in brackets."

5.2 Opportunities

The application of nanotechnologies and the use of nanomaterials can play an important role at all stages along the food chain. They can be applied and used either outside our food, for example as food contact materials in packaging, or inside, for example as nanoemulsions in ice creams and yogurt. The production of nanofoods starts in the field with the use of nanoagrochemicals and continues through processing, packaging, and finally to consumption, creating innovative opportunities, which range from providing a healthier and more diversified diet to ensuring better food safety and even allowing for more equality in food distribution (Figure 5.1).

5.2.1 Provide Better Quality and More Nutritious Food Options

It has been shown that regular consumption of functional foods or beverages enriched with bioactive agents, such as carotenoids, curcumin, flavonoids, coenzyme Q10, peptides, and ω-3 fatty acids, may help prevent heart disease, hypertension, and diabetes, and even have an impact on cognition by enhancing memory and concentration or improve brain development [8, 9]. However, there are challenges associated with the enrichment of foods with nutraceuticals such as chemical instability, poor food matrix compatibility, and low bioavailability. The use of nanodelivery systems can provide

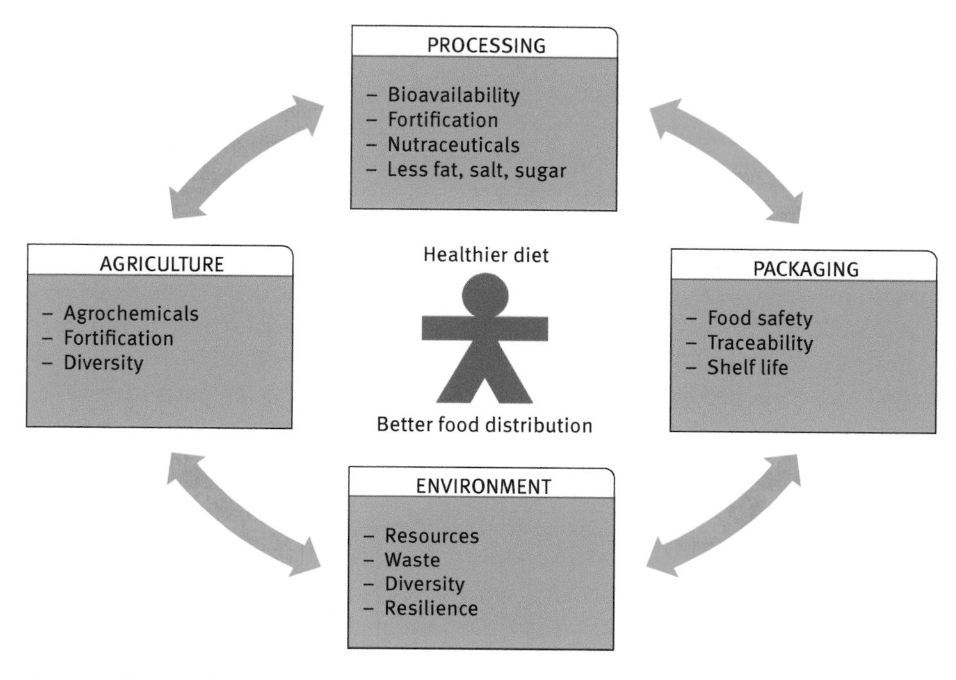

Figure 5.1: Nanofoods: opportunities.

new, adaptable solutions. According to the type of biomolecule that needs to be delivered (polar or non-polar bioactive molecules) or the target that wants to be reached (e.g., a specific region of the gastrointestinal tract [GIT]) [10], different nanodelivery systems, such as nanoliposomes, nanoemulsions, solid lipid nanoparticles, and nanostructured lipid carriers, can be chosen. Most of the above-mentioned delivery systems originate from primary food components such as proteins, polysaccharides, and lipids found in nature. They provide a highly tunable source of material for the production of nanostructures, which can be obtained through the application of many different methods, such as self-assembly (bottom-up approach) [11] and high-pressure homogenization (top-down approach) [12]. Nanoliposomes, for example, can self-assemble into bilayer structures able to trap either polar or non-polar bioactive molecules [13]. As an example of their potential use in food, soy phosphatidylcholine nanoliposomes containing ω-3 and ω-6 fatty acids, able to encapsulate vitamin C and vitamin D, were developed to be added to orange juice [14]. These carrier systems showed a good stability, in particular regarding the thermolabile vitamin C, and no change in taste, indicating their potential to deliver valuable bioactives. Another example is provided by nanoemulsions, which due to small droplet size impart unique rheological and textural properties to foodstuff, helping the production of low-fat nanostructured mayonnaise, spreads, and ice creams without compromising texture or taste, hence offering an alternative product to the consumer [15].

Nanodelivery systems can also be functionalized with intelligent nanosensors, able to release the bioactive component or the specific nutrient only when needed, and allowing for a dynamic and targeted response. This technology is in accordance with the development of a personalized, preventive nutrition and in combination with a personalized, preventive medicine.

Sensory improvement in terms of flavor/color enhancement and texture modification, as well as improvement of anticaking properties can be achieved through the use of engineered inorganic materials containing nanoparticles. While the European Food Safety Agency [16] observed that organic nanoparticles have been shown to present a low health risk, since they are probably metabolized as normal food, health hazards caused by inorganic nanomaterials still need to be investigated. The best known and studied such materials are amorphous synthetic silica and titanium dioxide. In the food industry, the main functions of amorphous synthetic silica are anticaking, carrier, spray drying aid, and milling aid. As anticaking agent silica significantly improves the flowability of powders and is therefore used in salt, sugar, coffee creamers, instant beverages, and soups. TiO_2, on the other hand, is a food additive belonging to the group of food colors, since it has a whitening effect. It is found in confectionary, dairy products' analogues (coffee creamers) and cheese, seasonings, and condiments. However, the highest content of TiO_2 was found in chewing gums [17]. The particle size distribution of raw powders of TiO_2 is from 30 to 400 nm [17, 18]. In E171 samples tested, the fraction of nanoparticles (<100 nm) ranged from less than 10% to 35%. It was hence always smaller than 50% [18–20] and, according to the EC definition [6], E171 should not be labeled as nanomaterial. However, recent studies showed that TiO_2 can be absorbed by the GIT and can bioconcentrate and bioaccumulate in the body [21]. This example shows that the current definition of nanomaterials may not always be appropriate and may cause uncertainties.

Specifically, in the processing of food, nanomaterials can be of use as coatings for food production machinery (biofilm formation) for reduced friction, antifouling, and self-cleaning. Furthermore, nanostructured sieves, filters, and membranes can be used in cold sterilization, separation process, or in the production of stable mixtures, while nanostructures can act as nanoscale adsorbents and catalysts. In all these applications, it is necessary to acquire sufficient basic knowledge about food structures on the micro- and nanoscale and the link between raw material, food processing, and food structures. It is also important to evaluate the extent of indirect migration to be able to quantify the potential uptake of nanoparticles through food contact materials.

5.2.2 Reduce Negative Environmental Impacts

Nanodelivery systems and nanosensors are finding use also in agriculture, providing advantages both to the production of food and to the quality of raw products. During the past years, there has been a remarkable development in nanoagrochemicals, and there have been increasing incentives in the scientific community to develop nano-products that are more efficient and less harmful to the environment compared to conventional agrochemicals. Current agrochemicals are responsible for increased pathogens and pest resistance, reduced biodiversity, diminished nitrogen fixation, and pollination decline. Preliminary results from the Global Food and Water System platform [22] show that an increase in crop yield productivity by at least 0.5% per year should be sufficient to meet food requirements of a crop-based food supply by 2050. However, this goal can be reached only with an increased use of fertilizers and water, which would put an unsustainable stress on our planet [23]. This stress is also increased by the fact that the production of nitrogen fertilizers is highly energy intensive and adds considerably to greenhouse gas emissions. Currently, the fertilizer nitrogen use efficiency by crops is not more than 30–50%, while the remainder is lost via volatilization, denitrification, leaching, and stabilization into soil organic matter. There is hence an urgent need to develop new agrochemicals, both pesticides and fertilizers, consistent with a more sustainable production and use of natural resources, in order to protect the ecosystem and ecosystem services. Focusing on plant protection, rapid detection of pathogens through nanobiosensors and nano-based kits allows a targeted use of inputs, while the improvement of pesticides efficiency and absorbance using nanodelivery systems will greatly reduce the dose levels required [24].

While researchers were originally interested in inorganic nanoagrochemicals, based on noble metals (Ag, Fe, Cu, TiO_2, ZnO, Ca, Mg, Se, and Au) due to their specific physicochemical and biological characteristics compared to the bulk counterpart, organic-based nanomaterials are now being intensively investigated. Equally interesting are nano-enabled formulations, for example, emulsions or microcapsules showing a well-defined nanopore network. Eventually, these new particles and formulations should allow the introduction of an essential functionality: the synchronization between crop demand and release of required inputs.

Multifunctional nanomaterials find use also in the remediation and purification of water. For example, nanocellulose is a renewable material that combines a high surface area with high strength and chemical inertness. It affects the adsorption behavior of important water pollutants, for example, heavy metal species, dyes, microbes, and organic molecules, combining a high removal efficiency with antifouling properties [25]. Oligodynamic metallic nanoparticles, nanoporous fibers, and nanoporous foams are being developed to be used in microbial disinfection, while nanocomposite membranes offer a low-energy alternative for desalination [26].

5.2.3 Increase Food Safety and Reduce Waste

Food waste through factors such as a lack of refrigeration and microbial spoilage is a tremendous global problem affecting food security and sustainability. Nanotechnologies can not only help in identifying contaminated or spoiled food, they can also provide tools to prevent contamination and spoilage. This is mainly achieved through the production of active and intelligent packaging and the introduction of new functionalities, able to monitor as well as react to data about product quality, presence of food contaminants, spoilage, supply chain and storage conditions, and food origin.

Nanocomposites with either antioxidant or antimicrobial properties are used in active packaging for food preservation. Nanoadditives are incorporated into the packaging matrix and provide a dynamic, rather than a passive protection of the food inside [27], absorbing ethylene, oxygen, carbon dioxide, moisture, odors, and other materials that release antioxidants, carbon dioxide, or antimicrobial agents [28]. Active materials containing nanosilver are the most used antimicrobial packaging materials [29] and some items are already on the market.

According to an EC report [30], it is estimated that globally inaccuracies in or misunderstanding of food date labels (including "best before," "sell by," and "display until") cause over 20% of waste of still-edible food. Innovative, intelligent packaging can be a solution for food freshness and safety monitoring. Nanoscale particles with their unique physical, chemical, and biochemical properties can provide the technological tools to enable this innovation. By adjusting the size, the structure, and the surface functionalization of the nanoparticle it is possible to tailor the response of the nanosensing system. Nanosensors can be incorporated into the packaging matrix, in labels or coatings. The response needs to be easily read (e.g., by consumers) and the nanoindicator system needs to be cheap (to match food cost), robust, and safe (no migration toward food). Nanoparticles, beside changing color in the presence of the targeted markers, can be integrated to radiofrequency identification tags, due to their electrochemical properties. This allows a continuous wireless tracking of quality and safety of packed food [31–32].

5.2.4 Encourage Local Food Production and Diversity

It has been recognized that a fundamental part of a healthy diet is provided by diversity. However, a diversified diet is possible only if diverse food is made available either through local production or through an adequate and responsible food distribution. Climate change, contributing among other things to greater weather variability, possible increase in the number of extreme events, habitat loss, and constraints in available water, is also posing a new threat to biodiversity and food production in general. Nanotechnologies are particularly suitable to address issues such as diversified needs as well as dynamically changing conditions, since they are per se inter- and transdisciplinary and therefore best suited to provide adaptable applications and solutions.

Local production can be supported by improving the quality of soil. This can be achieved through the application of intelligent nanoagrochemicals, which would avoid temporal overdose and reduces the amount of input needed, minimizing impact on environment and reducing waste. Furthermore, the quality of the crops could be enhanced by using intelligent delivery systems, for example nanocapsules, allowing for a targeted uptake of nutrients, such as Ca, Mg, S, Zn, and Fe, from roots and leaves. Applications of nanotechnology-enabled gene sequencing, with the use of solid-state nanopore devices, could contribute to the effective identification and utilization of plant trait resources, improving their capability to react against environmental stresses and diseases.

Nanosensors can promote precision agriculture, necessary in order to promptly adapt to extreme weather conditions (e.g., drought), and to produce more diverse crops in different regions and climates. Nanosensors also allow a better traceability of the food products, through the introduction of nanomarkers and sensors at field level, during processing, or on the packaging, ensuring not only better quality control but also supporting the development of a production and distribution network. This network, based on the data collected through the sensors, should allow an active response, supporting a food production better tailored to the needs of the populations. It should further encourage a more adaptable and intelligent food distribution, making food available where and when it is needed. This would also support biodiversity and a more diverse diet, while eventually supporting greater food security and reduction of waste.

5.3 Challenges

In the agrifood sector, engineered nanomaterials have without any doubts incredible innovative potential, and their applications will help address some of the key issues in the coming decades. However, there are still too many uncertainties regarding human safety, as well as their real environmental impact. It will be mandatory that all these uncertainties be addressed by researchers, industries, governmental bodies, and consumers' associations in a concerted and interdisciplinary way. In particular, long-term government funded programs to support innovation in the sector as well as R&D investments by large companies will be needed (Figure 5.2).

Responding to the need of a better overview, European Food Safety Authority (EFSA) published in 2014 an "Inventory of Nanotechnology applications in the agriculture, feed and food sector" [33]. The objective of the study was not only to prepare an inventory but also to review nanomaterials' regulations. Physicochemical characteristics, product names, suppliers, information on (eco)toxicity testing, targeted species, and exposure were considered. The inventory covers 55 different nanomaterials for agri/feed/food, and 12 different applications. In 2014, 276 nanomaterials were confirmed to be available on the market. They are very diverse, including metals, metal oxides, clay, and full-carbon materials and organic nanomaterials (nanoencapsulates and nanocomposites). A trend from inorganic toward organic materials can be detected in future applications, where novel foods, biocides, and pesticides are currently mostly at developmental stage.

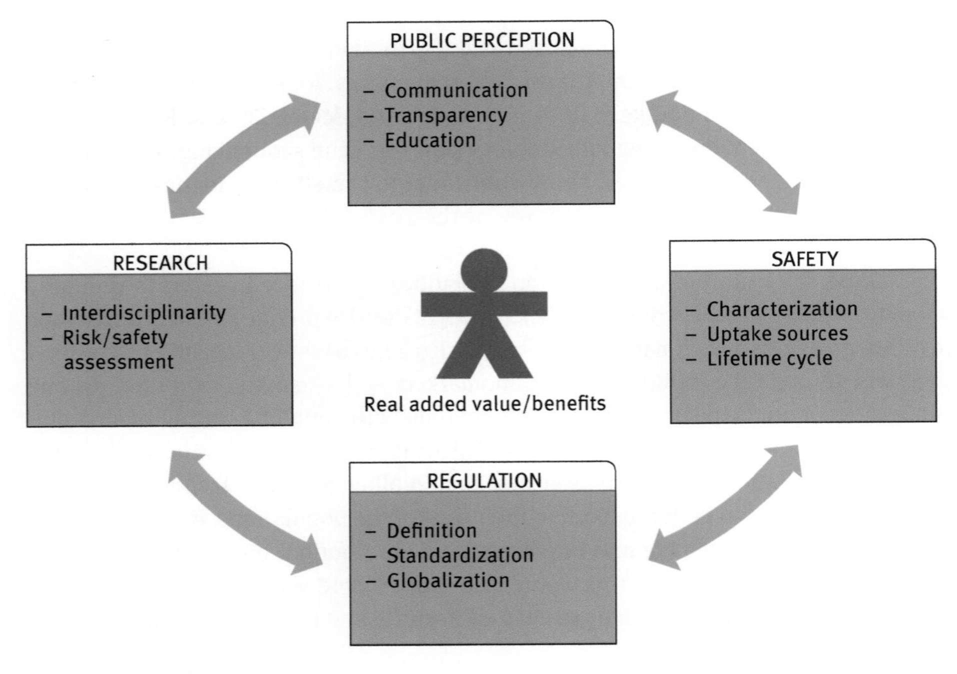

Figure 5.2: Nanofoods: challenges.

5.3.1 Safety Issues

Without always being aware of it, we have always been ingesting nanoparticles. These nanoparticles occur naturally in the environment and are generated through volcanic eruptions, fires, photochemical reactions, or simple erosion. Naturally occurring organic nanoparticle, such as proteins and carbohydrates, are also ubiquitous. However, more recently intentionally engineered nanoparticles have been developed, and while their added value has been recognized by many, the need to ensure a proper and safe use is mandatory. This validation process is not exclusive to the application of nanotechnologies or the use of new nanomaterials, it is required whenever a new technology is introduced. Nonetheless, with the unique properties of nanoparticles, new challenges and also new concerns arise. It is exactly these characteristics, which may induce new interactions with the biological systems [34–37] – both the human system and the environment – causing significant impacts on ecosystems and society and requiring adequate toxicological tests and life cycle assessments.

For humans, the main route of absorption of food nanoparticles is through the GIT after ingestion or inhalation; another minor route being through the epidermis after contact with nanoparticles containing materials. During the digestion process nanoparticles could undergo changes due to different pH, ionic strength, digestive enzymes, interaction with biomolecules, and so on [38]. When assessing the risk of nanoparticles, it is therefore important to keep in mind that the digested particles may be different from the ingested ones [39]. Properties such as size, aggregation, shape, surface properties, and corona influence the absorption by the GIT; on the other hand, it is exactly these properties that can change during the interaction with the gastrointestinal environment. It is therefore important to follow the fate of nanoparticles in the GIT, developing appropriate, qualitatively, and quantitatively sensitive methods.

It will also be critical to assess in the environment the life cycle of nanoparticles used in agriculture as fertilizers, pesticides, or sensors. Currently, there are only few studies, which consider the role that transformation or transport may have on these particles. Studies that replicate the soil conditions used in agriculture will be essential in understanding the benefits and the risks connected with the use of nanoagrochemicals.

5.3.2 Regulations and Policies

It is important to remark that up until now, no new toxicological effects have been attributed to nanoparticles. Most concerns connected to the use of such materials and particles in food may indeed be unfounded. Nonetheless, it is the responsibility and the interest of all parties concerned to combine their efforts in order to fill the knowledge gaps, and provide a better understanding on the fate, the interactions, and the effects of nanofoods, helping to remove doubts and conflicting results. Studies similar to the one undertaken by EFSA [33] are necessary and should be provided and updated also by other research and governmental bodies.

This would also support the development of clearer policies and regulations and diminish the uncertainties created by the lack of common definitions, standards, and control measures. Currently, apart from the European Union, who adopted a mandatory labeling regulation for nanofoods in 2011 [7], only few other countries have some regulation on nanotechnology, but not specifically on food. Many scientists agree that a specific "nanoregulation" may not be necessary, provided a common definition of nanomaterials can be agreed upon, and better methods for the characterization of nanoparticles can be developed. This is also necessary for a homogenous and comprehensive labeling, which should support consumers' decisions on the acquisition of nanofoods, and not just deliver a confusing signal of potential risk.

The implementation of safety regulations, however, could put an additional strain to poorer societies, since developing countries face larger barriers regarding the applications of nanotechnologies than do developed countries for several reasons, including lack of funding and human capacity. Policies can help avoid a potential "nano divide," which would result in shifting even further the focus of nanotechnologies' applications from the necessities of the poor. Primarily, policies related to the food sector should have as a key criterion the nutritional value of food and its assessment. They should hence support technologies and innovation that enhance the production and purchase of nutritious raw material, and recognize the benefits of food diversification and fortification. Highest standards in system integration of technologies as well as high standards of precaution and ethical validation should secure quality and increase transparency.

5.3.3 Public Perception

If we want to tackle all the challenges related to health, food security, climate change, and sustainability, it will be necessary to address them in a concerted way and develop a system approach, in particular a food system approach. At the center of this system are the human beings and their social, economic, and environmental needs. A technology becomes innovation only if it provides clear benefits and added value; otherwise, it remains an empty tool. This is certainly true also for nanotechnology and this paradigm should be applied and shared with the consumers through communication and transparency.

Inquiries showed that consumers are generally more willing to accept nanotechnological applications in the medical, rather than in the food sector. In a critical health situation, the risk/benefit balance is perceived in a different way than when we are addressing "just" nutrition and food. Furthermore, food is daily very close to each one of us, both literally and emotionally, and every possible risk factor causes immediate alarm. The prefix "nano" in combination with food and agriculture seems to be the opposite of the trendy terms of "natural," "organic," and "ecofriendly." Investments in communication and education should be a priority since they can greatly contribute to the understanding that organic does not necessarily equal healthy, and ecofriendly is not always sustainable. Knowledge is one of the best ways to empower consumers and allow them to make individual, knowledge-based decisions.

It is the responsibility of the scientific community to accept the challenge of sharing both newly acquired knowledge and open questions in nanoscience with as wide a community as possible, in order to build trust and encourage dialogue. It is the responsibility of the industries and the private sector to foster transparency to mitigate the fear that the reason why nanofood is produced is only to achieve a financial profit. It is finally the responsibility of the policy makers, providing clear regulations regarding the use of nanofoods, to contribute to a climate of confidence, reassuring the public, that their safety and their interests are looked after.

5.4 Conclusions

Nanotechnologies applied to food and agriculture have the potential to disruptively address many of the great challenges that mankind will be facing in the coming decades. By definition, nanomaterials will exploit new properties not shown by their macrocounterparts, and these characteristics have shown to be extremely useful in providing innovative solutions, outlined by their adaptability, responsiveness, and possibility of functionalization.

In particular, throughout the food system nanosensors can be applied to control soil quality, water usage, food quality, food safety, traceability, and more; nanodelivery systems find use in agriculture reducing the amount of input needed, minimizing impact on environment and reducing waste; nanodelivery systems find also use in food fortification, supporting better bioavailability and stability of nutrients, and hence allowing to address malnutrition. Nanoemulsions can also contribute to a healthier diet, supporting the processing of food with fewer fats, sugar, and salt. Finally, nanomaterials, adopted primarily as food contact materials, greatly support food safety, less spoilage, and less waste.

While there is a new awareness among the public of the huge challenges that lay ahead and the need to urgently develop new approaches to address them, the public opinion is often not formed by knowledge, but is influenced by emotions, fears, uncertainties, as well as trends. It is mandatory that in the future all actors engage in an open dialogue, based on the transparent exchange of knowledge and data regarding benefits, risks, safety, and life cycle of nanofoods. It is further essential to develop educational tools at all levels, to empower every individual to make decisions and take responsibility for their actions. Codesign and social innovation are further tools that need to be supported by private and public bodies alike, if we care that a new technology such as nanotechnology be embraced by all stakeholders alike.

While nanotechnology is not a panacea and should not be presented as such, it is a most valuable tool. Nanofoods are already part of our reality and they will certainly play an even greater role in the future. It is only by fostering a climate of trust and respect that farmers, researchers, industries, governments, consumers, and nongovernmental organizations will be able to work together, using a forward-looking approach that in the long term minimizes risks, while maximizing benefits in an equitable and sustainable way.

Acknowledgments: The author wishes to acknowledge the support of the Science and Technology Foresight Project, an Interdepartmental Project of the National Research Council of Italy, and Francesco Verginelli for the illustrations.

References

1. Global Panel on Agriculture and Food Systems for Nutrition. Food systems and diets: facing the challenges of the 21st century. London, UK, 2016.
2. Global Nutrition Report, Nourishing the SDGs. Available at: https://www.globalnutritionreport.org/files/2017/11/Report_2017.pdf. Accessed: 13 November 2017.
3. FAO: SAVE FOOD: Global Initiative on Food Loss and Waste Reduction. Available at: http://www.fao.org/save-food/resources/keyfindings/en/. Accessed: 27 November 2017.

4. Food Systems and Natural Resources. Available at: http://www.resourcepanel.org/reports/food-systems-and-natural-resources. Accessed: 13 November 2017.
5. Chaudry Q, et al. Applications and implications of nanotechnologies for the food sector. Food Addit Contam 2008;25(3):241–58.
6. European Commission, Commission Recommendation of 18 October 2011 on the definition of nanomaterial. Off J Eur Union 2011;L275/38.
7. Regulation (EU) No 1169/2011 of the European Parliament and of the Council of 25 October 2011. Off J Eur Union 2011;L304/18.
8. Ramaa CS, Shirode AR, Mundada AS, Kadam VJ. Nutraceuticals: an emerging era in the treatment and prevention of cardiovascular disease. Curr Pharm Biotechnol 2006;7 (1):15–23.
9. Espin JC, Garcia-Conesa MT, Tomas-Barberan FA. Nutraceuticals: facts and fiction. Phytochemistry 2007;68(22–24):2986–3008.
10. McClemens DJ. Nanoparticle- and microparticle-based delivery systems encapsulation, protection and release of active compounds. Boca Raton: CRC Press, 2014.
11. Raynes JK, Gerrard JA. Amyloid fibrils as bionanomaterials. In: Bionanotechnology: biological self-assembly and its applications. Ed. Rehm. Norfolk, UK: Caister Academic Press, 2013:85–106.
12. Jesorka A, Orwar O. Liposomes: technologies and analytical applications. Ann Rev Anal Chem 2008;1(1):801–32.
13. Mozafari MR, Johnson C, Hatziantoniou S, Demetzos C. Nanoliposomes and their applications in food nanotechnology. J Liposome Res 2008;18(4):309–27.
14. Marsanasco M, Piotrkowski B, CalabròV et al, Bioactive constituents in liposomes incorporated in orange juice as new functional food: thermal stability, rheological and organoleptic properties. J Food Sci Technol 2015;52(12):7828–38.
15. Chaudhry Q, Scotter M, Balckburn J, Ross B, Boxall A, Castle L, et al. Applications and implications of nanotechnologies for the food sector. Food Addit Contam: Part A 2008;25(3):241–58.
16. RIKILT and JRC. Inventory of nanotechnology applications in the agricultural, feed and food sector. Available at: http://www.efsa.europa.eu/fr/supporting/pub/621e Accessed: 17 November 2017.
17. Weir A, Westerhoff P, Fabricius I, Hritovski K, Goetz N. Titanium dioxide nanoparticles in food and personal care products. Environ Sci Technol 2012;46(4):2242–50.
18. Faust JJ, Doudrick K, Yang Y, Westerhoff P, Capco DG. Food grade titanium dioxide disrupts intestinal brush border microvilli *in vitro* independent of sedimentation. Cell Biol Toxicol 2014;30(3):169–88.
19. Peters RJB, van Bemmel G, Herrera-Rivera Z, Helsper HP, et al. Characterization of titanium dioxide nanoparticles in food products: analytical methods to define nanoparticles. J Agric Food Chem 2014;62(27):6285–93.
20. Yang Y, Doudrick K, Bi X, Hristovski K, et al. Characterization of food-grade titanium dioxide: the presence of nanosized particles. Environ Sci Technol 2014;48(11):6391–400.
21. Jovanovic B. Critical review of public health regulations of titanium dioxide, a human food additive. Inter Environ Assess Manage 2015;11(1):10–20.
22. Quentin Grafton R, Williams J, Jiang Q. Food and water gaps to 2050: preliminary results from the global food and water system (GFWS) platform. Food Secur 2015;7(2):209–20.
23. Rockström J, Falknemark M, Allan T, Folke, C, Gordon L, Jägerskog A, et al. The unfolding water drama in the anthropocene: towards a resilience-based perspective on water for global sustainability. Ecohydrology 2014;7:1249–61.
24. Rai M, Ingle A. Role of nanotechnology in agriculture with special reference to management of insect pests. Appl Microbiol Biotechnol 2012;94(2):287–93.
25. Voisin H, Bergström L, Liu P, Mathew AP. Nanocellulose-based materials for water purification. Nanomaterials 2017;7(3):7–25.

26. Rashid MH, Ralph SF. Carbon nanotubes membranes: synthesis, properties and future filtration applications. Nanomaterials 2017;7(5):99–126.
27. Lim L. Active and intelligent packaging materials. Compr Biotechnol 2011;1:629–44.
28. Vermeiren I, Devlieghere F, Van Beest M, De Kruijf N, Debevere J. Developments in the active packaging of foods. Trends Food Sci Technol 1999;10(3):77–86.
29. Rai M, Yadav A, Gade A. Silver nanoparticles as a new generation of antimicrobials. Biotechnol Adv 2009;27(1):76–83.
30. Preparatory Study on Food Waste across Europe. Technical Report, 2010-054. Available at: http://ec.europa.eu/environment/eussd/pdf/bio_foodwaste_report.pdf. Accessed: 24 November 2017.
31. Realin CE, Marcos B. Active and intelligent packaging systems for a modern society. Meat Sci 2014;98(3):404–19.
32. Martinez-Olmos A, Fernandez-Salmeron J, Lopez-Ruiz N, Rivadeneyra Torres A, Capitan-Vallvey LF, Palma AJ. Screen printed flexible radiofrequency identification tag for oxygen monitoring. Anal Chem 2013;85(22):11098–105.
33. EFSA supporting publication 2014: EN-621. External Scientific Report: Inventory of Nanotechnology applications in the agricultural, feed and food sector. Available at: https://www.efsa.europa.eu/en/supporting/pub/en-621. Accessed: 26 November 2017.
34. Oberdörster G, Stone V, Donaldson K. Toxicology of nanoparticles: a historical perspective. Nanotoxicology 2007;1:2–25.
35. Beer C, Foldbjerg R, Hayashi Y, Sutherlan DS, Autrup H. Toxicity of silver nanoparticles – nanoparticle or silver ion? Toxicol Lett 2012;208:286–92.
36. EFSA. Scientific Opinion of the scientific committee: the potential risks arising from nanoscience and nanotechnologies on food and feed safety. (eds. Barlow S, Chesson A, Collins JD, Flynn A, Hardy A, Jany KD, Knaap A, Kuioer H, Larsen JC, and Le Neindre P, et al. Eur Food Saf Authority 2009:1–39.
37. Nel A, Xia T, Madler L, Li N. Toxic potential of materials at the nanolevel. Supramol Sci 2006;311:622–27.
38. Böhmert L, Girod M, Hansen U, Maul R, Knappe P, Niemann B, Weidner SM, Thunemann AF, Lampen A. Analytically monitored digestion of silver nanoparticles and their toxicity on human intestinal cells. Nanotoxicology 2014;8:631–42.
39. Lefebvre DE, Venema K, Gombau I, Valerio Jr LG, Raju J, Bondy GS, Bouwmeester H, Singh RP, Clippinger AJ, Collnot AM, et al. Utility of models of the gastrointestinal tract for assessment of the digestion and absorption of engineered nanomaterials released from food matrices. Nanotoxicology 2015;9:523–42.

Oluwatosin Ademola Ijabadeniyi and Abosede Ijabadeniyi

6 Governance of nanoagriculture and nanofoods

6.1 Nanoagriculture and Nanofoods: Background and Definition

Nanotechnology is a multidisciplinary field that encompasses understanding and control of matter at about 1–100 nm, leading to development of innovative and revolutionary applications. Nanotechnology encompasses nanoscale science, engineering, and technology in addition to imaging, measuring, modelling, and manipulating matter [1]. The International Standards Organization has also recently defined nanotechnology as "understanding and control of matter and processes at the nanoscale, typically, but not exclusively, below 100 nm in one or more dimensions where the onset of size-dependent phenomena usually enables novel applications, where one nanometer is one thousand millionth of a metre" [2].

The Food Safety Authority of Ireland (FSAI) [3] defined nanotechnology as the use of engineered nanomaterial that has been intentionally synthesized or incidentally produced to exploit functional properties exhibited on the nanoscale. The European Commission has recognized nanotechnology as one of the six key enabling technologies [4], and several developing countries such as India, Brazil, South Africa, Thailand, the Philippines, Chile, Argentina, and Mexico invested millions of dollars in a number of nanotechnology initiatives and research. China, the United States, and Japan are however the countries with the highest number of nanotechnology patent applications in 2005 [5].

Nanomaterials and nanoparticles were invented and have been applied before the twentieth century with examples found throughout the fourth to seventeenth centuries [6]. Examples include gold nanoparticles; silver or copper nanoparticles were used to give lustre to ceramics in the Islamic world. Carbon nanotubes and cementite nanowires were also reported to be present in the famous Damascus sabre blades [6]. Nanotechnology has applications in a number of fields. According to Shea et al. [7], nanotechnology is a general-purpose technology with wide applications in different industries and fields. Table 6.1 shows non-food and agricultural applications of nanotechnology.

Nanotechnology has numerous applications in agriculture (i.e. nanoagriculture) and food industry. For example, food products derived when nanotechnology or nanomaterials are used directly or indirectly from farm to fork are referred to as nanofood. Garber [17] and Scrinis [18] have also defined nanofood. They both agreed that nanofood is not only a product of food processing rather they can also be derived during food production or cultivation and food packaging.

https://doi.org/10.1515/9783110547221-006

Table 6.1: Non-food and agricultural applications of nanotechnology

Representative nanomaterials	Applications	Field/industry	References
Gold nanoparticles	Laser applications	Medicine	Pustovalov and Babenko [8]
Carbon nanotubes, nanofibres, nano-Ag	Water and wastewater treatment (adsorption, membranes and membrane processes)	Water	Xiaolei et al. [9]
Engineered colloidal/ nanoparticles	Drug targeting, drug delivery	Pharmacy	Davis [10]
Nanocomposites	Nanofillers	Dentistry	Mitra et al. [11]
Nano-tex, nano-silver, nanosilica	Water repellence, UV protection, antibacterial, wrinkle resistance	Textile	Wong et al. [12]
Nanoemulsions, nanotubules	Vaccines development, adjuvants, immunomodulatory drugs	Immunology	Smith et al. [13]
Gold nanoparticles, nanobarcodes, nanopores	Microfluidic/lab-on-a-chip technology	Molecular diagnostics	Jain [14]
Carbon nanotubes, nanoscale	High-performance cementitious composites – improved macromechanical properties of cement paste	Construction industry	Konsta-Gdoutos et al. [15]
Nanoparticles/materials	AC Electrokinetics (dielectrophoresis and electrorotation)	Electricity/ power generation	Hughes [16]
Different nanoparticles	General-purpose technology	Across industry and technology sectors	Shea et al. [7]

6.2 Application of Nanotechnology in Agriculture and Agroprocessing Industry

Nanotechnology has scores of applications in the agrifood sector at the moment and it may likely change the sector within a short time. Furthermore, food nanotechnology has the potential to bring about golden opportunities for the food industry [19].

Agricultural nanotechnology, i.e. nanopesticides or nanofertilizers, can be used to increase crop yield while the technology could assist in biosecurity (e.g. sensors for detecting pathogens along the whole food chain from the farm to fork) [20].

FSAI [3] grouped the global food applications of nanotechnology into six categories: sensory improvements (flavour/colour enhancement and texture modification), increased absorption and targeted delivery of nutrients and bioactive compounds, stabilization of active ingredients such as nutraceuticals in food matrices, packaging and product innovation to extend shelf-life, sensors to improve food safety, and antimicrobials to kill pathogenic bacteria in food. It can be inferred from the foregoing categories that nanotechnology has the potential to improve food processing, packaging, and safety; enhance flavour and nutrition; produce more functional foods from everyday foods with added medicines and supplements, which, as a consequence, has a multiplied effect on food production and productivity.

In packaging, for example, bionanocomposites, which are hybrid nanostructured materials with improved mechanical, thermal, and gas properties, may be used to package food, increase its shelf-life, and also provide a more environmentally friendly solution because of reduction on reliance on plastics as packaging materials [21, 22]. An example of such bionanocomposites is zein, a prolamin and the major component of corn protein. When dissolved in ethanol or acetone, biodegradable zein films with good tensile and water barrier properties can be derived [22]. Also, Emamifar et al. [23] showed that packaging materials made from nanocomposite film containing nano-silver and nano-zinc oxide were significantly able to reduce microorganisms that could cause spoilage in orange juice.

Novel food packaging technology may in fact be the most promising benefit of nanotechnology in the food industry in the near future, and food companies are said to have started producing packaging materials based on nanotechnology that are delaying spoilage and improving microbial food safety [17].

Another area of interest and equally of great importance for the food industry is food safety and preservation. Food pathogens could be detected with the aid of nanosensors that had been placed directly into the packaging material serving as "electronic tongue" or noses by detecting chemicals released during food contamination and spoilage [24]. According to Lilie and Cantini [25], nanosensor could also help to recover one colony-forming unit of *Escherichia coli* in ground beef. Another advantage of nanosensor is that it can measure and ensure food safety at real time; and the procedures are quick, sensitive, and less labour intensive [26].

In agriculture, nanotechnology could be used to enhance productivity, agricultural water quality management, product processing, storage, and quality control with nanosensor [27]. According to Sastry et al. [28], nanotechnology has been reported to have the potential to enhance future food security measures such as agricultural productivity, soil health, water resources, and food packaging in India.

Examples of nanofood applications are presented in Figure 6.1 [29].

 Agriculture	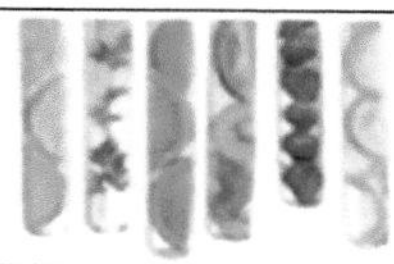 Food processing	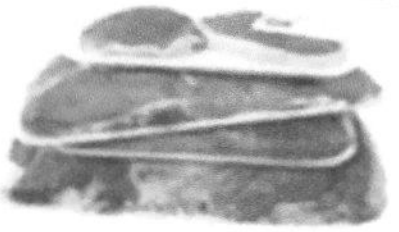 Food packaging	 Supplements
• Single molecule detection to determine enzyme/substrate interactions • Nanocapsules for delivery of pesticides, fertilizers and other agrichemicals more efficiently • Delivery of growth hormones in a controlled fashion • Nanosensors for monitoring soil conditions and crop growth • Nanochips for identity preservation and tracking • Nanosensors for detection of animal and plant pathogens • Nanocapsules to deliver vaccines • Nanoparticles to deliver DNA to plants (targeted genetic engineering)	• Nanocapsules to improve bioavailability of neutraceuticals in standard ingredients such as cooking oils • Nanoencapsulated flavor enhancers • Nanotubes and nanoparticles as gelation and viscosifying agents • Nanocapsule infusion of plant-based steroids to replace a meat's cholesterol • Nanoparticles to selectively bind and remove chemicals or pathogens from food • Nanoemulsions and -particles for better availability and dispersion of nutrients	• Antibodies attached to fluorescent nanoparticles to detect chemicals or foodborne pathogens • Biodegradable nanosensors for temperature, moisture and time monitoring • Nanoclays and nanofilms as barrier materials to prevent spoilage and prevent oxygen absorption • Electrochemical nano-sensors to detect ethylene • Antimicrobial and antifungal surface coatings with nanoparticles (silver, magnesium, zinc) • Lighter, stronger and more heat-resistant films with silicate nanoparticles • Modified permeation behavior of foils	• Nanosize powders to increase absorption of nutrients • Cellulose nanocrystal composites as drug carrier • Nanoencapsulation of neutraceuticals for better absorption, better stability or targeted delivery • Nanocochleates (coiled nanoparticles) to deliver nutrients more efficiently to cells without affecting color or taste of food • Vitamin sprays dispersing active molecules into nanodroplets for better absorption

Figure 6.1: Examples of nanofood applications

6.3 Development of Nanofood with Consumer Safety and Well-Being in Mind

It is important to be transparent about possible risks that may arise from exposure to different types of nanoparticles, as well as how risks can be eliminated or reduced to the level which will not be hazardous to humans. This is necessary before the potential benefits of nanotechnologies to food industry and agriculture can be harnessed.

It is equally important to be proactive in communicating the benefits of nanofoods to consumers, considering the low level of nanotechnology awareness and growing animosity towards nanofoods. Handford et al. [30] interviewed 14 agrifood stakeholders after 88 agrifood stakeholders had responded to an online questionnaire. They found that the current awareness of nanotechnology applications in the agrifood sector was inadequate and that respondents were indifferent to agrifood applications of nanotechnology.

6.4 Regulations of Nanoagriculture and Nanofood

It is without question that nanotechnology has the potential to revolutionize the whole agrifood sector, with the potential to increase agricultural productivity, food security, and economic growth; however, without focusing on the safety concerns and coming up with proper regulations, the consumers may not really accept the technology and its products [31]. Furthermore, it may suffer the same fate as genetically modified foods.

Apart from the European Union (EU) who in 2011 adopted a mandatory labelling regulation requiring food ingredients to be listed as "nano" if they fit with their definition of engineered nanomaterials. Few countries only have some regulations for nanotechnology; however, there is no regulation focusing on foods [32–34]. According to Sekhon [19], there is little or no EU/global legislation for regulation of nanotechnology in food. This however must change. Regulatory development for nanofoods should be carried out through a three-phase process described by Jones [35], which include (1) the use of research and development database to assess applications of nanotechnology to food and agriculture; (2) selecting particular products to assess and identify the risks and benefits; and (3) extrapolating to analyse appropriate regulatory or non-regulatory governance systems for the applications of nanotechnology in foods. Furthermore, criteria such as particular size range and measure, physical and chemical properties, processing, and safety concerns are needed to be considered for the development of the standard, definition, control measure, and regulation of nanofood [36].

According to Chau et al. [36], only a few government agencies or organizations from different countries have established standard and regulation to coordinate and control the use of nanotechnology. The US Food and Drug Administration (FDA) in 2006 formed an internal FDA Nanotechnology Task Force for determining regulatory approaches that encourage the continued development of innovative, safe, and effective FDA-regulated products that use nanotechnology materials [37]. This type of initiative and approach should be emulated by other regulatory bodies.

The Institute of Food Science and Technology in 2006 warned about the shortcomings in regulations concerning the impact of nanotechnology on food and packaging [38]. It is therefore important to develop an adequate regulatory system that addresses the definition and standard as well as regulates the labelling and safety assessment of nanoscale materials used in various food applications [36].

Valeria [39] identified three obstacles to the nanofood regulations. The first is the difficulty in determining an appropriate risk assessment for novel nanomaterials. The second is political, i.e. neoliberalism which expects states not to participate in the economy and thereby promoting deregulation, privatization, and liberalization. The third is soft regulation governance approach that does not make regulation effective [39]. Example of a soft regulation is the new EU regulation no. 2015/2283, which specifies that engineered nanomaterials require a novel food authorization before being used in food stuffs. Although it is welcome development in the nano-food regulatory process, it is yet still a soft regulation approach [40].

In concluding this section, it must be stated that present food laws and legislations are insufficient to combat the different nanorisks and control the nanofoods entering the marketplace [41–44].

6.5 Governance of Nanoagriculture and Nanofood

There is a need for risk governance of nanomaterials in the agro- and food industries. There should be proper risk–benefit assessment and risk management. It is also important that nanotechnology innovation does not proceed ahead of nanoagriculture and nanofood policy and regulatory system [45]. Different governmental and international organizations therefore have roles to play in nanotechnology governance. They must ensure that the nanofoods are produced, developed, and marketed in an atmosphere of adequate and unrestrictive regulation. A restrictive regulation discourages innovation and development, thereby negatively affecting the nanofood industry.

There is a need for an international cooperation (among states within the international organizations such as Organization for Economic Cooperation and Development, EU, African Union, World Health Organization, Food and Agricultural Organization, International Organization of Standardization [ISO], US Food and Drug Administration, and Codex) to develop a proper risk governance for nanofood and nanoagriculture. ISO's contribution towards the governance of food nanotechnology cannot be overemphasized. Current initiatives in ISO relevant for governance of nanotechnology include the technical committee on nanotechnologies (TC229) and the guidance on social responsibility (ISO 26000) [46].

Government must have political will and do away with neoliberal attitude such as in current EU policies. Regulatory frameworks based on true democratic participatory procedures (that do not think only about profit or business but also about the people, the environment, and the society large) are crucial for the development and sustainable of nanofood and nanoagriculture. Direct forms of regulation should be implemented, including mandatory labelling and the establishment of a public register of products and producers [40]. The registration of nanomaterials and development of nano-enabled product registers may also play an important role in enhancing the governance of nanoagriculture and nanofoods [47].

The need to collaboratively use both hazard- and risk-based approaches during regulations and governance has been defended by Barlow et al. [48]. In hazard-based approach, potentially harmful agent's detectable level in food is used as a basis for legislation and/or risk management action. Risk-based approaches, on the other hand, access the extent to which exposure may pose health risks to consumers [48].

Furthermore, an ongoing and worthwhile interaction between risk assessors, legislators, risk managers, risk communicators, food producers, food retailers, and the general public as consumers is needed. Such a dialogue will ensure the development of nanofood and at the same time allays consumers' fears and doubts. However, the believability of social dialogues and governance values leans on perceived authenticity of corporate social responsibility (CSR) initiatives and corporate identity communication. Morally conscious corporate identity communication and the demonstration of good corporate citizenship therefore play a key role in redressing the misconceptions surrounding the applications of nanotechnology to agribusiness.

6.6 Corporate Social Responsibility, Corporate Identity, Governance, and Nanotechnology

CSR is a business model that seeks to create value to stakeholders, through a relational approach which extends beyond legal compliance [49]. Not surprisingly, CSR has been posed as an important aspect of nanotechnology [50]. However, companies have been criticized for their inability to adapt to new standards on nanotechnology applications. It has also been observed that there is a need for companies to adopt innovative approaches to socially responsible supply chain practices, in order to foster the receptiveness of nanofoods and militate against consumers' scepticism of nanotechnology [51]. Furthermore, inadequate CSR communication and reporting have also been identified as factors that hamper the adoption of sustainable approaches to nanotechnology, which can be instrumental for advancing the shift from a "do no harm" to a "positive social force" model [52]. Since core organizational attributes are those that are central, distinctive, and enduring [53], it is apparent that the interrogation of the identity perspectives of CSR is pivotal for driving this shift. Of particular importance is consumers' receptiveness and perceived authenticity of the motives behind this shift. The foregoing section explicates how the model, as shown in Figure 6.2, demonstrates the interplay between CSR and how corporate governance can yield new insights into the identity perspectives of nanotechnology and agribusiness.

The adoption of a model that foregrounds the antecedents and aspirational notions of the identity perspectives of CSR (hereafter CSR corporate identity) is paramount for such an interrogation to drive a positive social force. While corporate identity has been defined as the impressions, image, and the personality of an organization [54], it is noteworthy that corporations are active players in a dynamic landscape of citizenship institutions [55], as they play a role in determining the configurations of corporate citizenship, which extends beyond the legal status [56]. It follows then that matters relating to corporate responsibilities and citizenships are inherently

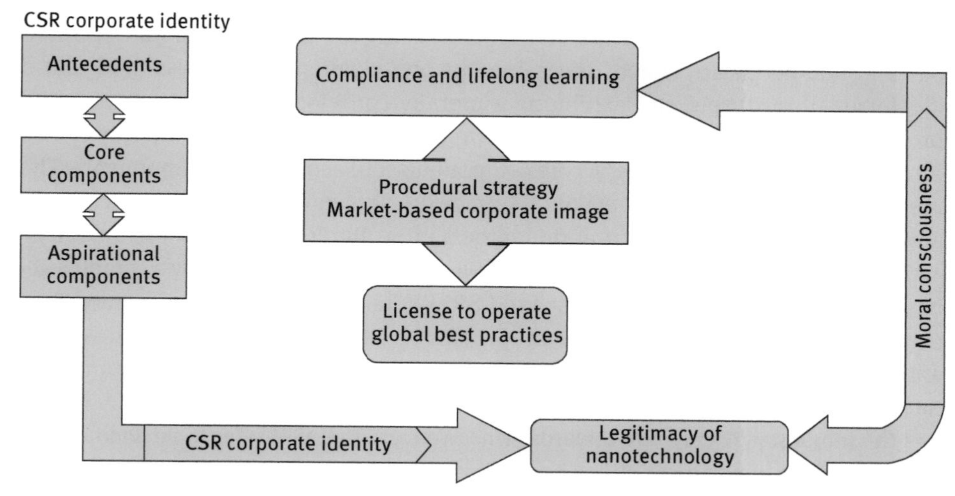

Figure 6.2: CSR Corporate Identity and the Legitimacy of Nanotechnology

contestable. Hence, there is a need for companies operating in the nanotechnology agribusiness industry to prioritize a communicative approach which incorporates the dynamism in the configurations of societal governance, in order to legitimize nanotechnology and nanofoods. Corporate responsibilities and citizenships are therefore premised on the key role corporations play in societal governance. In addition, a focused strategy to the identification of specific stakeholder issues [57] is equally instrumental for sustaining a communicative approach to legitimacy in the applications of nanotechnology.

In light of the above, we propose a model that addresses issues relating to foundational, prevailing, and aspirational notions of the identity perspectives of CSR, namely, the antecedents, core, and aspirational components of CSR corporate identity. The model as shown in Figure 6.2 illustrates the outcomes of the CSR corporate identity model of a Johannesburg Stock Exchange (JSE), Financial Times Stock Exchange (FTSE) listed food manufacturing company in South Africa, named Brand Empire. Brand Empire was one of three JSE FTSE listed food manufacturing companies which participated in a multiple case study of the implications of the antecedents and components of CSR corporate identity on corporate sustainability. Using Brand Empire as a hypothetic example of a company operating in the nanotechnology agribusiness industry, we offer scenarios of potential outcomes of the adoption of CSR corporate identity in the governance of nanotechnology and nanofoods.

The conceptual framework for CSR corporate identity, governance, and nanotechnology is presented in Figure 6.2 (adapted from Ijabadeniyi [58]).

The model demonstrates that the nature of and forces behind the antecedents of CSR corporate identity can be instrumental for the manifestations of core and aspirational corporate governance values. The implications of the antecedents and components of CSR corporate identity for enhancing the legitimacy of nanotechnology and nanofoods are largely determined by organizational adeptness to localized ideologies of nurturing and exhibiting a morally conscious CSR corporate identity profile. As shown in Figure 6.2, the core components of CSR corporate identity can be further divided into two, namely, core components and aspirational components. The multidimensionality of the conceptual framework shown in Figure 6.2 offers a novel and operational assessment criterion for the nanotechnology agribusiness industry.

The model shows that Brand Empire's CSR engagements are driven by a compliance-based and lifelong learning approach to CSR. While keeping abreast with prevailing standards on nanotechnology has been advocated in literature [51], a compliance-based approach to CSR is not sufficient enough to foster the legitimacy of the applications of nanotechnology in agribusiness. Furthermore, a compliance-based approach to CSR has the tendency to gravitate towards what Weaver et al. [59] refer to as decoupling, a phenomenon which exemplifies how companies gear conscientious efforts towards improving corporate image and reputation by developing coherent organizational identities through corporate rhetoric.

Drawing from the study conducted by Surroca et al. [60], CSR engagements can be driven by intangible resources such as innovation and human capital, we propose that the notion of a compliance-based approach to CSR, which has gained widespread popularity in theory and practice [57, 61], should normatively be driven by a morally conscious lifelong learning culture. It is on this note that moral consciousness is incorporated into the model, as shown in Figure 6.2. Nevertheless, a lifelong learning culture which is driven by a compliance-based approach to CSR may have deleterious effects on the legitimacy of the applications of nanotechnology in agribusiness, on the account that such an approach can be too legalistic to drive a morally conscious culture. Therefore, there is evidence suggesting that companies that are driven by a compliance-based culture may not be as actively involved in CSR, in the absence of industry charters and reporting standards. This phenomenon therefore calls for a balanced approach to CSR as advocated by Carroll [62], which, as a consequence, fosters good corporate governance values and the legitimacy of the applications of nanotechnology in agribusiness.

Similarly, the core components of Brand Empire's CSR corporate identity are premised on a procedural strategy, which draws from the company's antecedent of a compliance and lifelong learning-based approach to CSR. While such an approach can yield reputational benefits, there is evidence suggesting that it could be perceived as a promotional approach to CSR. In addition, the aspirational components of Brand Empire's CSR corporate identity relate to the acquisition of the license to operate and conformity to global best practices. The overarching role of the antecedents of CSR corporate identity is further demonstrated in the company's aspirations.

While the adoption of aspirational benchmarking strategies is commendable, caution is advised when adapting global best practices to local strategies given that CSR is a socially constructed concept [63]. We also propose that a CSR model which justifies a company's legitimate ownership of the license to operate can foster consumers' receptiveness and perceived authenticity of the shift from a do no harm to a positive social force, in the applications of nanotechnology in agribusiness. Moral consciousness, as shown in Figure 6.2, plays an overarching role in the operationalization of the model. The model explicates that CSR corporate identity communication and the ability to nurture and foster good corporate governance values play a vital role in fostering consumers' receptiveness and loyalty towards nanotechnology and nanofoods.

6.7 Conclusion

Application of nanotechnology in the food industry coupled with responsible governance may result in agrivalue chain that is ethical and sustainable with potential to enhance food security. For this to be achieved, nanoagriculture and nanofoods must be protected from nanomaterials that may cause harm to human health. Government in collaboration with international organizations and relevant stakeholders should develop nanotechnology regulatory frameworks that focus not only on profit but also on the people, the environment, and the society large. Also, there is a need for researchers, developers, and institutions to foster transparency by institutionalizing morally conscious CSR approaches, CSR approaches, in order to redress the negative antecedents of the historical applications and communications of nanotechnology or other emerging technologies.

References

1. NRC (National Research Council). A matter of size: triennial review of the national nanotechnology initiative. Washington, DC: National Academies Press, 2006.
2. ISO TC 229. Draft standard on nanotechnologies- terminology and definitions for nanoparticles. 2008. Available at: http://www.iso.org/iso/iso_catalogue/catalogue_tc/catalogue_detail.htm?csnumber=44278. Accessed: 17 June 2011.
3. FSAI. The Relevance for food safety of applications of nanotechnology in the food and feed industries. Edited by Food Safety Authority of Ireland Abbey Court, Dublin. 2008, 82pp.
4. European Commission. Commission of the European Communities preparing for our future: developing a common strategy for key enabling technologies in the EU. COM/30.09.2009/512 final, Brussels, 2009.
5. Salamanca-Buentello F, Persad DL, Court EB, Martin DK, Daar AS, Singer PA. Nanotechnology and the developing world. PLoS Med 2005;2(5):e97. Available at: https://doi.org/10.1371/journal.pmed.0020097. Accessed: 31 October 2017.
6. Bartolucci C. Nanotechnologies for agriculture and foods: past and future. In: Monique Axelos AV, Van de Voorde Marcel, editors. Nanotechnology in agriculture and food science. Germany: WILEY-VCH, 2017:3–14.

7. Shea CM, Grinde R, Elmslie B. Nanotechnology as general-purpose technology: empirical and implications. Technol Anal Strat Manage 2011;23:175–92.
8. Pustovalov VK, Babenko VA. Optical properties of gold nanoparticles at laser radiation wavelengths for laser application in nanotechnology and medicine. Laser Phys Lett 2004. Available at: http://iopscience.iop.org/article/10.1002/lapl.200410111/meta. Accessed: 30 October 2017.
9. Xiaolei Q, Pedro JA, Qilin L. Applications of nanotechnology in water and waste water treatments. Water Res 2013;47:3931–46.
10. Davis SS. Biomedical applications of nanotechnology – implications for drug targeting and gene therapy. Trends Biotechnol 1997;15:217–24.
11. Mitra SB, Wu D, Holmes BN. An application of nanotechnology in advanced dental materials. J Am Dent Ass 2003;134:1382–90.
12. Wong YW, Yuen EW, Leung MY, Ku SK, Lam HL Selected applications of nanotechnology in textile. AUTEX Res J 2006;6:1473–81.
13. Smith DM, Simon JK, Baker JR. Applications of nanotechnology for immunology. Nat Rev Immunol 2013;13:592–605.
14. Jain KK. Nanodiagnostics: application of nanotechnology in molecular diagnostics. Exp Rev Mol Diagn 2014;3:153–61.
15. Konsta-Gdoutos MS, Metaxa ZS, Shah SP. Highly dispersed carbon nanotube reinforced cement based materials. Cem Concr Res 2010;40:1052–9.
16. Hughes MP. AC Electrokinetics: applications for nanotechnology. Nanotechnology 2000;11. Available at: http://iopscience.iop.org/article/10.1088/0957-4484/11/2/314/meta. Accessed: 30 October 2017.
17. Garber C. Nanotechnology food coming to a fridge near you. 2007. Available at: http://www.nanowerk.com/spotlight/spotid=1360.php. Accessed: 11 June 2011.
18. Scrinis G. Nanotechnology: transforming food and the environment. 2010. Available at: http://www.foodfirst.org/en/node/2862. Accessed: 15 June 2011.
19. Sekhon BS. Food nanotechnology – an overview. Nanotechnol Sci Appl 2010;3:1–15.
20. Houllier F. Foreword. In: Nanotechnology in agriculture and food science. Axelos Monique AV, Van de Voorde Marcel. Germany: WILEY-VCH, 2017:XXI–XXVI.
21. Perch H. How is Nanotechnology being used in Food Science? 2007. Available at: http://www.understandingnano.com/food.html. Accessed: 11 June 2011.
22. Sozer N, Kokini JL. Nanotechnology and its applications in the food sector. Trends Biotechnol 2009;27:82–9.
23. Emamifar A, Kadivar M, Shahedi M, Zad-Soleimanian S. Effect of nanocomposite packaging containing Ag and ZnO on inactivation of Lactobacillus plantarum in orange juice. Food Control 2011;22:408–13.
24. Bhattacharya A, Datta PS, Chandhmi P, Barik BR. Nanotechnology – a new frontier for food security in socio economic development. 2011. Available at: http://disasterresearch.net/drvc2011/paper/fullpaper_22.pdf. Accessed: 15 June 2011.
25. Lilie M, Anna Cantini A. Nanotechnology in agriculture and food processing. Conference proceedings, University of Pittsburgh, Eleventh Annual Freshman Conference, 9 April 2001, 1–9
26. Das M, Saxena N, Dwivedi PD. Emerging trends of nanoparticles application in food technology: Safety paradigms. Nanotoxicology 2009;3:10–18.
27. Dasguta N, Ranjan S, Mundekkad D, Ramalingam C, Shanker R, Ashutosh K. Nanotechnology in agro-food: from field to plate. Food Res Int 2015;69:381–400.
28. Sastry RK, Rashmi HB, Rao NH. Nanotechnology for enhancing food security in India. Food Pol 2011;36:391–400.

29. Nanowerk. What's happening with nanofoods?, 2012. Available at: https://www.nanowerk.com/spotlight/spotid=24155.php. Accessed: 28 November 2017.
30. Handford CE, Dean M, Spence M, Henchion M, Elliott CT, Campbell K. Awareness and attitudes towards the emerging use of nanotechnology in the agri-food sector. Food Control 2015;57:24–34.
31. Handford CE, Dean M, Spence M, Henchion M, Elliott CT, Campbell K. Implications of nanotechnology for the agri-food industry: opportunities, benefits and risks. Trends Food Sci Technol 2014;40:226–41.
32. European Parliament and the Council. Regulation (EU) No 1169/2011 of 25 October 2011. Official Journal of the European Union. L304/18, 2011. Available at: http://faolex.fao.org/docs/pdf/den108120.pdf. Accessed: 27 May 2016.
33. Gruere GP. Implications of nanotechnology growth in food and agriculture in OECD countries. Food Policy 2012;37:191–8.
34. Marrani D. Nanotechnologies and novel foods in European law. Nanoethics 2013;7:177–88.
35. Jones PB. ISB News Report. A nanotech revolution in agriculture and the food industry. 2006. Available at: http: www.isb.vt.edu/news/2006/artspdf/jun0605.pdf. Accessed: 13 April 2016.
36. Chau C, Wu S, Yen G. The development of regulations for food nanotechnology. Trends Food Sci Technol 2007;18:269–80
37. FDA. FDA forms internal nanotechnology task force. 2006. Available at: http://www.fda.gov/bbs/topics/NEWS/2006/NEW01426.html. Accessed: 30 October 2017.
38. IFST. Institute of Food Science and Technology Nanotechnology. 2006. Available at: http://www.ifst.org/uploadedfiles/cms/store/ATTACHMENTS/Nanotechnology.pdf. Accessed: 30 October 2017.
39. Valeria S. European Union tackling nanofood risk: current situation and open issues. Bucharest 2016;17:199–204 .
40. Sodano V, Gorgitano MT, Quaglietta M, Verneau F. Regulating food nanotechnologies in the European Union: open issues and political challenges. Trends Food Sci Technol 2016;54:216–66.
41. Sodano V, Hingley MK. Conflicting interests and regulatory systems of new food technologies: The case of nanotechnology. In: Lindgreen A, Hingley MK, Angell RJ, Memery J, Vanhamme J, editors. A stakeholder approach to managing food local, national, and global issues. United Kingdom: Routledge, 2016.
42. Ehnart T. The legitimacy of new risk governance – a critical review in light of the EU's approach to nanotechnologies in food. Eur Law J 2015;21:44–67.
43. Salvi L. EU's soft reaction to nanotechnology regulation in the food sector. Eur Food Feed Law Rev 2015;3:186–93
44. Sodano V. Regulating food nanotechnologies: ethical and political challenges. In: Dumitras DE, Jitea IM, Aerts S, editors. Know your food. Netherlands: Wageningen Academic Publishers, 2015:36–41.
45. Renn O, Roco MC. Nanotechnology and the need for risk governance. J Nanopart Res 2006;8:153–91.
46. Forsberg E. The role of ISO in the governance of nanotechnology. 2010. Available at: TheroleofISOinthegovernanceofnanotechnologyForsberg.pdf. Accessed: 31 October 2017.
47. Munir AB, Karim ME, Yasin SH, Muhammad-Sukki F. Registration of nanomaterials and nano-enabled products: solution in regulation and governance or new challenge. In the proceedings of the international seminar on nanoscience and nanotechnology (NANOSciTech2016), 26–29 February 2016, Shah Alam, Malaysia. Shah Alam: NANO-SciTech Centre, pages 10–11. https://openair.rgu.ac.uk/handle/10059/1744. Accessed: 28 October 2017.
48. Barlow SM, Boobis AR; Bridges J, Cockburn A, Dekant W, Hepburn P, Houben GF, Konig J, Nauta MJ, Scheurmans J. The role of hazard-and risk-based approaches in ensuring food safety. Trends Food Sci Technol 2015;46:176–88.

49. UN Global Compact. Global compact international yearbook. Germany: United Nations Publications, 2014.
50. Kuzma J, Kuzhabekova A. Corporate social responsibility for nanotechnology oversight. Med Health Care Philos 2011;14(4):407.
51. Pelle S, Reber B. Responsible innovation in the light of moral responsibility. J Chain Network Sci 2015;15(2):107–17. Available at: https://doi.org/10.3920/JCNS2014.x017. Accessed: 10 October 2017.
52. Groves C, Frater L, Lee R, Stokes E. Is there room at the bottom for CSR? Corporate social responsibility and nanotechnology in the UK. J Bus Ethics 2011;101(4):525–52.
53. Albert S, Whetten DA. Organizational identity. Greenwich, CT: JAI Press, 1985.
54. West D. Strategic marketing: creating competitive advantage, 2nd ed. United Kingdom: Content Technologies Publications, 2014.
55. Crane A, Matten D. The emergence of corporate citizenship: historical development and alternative perspectives. In: Scherer AG, Palazzo G, editors. Handbook of research on global corporate citizenship. Cheltenham: Edward Elgar, 2008:25–49.
56. Isin EF, Turner BS. Citizenship studies: an Introduction. In: Isin EF, Turner BS, editors. Handbook of citizenship studies. London: SAGE, 2002:1–10.
57. Kleyn N, Abratt R, Chipp K, Goldman M. Building a strong corporate ethical identity. Calif Manage Rev 2012;54(3):61–76. DOI:10.1525/cmr.2012.54.3.6. Accessed: 10 May 2015.
58. Ijabadeniyi A. 2018. Exploring corporate marketing optimisation strategies for the KwaZulu-Natal manufacturing sector: a corporate social responsibility perspective. PhD Thesis, Durban University of Technology.
59. Weaver GR, Trevino LK, Cochran PL. Integrated and decoupled corporate social performance: Management commitments, external pressures, and corporate ethics practices. Acad Manage J 1999;42(5):539–52. Available at: https://doi.org/10.2307/256975. Accessed 10 August 2016.
60. Surroca J, Tribó JA, Waddock S. Corporate responsibility and financial performance: The role of intangible resources. Strat Manage J 2010;31(5):463–90.
61. Smit A. What hinders successful CSR implementation in African companies? 2010. Available at: https://www.google.co.za/webhp?hl=en&sa=X&ved=0ahUKEwjJ5bzo0NfRAhVGOJoKHb7tB g0QPAgD#hl=en&q=what+hinders+successful+csr+implementation+in+African+ companies)&spf=1500986202637. Accessed: 25 July 2017.
62. Carroll AB. Carroll's pyramid of CSR: taking another look. Int J Corporate Social Responsibility 2016;1(3):1–8. Available at: https://doi.org/10.1186/s40991-016-0004-6. Accessed: 5 December 2016.
63. Dartey-Baah K, Amponsah-Tawiah K. Exploring the limits of Western corporate social responsibility theories in Africa. Int J Bus Soc Sci 2011;2(18):126–37. Available at: ijbssnet.com/ journals/Vol_2_No_18_October_2011/18.pdf. Accessed 14 March 2016.

Martina C. Meinke, Jürgen Lademann, Fanny Knorr,
Alexa Patzelt and Silke B. Lohan

7 Nanocosmetics

7.1 Introduction

What are nanoparticles (NPs) and why are they used in cosmetics? These questions are addressed by highlighting recent developments in the field of nanotechnology that have increased the potentially skin exposure to NPs. Trends emerging from recent literature suggest that the positive benefits of engineered NPs for use in cosmetics outweigh potential toxicity concerns. This chapter begins with an overview about the nanostructures that are applied in cosmetics and why. It is followed by a discussion of the current state of understanding of the skin penetration of NPs, the methods that can study the distribution, and the risk factors induced by the nanostructures. The chapter focuses on NPs used in formulation.

7.2 What Are Nanocosmetics?

7.2.1 Where and Why Are They Used?

Strictly speaking, NPs are only 1–100 nm in size. Thus, they are 1,000 times smaller than the diameter of a human hair. The name "nano" is derived from the Greek word "nanós" – dwarf. Their small size lends them special properties. For this reason, NPs are already widely used in numerous industries such as electronics, chemistry, medicine, and cosmetics.

In many cases, cosmetic products contain nanosized (50–5,000 nm) components such as nanoemulsions, nanocapsules, nanosomes, niosomes, or liposomes that comprise traditional cosmetic materials [1]. Several sunscreens, deodorants, and toothpastes contain NPs. The small size is responsible for specific characteristics, which can affect their interactions with the skin [2]. The reasons why NPs are used are diverse. Nanosized paint particles, for example, ensure long-lasting effects in eyeliner pencils and mascara. They can function as reflective agents in sunscreens, as depots or penetration enhancers of actives in creams, or they protect the actives or colors against degradation. Silver NPs have an antibacterial or preservative function and are used in deodorants and hand creams. Furthermore, NPs may be used to impart stability to formulations that contain ingredients, which can decompose as a result of oxidation or other causes [3]. For this application, liposomes or cellulose can be used [4, 5]. This is important if the actives are instable

https://doi.org/10.1515/9783110547221-007

due to the biological environment or if they should be photoprotected. One new and challenging aspect is the penetration enhancement by NPs. The main reason to use NPs is the fact that most of the actives do not penetrate the *stratum corneum* (*SC*) barrier. Only 0.1–1% can pass this barrier if the molecular mass is below 500 Dalton. Penetration enhancers including chemicals such as dimethyl sulfoxide and ethanol and mechanical perforation such as microneedles greatly disturb the *SC*. For most NPs, the mechanisms of active penetration enhancement are not yet clear. Nevertheless, it could be shown that several carriers significantly enhance active delivery [6–10].

Several skin penetration pathways for topically applied substances are known to date: Diffusion through the intercellular substance of the *SC*, the transcellular route involving the corneocytes, and penetration along the adnexae of the skin, in particular the hair follicles (HFs). In most cases, topically applied drugs penetrate into the upper corneocyte layers of the *SC*, which are subject to rapid desquamation and thereby depletion of the drug reservoir. Recently, studies have shown that the penetration of drugs and specific skin care products into HFs can also be effectively exploited for the localized treatment of skin diseases with sustained drug release and reduced drug side effects. Modern concepts in the field of targeted pharmacotherapy are based on the use of NPs, because these particle systems show a tendency to aggregate in HFs. Although NPs preferentially accumulate in HFs compared to nonparticulate substances, an additional massage during application can enhance and improve the follicular penetration [11]. The HFs display a long-term reservoir for NPs, and only hair growth, sebum production, or degradation of the NPs will result in depletion of the reservoir [12]. The transfollicular absorption of small molecules diffusing from NPs into the blood vessels which surround the HFs has also been reported [13, 14]. The risk of systemic side effects caused by NPs themselves is minimal, as tight junctions are present in HFs [15], preventing their penetration into cells of the viable skin [16]. Patient compliance may be increased by reducing the frequency of drug applications necessary. For follicular penetration, versatile applications of NPs are possible: Depending on the question of interest, NPs with modified surfaces and various chemical and physical properties can be used to promote an active targeting of selected cell populations. But not only HF-associated diseases can be addressed, also gene- or immunotherapy can be performed [17].

7.2.2 Types of NP in Dermal Formulations

NP systems are developed mainly for the delivery of drugs to target structures where a controlled release at the site of action can take place as well as for the protection of active ingredients. Besides the NP size, the selection of the vehicle itself has a strong effect on the dermal/follicular penetration and not every particle system can be used for every treatment goal.

7.2.2.1 Solid Particles

Solid NPs such as titanium dioxide (TiO_2) and zinc oxide (ZnO) act as physical ultraviolet (UV) filters in sunscreens. They absorb UV and scatter the light in the whole spectral range of the sunlight. For TiO_2, this occurs most effectively at a size range of 60–120 nm [18], whereby ZnO is generally used in the form of particles sized 30–200 nm. Inert coating materials including aluminum oxide or silicon oils are often used to treat these NPs [1]. The scattering effect decreases with increasing wavelength and also in the near-infrared region, a protection effect could be determined [19]. In the UV range, synergistic effects could be observed if both chemical (soluble) UV filter and particle (nonsoluble) filters were used together [20]. Some sunscreens, particularly those for children, contain only particle filters because they do not cause cutaneous adverse health effects. To provide a convenient application, the number of particles is limited depending on the size of the added NPs to about 10%. The particle size of the metal oxide in sunscreens is about 40 nm.

7.2.2.2 Nonsolid Particles

Nonsolid NPs are used to transport and/or stabilize actives in the skin. There are many different NPs that can act in different ways.

Liposomes are globular vesicles sized 25–5,000 nm [1] that have been widely used in cosmetics for several years with the purpose of increasing the delivery of ingredients into the skin. They are concentric bilayered particulates that enclose an aqueous volume via a lipid bilayer that is comprised of natural or synthetic phospholipids. These are mostly regarded as being safe. Other vesicular particulates including invasomes [21], transferosomes [22, 23], niosomes [24, 25], and ethosomes [26] are also implemented for enhancing skin penetration. Cevc et al. [27] demonstrated that transferosomes (flexible liposomes) were more effective than rigid liposomes in the delivery of ingredients through the skin. It is presumed that their ability to change shape as they pass through small spaces in the skin may be beneficial in their ability to promote absorption [28].

Nanoemulsions are used in cosmetic products such as conditioners or lotions. Traditional cosmetic ingredients such as water, oils, and surfactants are combined in a two-phase system in which droplets sized 50–100 nm are dispersed in an external aqueous phase [1]. Their smaller particle size provides higher stability and better suitability to carry actives [29], and renders nanoemulsions transparent and smooth to the touch [29].

Nanocapsules are vesicles in which a drug is encapsulated in a cavity of an inner liquid core that is surrounded by a polymeric membrane [30]. In the cavity, the drug can be in liquid or solid form or as a molecular dispersion [31]. The drugs can also be bound onto their surfaces or within the polymeric membrane [32]. Hwang et al. [33] reported that the use of nanocapsules decreased the penetration of UV filter octyl methoxycinnamate in porcine skin when compared to conventional emulsions.

Solid lipid NPs are comprised of a solid lipid core that can solubilize lipophilic molecules. The core is stabilized by surfactants. Nanostructured lipid carriers are prepared with a mixture of solid and liquid lipids [34]. They both have the advantage of high stability. On the skin surface, such lipid NPs form a lipid nanolayer film, resulting in less evaporation of water and thereby increased skin hydration. Encapsulated drugs are protected from degradation. As they enhance the penetration of active substances into the *SC*, they are used for the sustained delivery of cosmetics to the skin [35].

Nanocrystals are NPs with a crystalline character composed completely of drug. There is no carrier material such as in polymeric NPs. The increased solubility and dissolution velocity lead to an acceptable bioavailability. Furthermore, nanocrystal technology enables formulations to be developed without the use of surfactants. They enable a fast onset of action, as the drug is absorbed quickly due to their fast dissolution [36].

Dendrimers and other polymers are used as a method of drug delivery into the skin. Polymers are relatively small and can be soluble or insoluble particles depending on their structure. Soluble polymer NPs are stable and do not decompose on the skin surface. Chauhan et al. [37] found enhanced skin absorption of indomethacin by the addition of polyamidoamine dendrimers to aqueous formulations administered to rat skin. Core multishell nanocarriers consist of a hydrophilic core, an intermediate lipophilic shell, and an outer hydrophilic shell, enabling them to incorporate hydrophilic and lipophilic substances [38, 39]. Dendritic core multishell neurotransmitter NPs increased the penetration of Nile red dye into the *SC* and viable epidermis considerably in comparison to the cream formulation [40].

Cubosomes are naturally occurring capsules that can distribute medicinal active ingredients or even nutrients throughout the human body. In general, these small capsules are formed from lipid molecules and water and can contain active substances. These NPs share similarities with liposomes. Both liposomes and cubosomes are composed of fat molecules (lipids), but the former are spherical, while the latter have a cubic structure [41]. During the chemical synthesis, they are formed by the self-assembly of liquid crystalline particles of certain surfactants when mixed with water and a microstructure at a certain ratio. Cubosomes offer a large surface area, low viscosity, and can exist at almost any dilution level. They have high heat stability and are capable of carrying hydrophilic and hydrophobic molecules [42]. The cubosomes show the potential for controlled drug release through functionalization and thus display an attractive choice for cosmetic applications as well as drug delivery processes.

7.3 Where Are We?

While NPs have been used in cosmetics for several years, their presence in a product must be labeled on the package since 2012 (Regulation [EC] No. 1223/2009 of the

European Parliament and the Council of 30 November 2009 on Cosmetic Products L 342/59). This has led to an uncertainty among the consumers. Several methods were developed to answer the main relevant questions: Do NPs really make the product better? Is the product safe? What are the risks of NPs particularly when used in cosmetic formulations?

First, we will focus on the techniques to study the efficiency of NPs. This will be followed by a risk assessment and a description of the methods that can be applied to estimate the safety of the used NPs and cosmetic product.

7.3.1 Techniques to Determine NP Distribution and Effects

The penetration of NPs can be investigated using several in vitro methods on membranes or reconstructed human skin, ex vivo on excised human or porcine skin, and in vivo on mice or, if all cosmetic ingredients are approved to be applicable, also on human volunteers. Franz cell experiments are used to investigate the permeation, because the receptor fluid can be analyzed with various analytical methods such as high-performance liquid chromatography (HPLC), mass spectroscopy, and high-performance thin-layer chromatography (HPTLC) depending on the substance. If the NPs are found in the receptor fluid, the viable cells can potentially be reached and a further risk assessment is necessary. Often no permeation is found, especially if the NPs are designed to remain on the surface.

In order to study dermal penetration, resorption, and metabolic processes of active substances, dermal microdialysis can be used. This method is based on diffusion processes that take place through a semipermeable membrane, which is inserted into the dermis via a guide needle. It can expand over a distance of one to several centimeters. A carrier solution (perfusate) is perfused at a constant flow rate through the tissue, which is maintained by a microdialysis pump, allowing the collection of the interstitial fluid from excised skin as well as from humans. Due to the concentration gradient between the interstitial fluid and the perfusate, endogenous molecules will diffuse through the probe membrane into the perfusate, which thus becomes the dialysate. Finally, the obtained dialysate will be analyzed concerning the molecules of interest [43, 44]. In the field of dermatology, the microdialysis technique is currently used mainly in research regarding the skin barrier function, penetration and dermal metabolism of drugs and chemical agents, as well as the release of mediators in inflammation [45].

Radioactive-labeled NPs are predominantly applied in the medical and pharmaceutical fields for studying therapeutic and toxicological effects as well as for the visualization of possible penetration pathways. The radioactive labeling of the outer shell of NPs can be achieved by using radioactively labeled biomolecules, such as antibodies, which are covalently linked to the surface of the NPs. But isotopes are also often used, for example gold isotopes [46]. A dual labeling of the core and the outer shell of NPs enables the stability of such a construct as well as the distribution within the body to be studied. After the injection of radioactive gold NPs into mice, it could be demonstrated that the supposed stable NP conjugates changed their original structure and thus their properties within the body and an accumulation in different organs could be visualized [47]. Such studies are very helpful and thus have a strong influence on the medical applications of NPs and their risk assessment.

Laser scanning microscopy (LSM) can be used if the particle and/or the active is fluorescent or fluorescently labeled. It can be applied ex vivo on a biopsy or on cryosections or in vivo to visualize the penetration of the carrier or the active (Figure 7.1). Penetration depths into HF can be estimated. The lateral resolution of the method is 1–5 μm. Patzelt et al. [48] demonstrated that NPs with a size of approximately 650 nm penetrated the deepest. The fluorescence technique is very sensitive. Fluorescence lifetime imaging can further improve the findings because it monitors the microenvironment and if the active is fluorescent it can distinguish different surroundings [49].

Multiphoton microscopy (MPM) can also be applied ex vivo and in vivo. It offers to track second harmonic generation signals from order structures such as ZnOs. Darvin et al. [50] could prove that ZnO did not penetrate the *SC* as a whole compact particle. MPM has a high resolution of 0.5 μm lateral and 1 μm vertical.

Confocal Raman spectroscopy can be applied ex vivo and in vivo if the chemical contrast between the particle and the skin is sufficient and the concentration is high enough [51]. The advantage here is that it is a label-free method, and no alterations are necessary. Related highly sophisticated methods have been developed such as stimulated Raman spectroscopy, which are much more sensitive [52].

Electron paramagnetic resonance (EPR) spectroscopy can be used to quantify penetrated actives or carriers if they provide paramagnetic structures. This can be

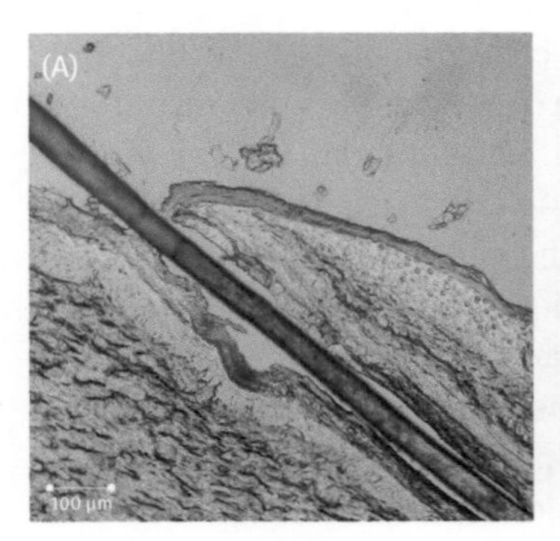

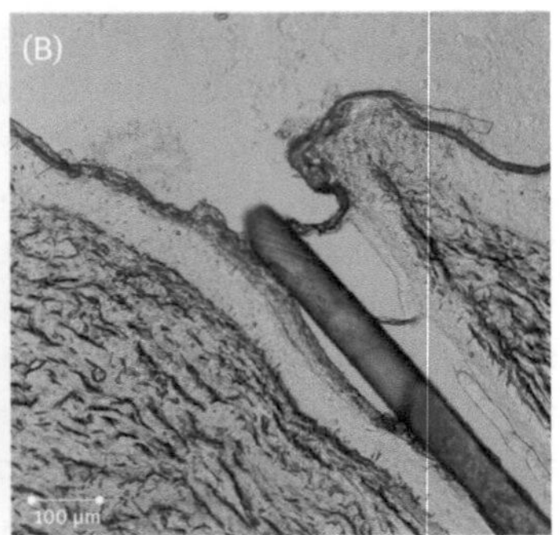

Figure 7.1: Laser scanning images of porcine ear skin sections after the penetration of the fluorescent dye, Nile red, incorporated into NP (A) and a cream (B). The penetration into the HF is clearly visible if applied in NP. Furthermore, the transfollicular penetration of the dye into the living epidermis is facilitated.

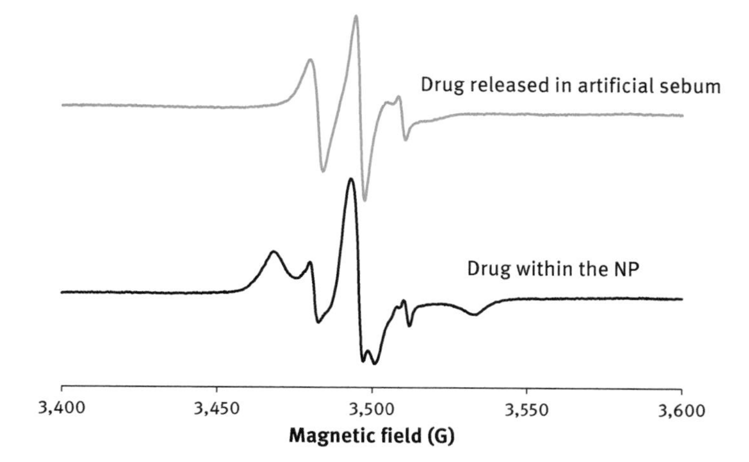

Figure 7.2: EPR spectra of spin-labeled drug before and after the release from a NP which is dispersed in an aqueous solution. The shape of the spectrum changes from a strong immobilized spectrum within the NP with little amount of free mobile drug in the aqueous solution to a moderate immobilized nitroxid spectrum within artificial sebum.

reached by labeling the substances with spin probes, which have the advantage to be small compared to fluorescence labels. The shape of the EPR signal gives information about the microenvironment and a possible release can be easily monitored (Figure 7.2). EPR is mostly used as an integrative method because EPR can easily give quantitative values of the EPR active substances, and it can be combined with the imaging methods that do not provide quantitative data [6, 7, 39, 53].

7.3.2 Risk Assessment

The small size of NPs alters their physicochemical properties and may result in increased uptake and interactions with biological tissues, including the production of reactive oxygen species (ROS) such as free radicals. These can lead to oxidative stress, inflammation, as well as protein and DNA damage [54]. Carbon nanotubes have been shown to cause kidney cell death and to inhibit further cell growth [55]. About 20 nm TiO_2-NPs are capable of causing complete destruction of super-coiled DNA, even at low doses and in the absence of exposure to UV [56].

Some of these effects are desired under controlled conditions such as the application of silver NPs for disinfection. The main question is: Do NPs stay on the surface within the *SC* or could they pass the barrier? The second question is the stability of the NPs. If they do not pass the barrier as the whole particle, do they metabolize, or decay? If particle fractions penetrate the *SC*, are they toxic? Therefore, the ability of penetration must be investigated. This is dependent on the size of the particle. The stability in a biologically relevant environment must be studied and the toxicity to cells should be investigated because the formulation could be applied on lesions where the penetration is facilitated.

7.3.3 Evaluation of the Toxic Potential In Vitro

For the evaluation of the toxic potential of such nanomaterials, a comprehensive characterization of their physicochemical properties as well as their possible interference with various in vitro test systems is necessary (Figure 7.3). In comparison to larger chemical constructs, nanomaterials show a largely enhanced ratio of surface to volume, whereby they are much more reactive and can interact more strongly with biological systems and their components. The toxic potential of nanomaterials can differ depending on their underlying basic material. However, the same chemical composition can also cause different cellular responses. Furthermore, aggregation/ agglomeration, the shape, and surface charges can have a strong influence on biological systems [57–59].

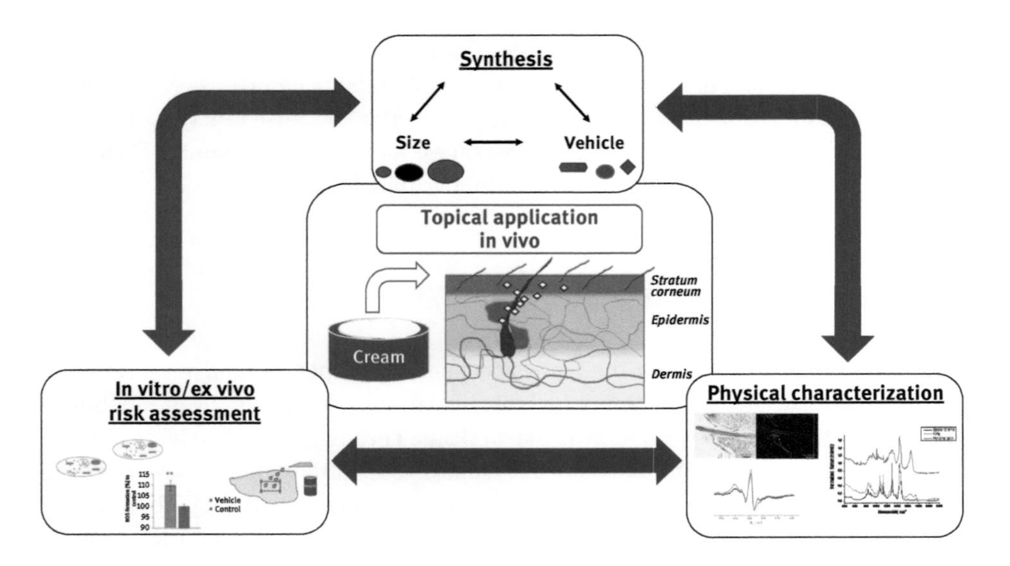

Figure 7.3: Nanocosmetics – from synthesis to application. Before cosmetic products containing NPs are allowed to be administered in vivo, a physicochemical characterization and investigations of the toxicity in in vitro and ex vivo systems have to be performed. Particle size and vehicle shape of the NPs depend on the field of application, their toxicity potential, and physical and chemical properties. For the topical application on skin, the HFs display one of the most important pathways. The illustrations are partly taken from Refs. [6, 39, 60].

7.3.3.1 Physicochemical Characterization

A physical characterization of nanomaterial is essential and should be performed in the crystalline (powder) and in the dissolved form. The particle size and shape can be examined with transmission electron microscopy. X-ray diffraction is a suitable method for the analysis of crystallinity and crystal modification [61]. The electrical potential of the surface (zeta potential), the solubility, aggregation, and agglomeration potential can be investigated for aqueous solutions and the respective test medium by electrophoreses and dynamic light scattering measurements [62]. Surface modifications (protein corona) can be identified by X-ray photoelectron spectroscopy, energy-dispersive X-ray spectroscopy, and/or mass spectrometry [59, 63]. Furthermore, structure and stability of soft NPs can be investigated using EPR spectroscopy. Therefore, part of the substances are labeled or marker substances are added and the spectra are measured over the desired time frame [4, 53]. Comparable structure analysis can be performed using fluorescence lifetime microscopy [64].

7.3.3.2 Testing of Possible Interference of NPs with Various In Vitro Test Systems

Currently, the in vitro methods for the determination of the toxic potential of NPs are based on standard procedures, which have been established for conventional chemical compounds. The potential damaging effects are tested in a cell culture model system on cell lines that correspond to a possible in vivo exposure. For example, pulmonary epithelial cells are researched if the substance of interest can be absorbed via the inhalation route; skin cells are used as test material if the substance is applied on the skin [65, 66].

As a parameter for cytotoxic investigations and biocompatibility, NPs are analyzed by a cell viability assay, whereby a tetrazolium salt-based assay (MTT (3-(4,5-dimethylthiazol-2-yl)-2,5-diphenyltetrazolium bromide)/XTT ((2,3-bis-(2-methoxy-4-nitro-5-sulfophenyl)-2H-tetrazolium-5-carboxanilide)) is often used [67]. The stimulation of cellular apoptosis or necrosis is determined by the markers Annexin V, caspase-3, or lactate hydrogenase [68]. Changes in the membrane integrity are achieved using neutral red [69], trypan blue, or propidium iodide [70].

NPs are frequently used for drug delivery processes. But several studies have shown that NPs may promote the development of free radicals, especially ROS [71]. Different detection methods for oxidative stress development in cells are available, ranging from the determination of the total ROS concentration in tissue/cell culture (dichlorofluorescein [DCF] assay) [72] to the specific detection of oxidation products of various cell components, such as DNA components (Comet assay) [73]. The DCF assay is a highly sensitive indicator for oxidative stress in cells and thus mainly used for ROS detection [72]. The Comet assay is a technique of gel electrophoresis, which enables the detection of DNA damage in individual cells. DNA double-strand breaks as well as DNA single-strand breaks can be detected [74]. The EPR spectroscopy represents an alternative noninvasive method for the detection of radical formation in cells and tissue (in vivo, ex vivo, and in vitro) [75–77]. The main advantage of the EPR spectroscopy to fluorescence-based assays is that possible artifacts of particle systems will be overcome and the measurements cannot be affected by light scattering properties of the analyzed material, minimizing misinterpretations and incorrect results [78, 79]. EPR represents a more sensitive method versus the DCF assay, which could be confirmed by the treatment of secondary keratinocytes with silver NPs [79]. For the investigation of ROS development, tissue or a cell culture will be treated with the EPR probe TEMPO (2,2,6,6-tetramethylpiperidine-1-oxyl). Its amphiphilic character enables the uptake into cells, allowing the direct interaction with metabolic radicals and antioxidants. A decrease of the EPR signal of TEMPO after exposure to stress over time enables statements concerning the ROS formation, whereby it could be not excluded that the spin probe also interacts with endogenous antioxidants [75, 80, 81]. To analyze the endogenous redox system of a cell, glutathione (GSH) represents one of the main components. The determination of the ratio between reduced and oxidized GSH gives a good insight into the endogenous redox system and thus to the oxidative stress potential of applied nanomaterial [82, 83].

Enzyme-linked immunosorbent assay studies are very useful for detecting inflammatory biomarkers in cell culture. To estimate cell inflammation, the chemokines interleukin (IL)-8, tumor necrosis factor α, and IL-6 are useful biomarkers [84]. Especially for these studies, but also in general, a sterile production of NP is of great importance in order to prevent a change in the level of inflammatory markers by bacteria or other endotoxins. By EPR spectroscopy, it could be demonstrated that primary and secondary keratinocytes do have different thresholds for oxidative stress development after incubation with nanomaterials of different origin. In addition, different test methods show different sensitivities [65, 79].

Thus, nanomaterial has to be tested with more than one assay/method, and different cell lines should be used, to provide sufficiently statements regarding their toxic potential. Currently, in vivo studies are essential for toxicological assessments. Using standardized in vitro test methods of testing nanomaterials on their toxic potential, the number of necessary in vivo studies could be reduced.

7.4 Where Should We Go?

NPs have shown to enhance the desired effects in cosmetics and are indispensable in nanocosmetics. The main advantages are the penetration enhancement and the reservoir function for cosmetic substances. The exception is the solid NPs for sunscreens, which should not penetrate but reflect the sunlight and provide high sun protection factors in a safe manner. Nevertheless, it must be kept in mind that each new nanomaterial must be investigated toward its safety and beneficial effects. The size of nonbiodegradable NPs should not be smaller than 20 nm to ensure that the particles do not pass the *SC* barrier.

In the last years, stimulus-responsive delivery via NPs has become an effective experimental approach. The drug of interest is loaded or trapped on an NP and after initiating of a stimulus, that is pH, temperature, optical absorption, etc., the therapeutic agent is released at the site of action [85, 86]. It could be demonstrated that such an approach is more effective versus gradual diffusion or particle decomposition [87]. In this context, the pH gradient present in the follicular ducts was recently determined using pH-responsive dendritic polyglycerol nanogels. The pH values ranged from 6.50 on the skin surface to 7.44 at a depth of 530 µm [88]. pH-sensitive NPs that erode near the skin surface but dissolve in the HFs could enable controlled and specifically targeted drug delivery. Recently, the first successful ex vivo evaluation of NPs with pH-dependent targeting potential was conducted [89].

In any case, the beneficial effects of the application of NPs should be proven in cosmetic studies with reasonable control groups.

7.5 What Are the Social Science Aspects?

The implementation of NPs in cosmetic products has many advantages. The enhanced penetration efficacy and the increased stability will require smaller amounts of the active agent to be used as well as fewer applications, reducing the cost of the product. As a penetration of solid NPs above a size of 40 nm has thus far not been observed for intact skin, adverse effects are not very probable.

In the case of sunscreen products that use solid NPs, high sun protection factors can be realized due to the synergistic effect of chemical and physical filters. But the NPs not only reduced the transmittance of the light in the UV part, the scattering effect of the NPs is also effective in the visible and near-infrared part of the sunlight [90, 91]. This is important because the visible and infrared regions also contribute up to 50% of radical production induced by sunlight [92]. The incorporation of NPs into sunscreens can reduce the radical formation in the skin and thus can reduce the incident of skin cancer. This is of great relevance particularly for the health system because the treatment of skin cancer is very cost intensive.

Additionally to beneficial effects of NPs in cosmetics, adverse effects to the environment could appear by the exposure of nanomaterials through release into the water, air, and soil during the manufacture, use, or disposal of these materials. Antibacterial NPs, for example, could potentially interfere with beneficial bacteria in sewage and wastewater treatment plants and possibly contaminate water that is intended for reuse [93]. In another study, nano-TiO_2 used in personal care products reduced biological functions of bacteria after less than one hour of exposure. As the

TiO_2-NPs end up at municipal sewage treatment plants, this could eliminate microbes that play vital roles in ecosystems and help treat wastewater [94].

Therefore, also such adverse effects must be taken into account if NPs are used in cosmetics. The soft particles are mainly biodegradable and therefore harmless. The solid NPs such as TiO_2 are of such great importance for the skin cancer prevention that waiving of such NP is not an option.

The implementation of NPs in cosmetics should be controlled through guidelines, to ensure that only safe and effective products are marketed. Possible risks and benefits should be considered.

7.6 Conclusion

NPs represent highly effective carrier systems for actives that can be exploited in cosmetics. They facilitate the penetration of drugs deep into HFs, where these can be stored for up to 10 days. Solid and nonbiodegradable NPs in cosmetics should not be smaller than 20 nm to ensure that they do not penetrate into viable skin. Furthermore, each particle type and possible fragments or metabolites should be subjected to risk assessments and toxicity testing even if it was previously shown that the material is nontoxic on a macroscopic scale. NPs should only be incorporated if their benefits have been verified experimentally. EU regulations should be adapted to fit these aspects so that all doubts regarding safety are eliminated.

References

1. Nohynek GJ, Dufour EK, Roberts MS. Nanotechnology, cosmetics and the skin: is there a health risk? Skin Pharmacol Physiol 2008;21(3):136–49.
2. Honeywell-Nguyen PL, et al. The in vivo and in vitro interactions of elastic and rigid vesicles with human skin. Biochim Biophys Acta 2002;1573(2):130–40.
3. Ourique AF, et al. Tretinoin-loaded nanocapsules: preparation, physicochemical characterization, and photostability study. Int J Pharm 2008;352(1–2):1–4.
4. Haag SF, et al. Stabilization of reactive nitroxides using invasomes to allow prolonged electron paramagnetic resonance measurements. Skin Pharmacol Physiol 2011;24(6):312–21.
5. Suwannateep N, et al. Encapsulated curcumin results in prolonged curcumin activity in vitro and radical scavenging activity ex vivo on skin after UVB-irradiation. Eur J Pharm Biopharm 2012;82(3):485–90.
6. Lohan SB, et al. Investigation of the cutaneous penetration behavior of dexamethasone loaded to nano-sized lipid particles by EPR spectroscopy, and confocal Raman and laser scanning microscopy. Eur J Pharm Biopharm 2017;116:102–10.
7. Jager J, et al. Characterization of hyperbranched core-multishell nanocarriers as an innovative drug delivery system for the application at the oral mucosa. J Periodontal Res 2017, 53(1):57–65.
8. Rancan F, et al. Effects of thermoresponsivity and softness on skin penetration and cellular uptake of polyglycerol-based nanogels. J Controlled Release 2016;228:159–69.

9. Rancan F, et al. Skin penetration and cellular uptake of amorphous silica nanoparticles with variable size, surface functionalization, and colloidal stability. ACS Nano 2012;6(8):6829–42.
10. Vogt A, et al. Interaction of dermatologically relevant nanoparticles with skin cells and skin. Beilstein J Nanotechnol 2014;5:2363–73.
11. Lademann J, et al. Nanoparticles – an efficient carrier for drug delivery into the hair follicles. Eur J Pharm Biopharm 2007;66(2):159–64.
12. Otberg N, et al. Laser spectroscopic methods for the characterization of open and closed follicles. Laser Phys Lett 2004;1(1):46–9.
13. Blume-Peytavi U, et al. Follicular and percutaneous penetration pathways of topically applied minoxidil foam. Eur J Pharm Biopharm 2010;76(3):450–3.
14. Otberg N, et al. The role of hair follicles in the percutaneous absorption of caffeine. Br J Clin Pharmacol 2008;65(4):488–92.
15. Langbein L, et al. Tight junctions and compositionally related junctional structures in mammalian stratified epithelia and cell cultures derived therefrom. Eur J Cell Biol 2002;81(8):419–35.
16. Nohynek GJ, et al. Grey goo on the skin? Nanotechnology, cosmetic and sunscreen safety. Crit Rev Toxicol 2007;37(3):251–77.
17. Vogt A, et al. Follicular targeting – a promising tool in selective dermatotherapy. J Investig Dermatol Symp Proc 2005;10(3):252–5.
18. Popov AP, et al. Effect of size of TiO2 nanoparticles embedded into stratum corneum on ultraviolet-A and ultraviolet-B sun-blocking properties of the skin. J Biomed Opt 2005;10(6):064037.
19. Meinke MC, et al. Radical protection by differently composed creams in the UV/VIS and IR spectral ranges. Photochem Photobiol 2013;89(5):1079–84.
20. Lademann J, et al. Synergy effects between organic and inorganic UV filters in sunscreens. J Biomed Opt 2005;10(1):14008.
21. Haag SF, et al. Skin penetration enhancement of core-multishell nanotransporters and invasomes measured by electron paramagnetic resonance spectroscopy. Int J Pharm 2011;416(1):223–8.
22. Cevc G. Transferosomes, liposomes and other lipid suspensions on the skin: permeation enhancement, vesicle penetration, and transdermal drug delivery. Crit Rev Ther Drug Carrier Syst 1996;13(3–4):257–388.
23. Thong HY, Zhai H, Maibach HI. Percutaneous penetration enhancers: an overview. Skin Pharmacol Physiol 2007;20(6):272–82.
24. Uchegbu IF, Vyas SP. Non-ionic surfactant based vesicles (niosomes) in drug delivery. Int J Pharm 1998;172(1):33–70.
25. Balakrishnan P, et al. Formulation and in vitro assessment of minoxidil niosomes for enhanced skin delivery. Int J Pharm 2009;377(1–2):1–8.
26. Touitou E, et al. Ethosomes – novel vesicular carriers for enhanced delivery: characterization and skin penetration properties. J Controlled Release 2000;65(3):403–18.
27. Cevc G, Schatzlein A, Blume G. Transdermal drug carriers – basic properties, optimization and transfer efficiency in the case of epicutaneously applied peptides. J Controlled Release 1995;36(1–2):3–16.
28. Katz LM, Dewan K, Bronaugh RL. Nanotechnology in cosmetics. Food Chem Toxicol 2015;85:127–37.
29. Sonneville-Aubrun O, Simonnet JT, L'Alloret F. Nanoemulsions: a new vehicle for skincare products. Adv Colloid Interface Sci 2004;108:145–9.
30. Quintanar-Guerrero D, et al. Preparation techniques and mechanisms of formation of biodegradable nanoparticles from preformed polymers. Drug Dev Ind Pharm 1998;24(12):1113–28.

31. Radtchenko IL, Sukhorukov GB, Mohwald H. A novel method for encapsulation of poorly water-soluble drugs: precipitation in polyelectrolyte multilayer shells. Int J Pharm 2002;242 (1–2):219–23.
32. Khoee S, Yaghoobian M, An investigation into the role of surfactants in controlling particle size of polymeric nanocapsules containing penicillin-G in double emulsion. Eur J Med Chem 2009;44(6):2392–9.
33. Hwang SL, JC Kim. In vivo hair growth promotion effects of cosmetic preparations containing hinokitiol-loaded poly(epsilon-caprolacton) nanocapsules. Journal of Microencapsulation 2008;25(5):351–6.
34. Schafer-Korting M, Mehnert WG, Korting HC. Lipid nanoparticles for improved topical application of drugs for skin diseases. Adv Drug Deliv Rev 2007;59(6):427–43.
35. Muller RH, Radtke M, Wissing SA. Solid lipid nanoparticles (SLN) and nanostructured lipid carriers (NLC) in cosmetic and dermatological preparations. Adv Drug Deliv Rev 2002;54(Suppl 1):S131–55.
36. Junghanns JU, Muller RH. Nanocrystal technology, drug delivery and clinical applications. Int J Nanomed 2008;3(3):295–309.
37. Chauhan AS, et al. Dendrimer-mediated transdermal delivery: enhanced bioavailability of indomethacin. J Controlled Release 2003;90(3):335–43.
38. Kurniasih IN, Keilitz J, Haag R. Dendritic nanocarriers based on hyperbranched polymers. Chem Soc Rev 2015;44(12):4145–64.
39. Lohan SB, et al. Investigation of cutaneous penetration properties of stearic acid loaded to dendritic core-multi-shell (CMS) nanocarriers. Int J Pharm 2016;501(1–2):271–7.
40. Kuchler S, et al. Nanoparticles for skin penetration enhancement – A comparison of a dendritic core-multishell-nanotransporter and solid lipid nanoparticles. Eur J Pharm Biopharm 2009;71(2):243–50.
41. Barauskas J, et al. Cubic phase nanoparticles (cubosome): principles for controlling size, structure, and stability. Langmuir 2005;21(6):2569–77.
42. Spicer PT, et al. Bicontinuous cubic liquid crystalline phase and cubosome personal care delivery systems. In: Rosen M, editor. Personal Care Delivery Systems and Formulations. Berkshire: Noyes Publishing, 2003, 1–44.
43. Korinth G. et al. Percutaneous absorption and metabolism of 2-butoxyethanol in human volunteers: a microdialysis study. Toxicol Lett 2007;170(2):97–103.
44. Li Y, et al. Microdialysis as a tool in local pharmacodynamics. AAPS J 2006;8(2):E222–35.
45. Hersini KJ, et al. Microdialysis of inflammatory mediators in the skin: a review. Acta Derm Venereol 2014;94(5):501–11.
46. Kharisov BI, Kharissova OV, Berdonosov SS. Radioactive nanoparticles and their main applications: recent advances. Recent Pat Nanotechnol 2014;8(2):79–96.
47. Kreyling WG, et al. In vivo integrity of polymer-coated gold nanoparticles. Nat Nanotechnol 2015;10(7):619–23.
48. Patzelt A, et al. Selective follicular targeting by modification of the particle sizes. J Controlled Release 2011;150(1):45–8.
49. Zoubari G, et al. Effect of drug solubility and lipid carrier on drug release from lipid nanoparticles for dermal delivery. Eur J Pharm Biopharm 2017;110:39–46.
50. Darvin ME, et al. Safety assessment by multiphoton fluorescence/second harmonic generation/hyper-Rayleigh scattering tomography of ZnO nanoparticles used in cosmetic products. Skin Pharmacol Physiol 2012;25(4):219–26.
51. Ascencio SM, et al. Confocal Raman microscopy and multivariate statistical analysis for determination of different penetration abilities of caffeine and propylene glycol applied simultaneously in a mixture on porcine skin ex vivo. Eur J Pharm Biopharm 2016;104:51–8.

52. Klossek A, et al. Studies for improved understanding of lipid distributions in human skin by combining stimulated and spontaneous Raman microscopy. Eur J Pharm Biopharm 2017;116:76–84.
53. Saeidpour S, et al. Drug distribution in nanostructured lipid particles. Eur J Pharm Biopharm 2017;110:19–23.
54. Oberdorster G, Oberdorster E, Oberdorster J. Nanotoxicology: an emerging discipline evolving from studies of ultrafine particles. Environ Health Perspect 2005;113(7):823–39.
55. Magrez A, et al. Cellular toxicity of carbon-based nanomaterials. Nano Lett 2006;6(6):1121–5.
56. Donaldson K, Beswick PH, Gilmour PS. Free radical activity associated with the surface of particles: a unifying factor in determining biological activity? Toxicol Letters 1996;88 (1–3):293–298.
57. Yang L, Watts DJ. Particle surface characteristics may play an important role in phytotoxicity of alumina nanoparticles. Toxicol Lett 2005;158(2):122–32.
58. Garnett MC, Kallinteri P. Nanomedicines and nanotoxicology: some physiological principles. Occup Med (London) 2006;56(5):307–11.
59. Corbo C, et al. The impact of nanoparticle protein corona on cytotoxicity, immunotoxicity and target drug delivery. Nanomedicine (London) 2016;11(1):81–100.
60. Lohan SB, et al. ROS production and glutathione response in keratinocytes after application of ß-carotene and VIS/NIR irradiation. in preparation, 2017.
61. Lin PC, et al. Techniques for physicochemical characterization of nanomaterials. Biotechnol Adv 2014;32(4):711–26.
62. Bhattacharjee S. DLS and zeta potential – What they are and what they are not? J Controlled Release 2016;235:337–51.
63. Cedervall T, et al. Understanding the nanoparticle-protein corona using methods to quantify exchange rates and affinities of proteins for nanoparticles. Proc Natl Acad Sci USA 2007;104(7):2050–5.
64. Alexiev U, et al. Time-resolved fluorescence microscopy (FLIM) as an analytical tool in skin nanomedicine. Eur J Pharm Biopharm 2017;116:111–24.
65. Bahadar H, et al. Toxicity of nanoparticles and an overview of current experimental models. Iran Biomed J 2016;20(1):1–11.
66. Medina C, et al. Nanoparticles: pharmacological and toxicological significance. Br J Pharmacol 2007;150(5):552–8.
67. Riss TL, et al. Cell viability assays. In: Sittampalam GS, et al. editors. in assay guidance manual, 2004: Bethesda (MD).
68. Cummings BS, Wills LP, Schnellmann RG. Measurement of cell death in Mammalian cells. Curr Protoc Pharmacol 2012;Chapter 12:Unit12 8.
69. Fautz R, Husein B, Hechenberger C. Application of the neutral red assay (NR assay) to monolayer cultures of primary hepatocytes: rapid colorimetric viability determination for the unscheduled DNA synthesis test (UDS). Mutat Res 1991;253(2):173–9.
70. Jones KH, Senft JA. An improved method to determine cell viability by simultaneous staining with fluorescein diacetate-propidium iodide. J Histochem Cytochem 1985;33(1):77–9.
71. Manke A, Wang L, Rojanasakul Y. Mechanisms of nanoparticle-induced oxidative stress and toxicity. Biomed Res Int 2013;2013:942916.
72. Wang H, Joseph JA. Quantifying cellular oxidative stress by dichlorofluorescein assay using microplate reader. Free Radic Biol Med 1999;27(5–6):612–6.
73. Fang L, et al. Comet assay as an indirect measure of systemic oxidative stress. J Vis Exp 2015(99):e52763.
74. Bowman L, Castranova V, Ding M, Single cell gel electrophoresis assay (comet assay) for evaluating nanoparticles-induced DNA damage in cells. Methods Mol Biol 2012;906:415–22.

75. Lohan SB, et al. EPR Technology as sensitive method for oxidative stress detection in primary and secondary keratinocytes induced by two selected nanoparticles. Cell Biochem Biophys 2017;75(3-4):359–367.
76. Haag SF, et al. Determination of the antioxidative capacity of the skin in vivo using resonance Raman and electron paramagnetic resonance spectroscopy. Exp Dermatol 2011;20(6):483–7.
77. Lohan SB, et al. Ultra-small lipid nanoparticles promote the penetration of coenzyme Q10 in skin cells and counteract oxidative stress. Eur J Pharm Biopharm 2015;89:201–7.
78. Vogt, A., et al. Interaction of dermatologically relevant nanoparticles with skin cells and skin. Beilstein J Nanotechnol 2014;5:2363–73.
79. Ahlberg S, et al. Comparison of different methods to study effects of silver nanoparticles on the pro- and antioxidant status of human keratinocytes and fibroblasts. Methods 2016;109:55–63.
80. Fuchs J, et al. Electron paramagnetic resonance studies on nitroxide radical 2,2,5,5-tetramethyl-4-piperidin-1-oxyl (TEMPO) redox reactions in human skin. Free Radical Biol Med 1997;22(6):967–76.
81. Dikalov SI, Harrison DG. Methods for detection of mitochondrial and cellular reactive oxygen species. Antioxid Redox Signal 2014;20(2):372–82.
82. Aquilano K, Baldelli S, Ciriolo MR. Glutathione: new roles in redox signaling for an old antioxidant. Front Pharmacol 2014;5:196.
83. Kerksick C, Willoughby D. The antioxidant role of glutathione and N-acetyl-cysteine supplements and exercise-induced oxidative stress. J Int Soc Sports Nutr 2005;2:38–44.
84. Holub M, et al. Cytokines and chemokines as biomarkers of community-acquired bacterial infection. Mediators Inflammation 2013;2013:190145.
85. Mak WC, et al. Triggering of drug release of particles in hair follicles. J Controlled Release 2012;160(3):509–14.
86. Mak WC, et al. Drug delivery into the skin by degradable particles. Eur J Pharm Biopharm 2011;79(1):23–7.
87. Tran N, et al. Gradient-dependent release of the model drug TRITC-dextran from FITC-labeled BSA hydrogel nanocarriers in the hair follicles of porcine ear skin. Eur J Pharm Biopharm 2017;116:12–16.
88. Dimde M, et al. Synthesis and validation of functional nanogels as pH-sensors in the hair follicle. Macromol Biosci 2017, doi: 10.1002/mabi.201600505.
89. Sahle FF, et al. Formulation and in vitro evaluation of polymeric enteric nanoparticles as dermal carriers with pH-dependent targeting potential. Eur J Pharm Sci 2016;92:98–109.
90. Meinke MC, et al. Radical protection by sunscreens in the infrared spectral range. Photochem Photobiol 2011;87(2):452–6.
91. Souza C, et al. Radical-scavenging activity of a sunscreen enriched by antioxidants providing protection in the whole solar spectral range. Skin Pharmacol Physiol 2017;30(2):81–9.
92. Zastrow L, et al. The missing link – light-induced (280–1,600 nm) free radical formation in human skin. Skin Pharmacol Physiol 2009;22(1):31–44.
93. Potera C, Nanomaterials: transformation of silver nanoparticles in sewage sludge. Environ Health Perspect 2010;118(12):A526–7.
94. Cimitile M. Nanoparticles in sunscreen damage microbes. Available at: https://www.scientificamerican.com/article/nanoparticles-in-sunscreen/. Scientific American, 2009.

Part II: Nanotechnology in Information and Communication

Livio Baldi

8 Nanooptoelectronics

8.1 Introduction

Nanoelectronics permeates now countless aspects of the world economy, impacting sectors as varied as banking, retail, energy, transportation, education, publishing, media, and health, even if most people are unaware of it. In a perhaps less evident way, also nanophotonics has entered our lives: it is in the flat screen television set, in the display of our computers and smartphones, in our music CD and DVD readers, in the optical fibres that are bringing us high-speed Internet, and even less evidently in several medical examinations. Together, nanoelectronics and nanophotonics are at the basis of information and communication technologies that are transforming the ways social interactions and personal relationships are taking place.

8.2 A Tale of Two Sciences

The history of electronics and photonics goes back a long time, to when they were called simply light and electricity. Light was the first to be widely used by man: optical signals in the form of fires or flags were used to transmit information already at the beginning of history,[1] while electricity was little more than a parlour game until Volta discovered the way to generate it chemically at the beginning of the nineteenth century. After that, electricity replaced light also in communications with the invention of the telegraph and the telephone, and, with the invention of mechanical electricity generators and of the light bulb by Edison, it took the role of light generation, replacing gas lighting. At the end of the nineteenth century, science started the exploration of the atomic world and the two elementary particles behind electricity and light, electrons and photons were discovered, and things just became more confused. It was soon found that photons could behave like particles, with a well-defined energy and a momentum that could be exchanged with more "solid" particles (Compton effect), while electrons could behave as waves, a property currently exploited in electron microscopes. The invention of the triode by Forest in 1906 was the beginning of electronics and resulted in a variety of applications in communications, control systems, and computing, ending up in

---

1 In its tragedy, Agamemnon Aeschylus (fifth century BC) describes a chain of fires that bring news of the fall of Troy to Argos.

https://doi.org/10.1515/9783110547221-008

the ENIAC[2] machine in 1946, a 30 tonne monster that was also the first operational computer in the modern sense. The invention of the transistor in 1947 and of the integrated circuit in 1957 opened the way to microelectronics and started an explosive expansion of electronics in all fields of everyday life, following the exponential cost reduction of the Moore's law. In the early 2000, microelectronics has slowly evolved into nanoelectronics, with typical device feature size going below 100 nm. At the same time, the progress made in the field of understanding the operation of devices at the quantum level and in the technology for material manipulation on that scale has allowed the development and production of advanced optical devices that were available before only in the lab environment: it was the birth of nanophotonics. The two technologies are now completely intertwined: optical devices are relying on nanoelectronic components for power supply and control operations, while nanophotonics allows the interconnection of nanoelectronic devices among themselves and with the external world. Perhaps from now on we should speak only on nano-optoelectronics.

8.3 Small Devices, Big Impact

The man in the street is not aware in general of the presence of nanoelectronics and nanophotonics, but his world would collapse if they would be taken away. Communications, social life, transportations, banking, business, entertainment, medical diagnosis, and treatments all would be severely disrupted or totally disappeared. He would lose connections, appointments, access to his money, and the capability to book a plane or train ticket, or to reserve a hotel. Half of the appliances at home would stop functioning, as well as the ecological light bulbs.

Information and communication technologies are transforming the way social interactions and personal relationships are conducted, with fixed, mobile, and broadcast networks converging, and devices and objects increasingly connected to form the Internet of Things. The volume of data traffic on Internet has grown by a factor of 20 between 2005 and 2013, reaching the astonishing amount of more than 51 exabytes per month in 2013.[3] A report of the Strategic Policy Forum on Digital Entrepreneurship of the European Commission estimated that digitalization could induce a growth of 15–20% in European manufacturing and create 3.8 million jobs, 1.5 of which in Internet services. Big data technology and services alone were expected to reach USD 16.9 billion worldwide in 2015 at a compound annual growth rate of 40% [1].

---

2 ENIAC: acronym of Electronic Numerical Integrator and Computer. The first general purpose electronic computer of history, realized in 1946 at the Moore School of Electrical Engineering of the University of Pennsylvania

3 OECD data. 1 exabyte = 10^{18} bytes.

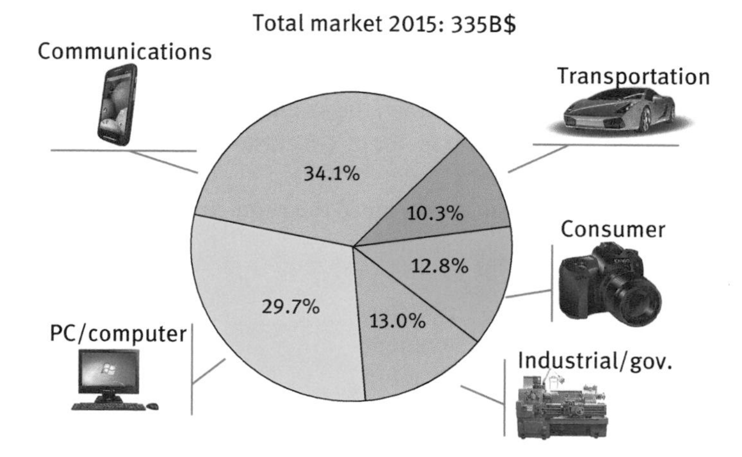

Figure 8.1: Semiconductor market 2015 by end user (data source: SIA Factbook 2016).

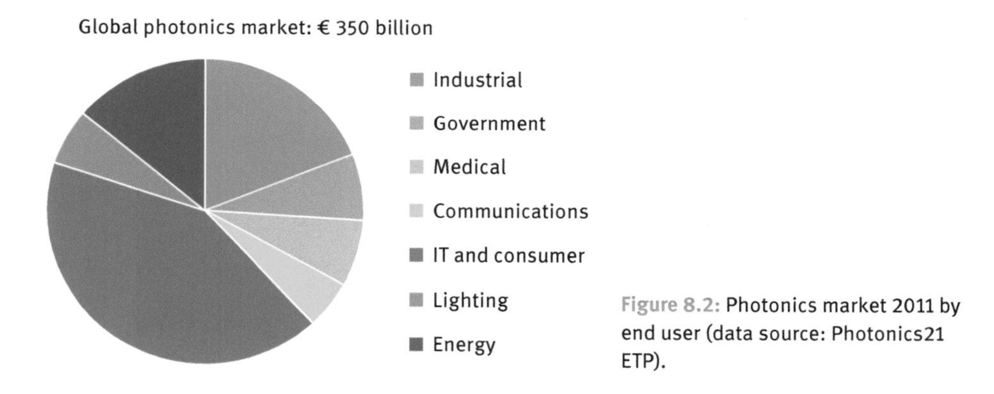

Figure 8.2: Photonics market 2011 by end user (data source: Photonics21 ETP).

An analysis of the end user markets for semiconductor devices (Figure 8.1) confirms the importance of communications for the nanoelectronics market [2]. It should be added also that with the appearance of tablets and smartphones, the distinction between communications, computers, and consumers has been progressively blurring.

The photonics market shows similar data (Figure 8.2). The large weight of IT (information technology) and consumers is related to the large market of flat panel displays, while typical of photonics are the markets of lighting (mostly light-emitting diode [LED]) and energy (photovoltaics) [3].

All in all it is an impressive market, even if the two values cannot be summed up, since several devices are listed both as semiconductors (based on the fabrication technology) and as photonics (based on the function), while the figures for the photonics market refer to the value of the final system rather than to the one of the discrete components.

Much more important however are the markets that nanoelectronics and nanophotonics are enabling: it is not only the matter of electronic devices, like cellular phones, computers, digital cameras, car automation, and similar, but also of the huge spectrum of services that have been enabled by mobile communications and Internet. A rough evaluation gives a ratio of 5 between the semiconductor market and the market of electronic systems, while another factor of 3–5 exists between the electronic systems and the services based on them.

The term "nano" is used here in a broad sense, since the geometrical features of some of the devices in nanoelectronics, and even more in nanophotonics, are far from the nanometre scale. According to the definition given by the US Patent Office, nanotechnology patents, classified under Class 977, must have the following characteristics:

- Related to research and technology development at the atomic, molecular, or macromolecular levels in the length of scale approximately 1–100 nm range in at least one dimension.
- That provide a fundamental understanding of phenomena and materials at the nanoscale and to create and use structures, devices, and systems that have novel properties and functions because of their size [4].

The devices considered under nanoelectronics and nanophotonics fall under the second conditions, since their operation is based on quantum properties of matter, and their manufacturing requires technologies (layer deposition, etching, etc.) that are controlled on the scale of hundreds, if not of tens of nanometres.

8.4 The Growth of Nanoelectronics

The basic device, which builds up most of nanoelectronics devices, is the MOS (Metal Oxide Semiconductor) transistor. As shown in Figure 8.3, almost 70% of the nanoelectronics market is made up of microprocessors or microcontrollers, logic devices, and memories, all based on CMOS (Complementary MOS) technology, and for a large part with typical feature size below 100 nm.

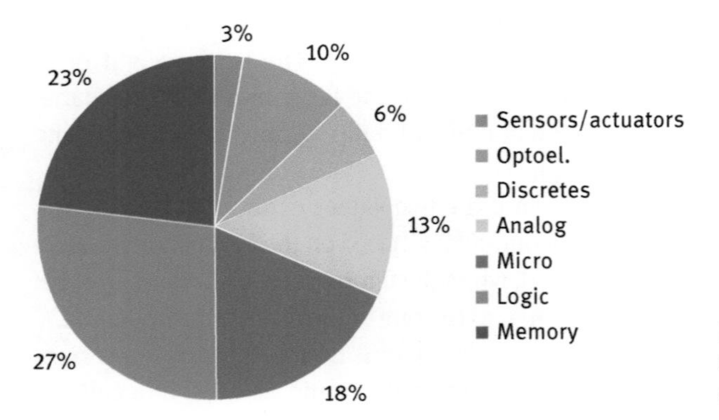

Figure 8.3: Nanoelectronics market by technology (data source: WSTS).

The MOS transistor is probably the simplest basic switching element, and its origin goes back to a patent of 1928 [5]. The main advantages of MOS transistor are the low intrinsic complexity of the transistor architecture, the very low power dissipation in standby, and above all, its intrinsic scalability. In simple terms, scalability, first described by Dennard in 1974 [6], means that two transistors of different size behave in the same way if their physical and geometrical parameters are scaled by the same factor. As a fringe benefit, if proper scaling rules are applied, main performances like speed and power dissipation are improving (save power density, but it was not considered a problem for a long time). Another strong point of microelectronics, and afterwards of nanoelectronics, has been the batch processing. Integrated circuits are printed in parallel on silicon substrates (wafers); therefore, the smaller the device size, the larger the number of produced devices per wafer and the lower the cost.

Since scaling laws reduce a physical problem to a "simple" technological one, the economic pressure has led to a continuous race towards smaller device size, exemplified by the so-called Moore's law. First enunciated by Gordon Moore in 1965 [7] (after only 8 years from the invention of the integrated circuit) and confirmed in 1975 [8], it stated that the "complexity of integrated circuits has approximately doubled every year since their introduction" and that "there is no present reason to expect a change in the trend". Even if with small adjustments to the rate of progress, which has gone down to doubling every 18 months, this prediction has hold true for 50 years!

The decreasing cost of CMOS-based logic devices and the invention of the microprocessor brought about a second revolution, leading to the digitalization not only of computation but also of data and processes that were eminently analogue in nature. Analogue process control gave place to digital process control, analogue communications to digital communications (GSM), disks and films to MP3, digital cameras, and Photoshop. The trend has been to reduce any problem to a computable one to be solved with brute force approach (number crunching), thanks to the always decreasing cost of computing power. Nanophotonics with the introduction of CMOS imagers and digital displays had an important role in this evolution.

The final result was the birth to the digital society in which we are living now, where all information is available in digital form over the Internet. For nanoelectronics, it had the effect of expanding exponentially the addressable market, transforming professional markets into a consumer-driven ones.

8.5 Nanophotonics Enters the Picture

Photonics is the science and technology of harnessing light. Photonics includes the generation, detection, and management of light through guidance, manipulation, and, most importantly, its utilization for the benefit of society.

The reasons for the wide range of application of photonics are in the specific properties that differentiate photons from electrons:
- Photons have no mass, and therefore can travel at the speed of light. Nothing can go faster in our universe (if Einstein is right).
- Photons, unlike electrons, can propagate in transparent media (air, water, and glass fibres) with minimal losses.
- Photons are bosons, following the Bose–Einstein statistics, which means that there is no limitation to the number of photons that can occupy the same quantum state. Laser beams are a practical example of this property.
- A pulse of photons can be as short as 100 billionth of a billionth of a second (around 70 attoseconds[4] being the record until now [9]), allowing the exploration of molecular and atomic reactions.
- Analysis of the emitted/reflected light (spectroscopy) provides a contact-free analytic tool for a large variety of materials.

Even if optoelectronic devices have been around for some time (detectors and photomultipliers), it was only with the development of solid-state technology, coming from nanoelectronics, that modern photonics came to the world.

8.6 The Devices of Nanophotonics

The fact that light can stimulate electron emission from materials is well known since the late nineteenth century and eventually led Einstein to the formulation of the quantum theory [10] and to the Nobel Prize in 1921. The basis of much of the nanophotonics is in the interaction between electrons and light in semiconductor materials, and more specifically in junction diodes.

Semiconductor materials are characterized by an energy gap that separates the allowed energy states. Junction diodes put in contact two materials with different energy levels (due to doping) and different electron populations. One side (n-doped) is rich of electrons in the conduction (high-energy) band, the other (p-doped) has a lack of electrons, conventionally represented as holes, in the valence (low-energy) band. Charge diffusion across the border produced a charge-depleted zone, in which an electric field exists that prevents further electron diffusion. This device can provide both functions of electricity generation by light absorption, and light generation by electron decay from a higher to a lower energy state as shown in Figure 8.4.

When light of the proper frequency hits the junction, couples of holes and electrons are generated (middle picture). If it happens in the depleted zone, the local

4 1 attosecond = 1×10^{-18} seconds

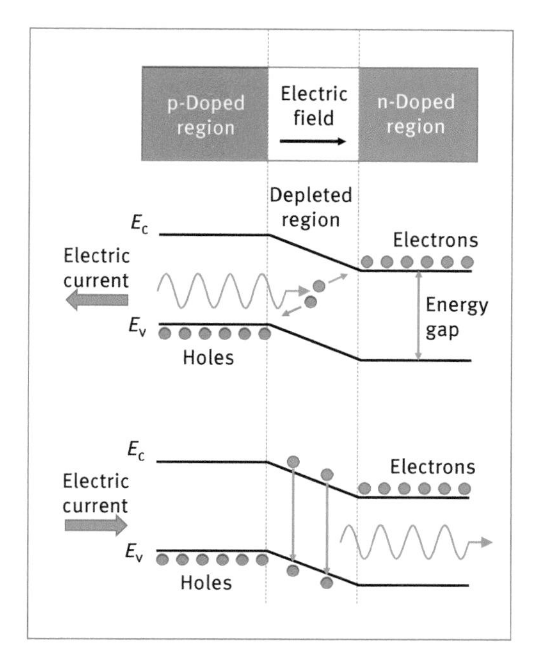

Figure 8.4: Schematic picture of light–electron interaction in a diode.

electric field separates the charges and sweeps them to the opposite regions of the junction, generating a current in an external circuit. This principle is at the basis of photovoltaic energy generation (solar cells), light detectors, and imagers. The energy of the light must be at least as large as the semiconductor energy gap. Radiation with higher energy can also be absorbed, but the extra energy is not creating more electrons and is merely dissipated as heat.

On the other side, if a current is forced inside the junction (lower picture) the light is generated by the recombination of electrons from the n-doped side with holes in the p-doped side. An external voltage can provide a continuous flow of electrons and holes and therefore a constant light emission. It is the principle of LEDs and of semiconductor lasers. The frequency of the emitted light is related to the energy gap by the well-known Planck relationship $E = h\nu$.

Silicon is a poor material for light emission, because electrons and holes usually recombine with a nonradiative transition, due to the band structure, while III–V materials such as GaAs, GaN, InP, and more complex ones are good emitters. However, silicon is efficient and cheap enough for photovoltaics.

The single most important photonic device is probably the laser (acronym of Light Amplification by Stimulated Emission of Radiation), which has applications in all fields of science and technology. The basic operation principle was already anticipated by Einstein in 1917 [11]. The emission of radiation from an electron in an excited state is a random event, with a well-defined decay time. However, if it is stimulated by an external radiation with the frequency associated with that state

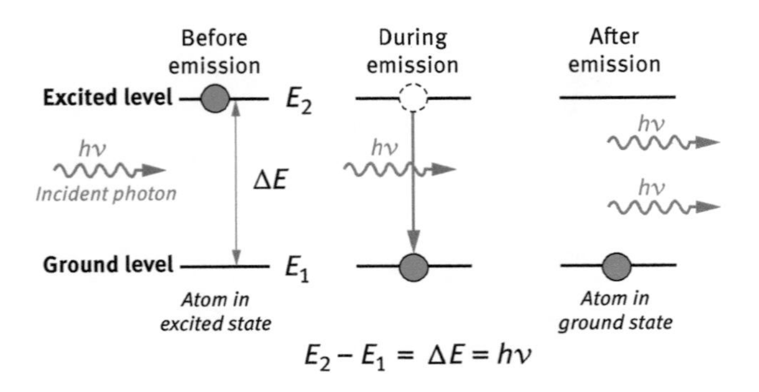

Figure 8.5: Schematic representation of laser effect (source: Wikimedia).

transition, the excited electron tends to decay immediately, releasing a photon that adds to the input signal, with the same wavelength, phase, and polarization (Figure 8.5).

This is possible because photons are bosons and any number of them can occupy the same quantum state. The result is therefore a strong intensification of the incident light. In practical terms, a laser consists in an optical cavity, filled with the material that can be excited to emit light, put between two mirrors that reflect back and forth the radiation to intensify it. Semiconductor lasers rely on the consolidated manufacturing technology of microelectronics. Similarly to LEDs, the light emission is associated with the recombination of electrons and holes within a junction, and the presence of mirrors on both ends induces the stimulated emission. Performances can be further enhanced by inserting a thin undoped layer, where recombination takes place, between the n and p sides of the junction. If the layer is thin enough, energy states in it are quantized, enhancing the concentration of electrons in the allowed states and intensifying the laser effect. These devices are known as quantum well lasers and represent the leading edge of nanophotonics. Semiconductor lasers are used for common applications like fibreoptic communications, barcode readers, laser pointers, CD/DVD reading and recording, laser printing, and laser scanning.

Another important component of nanoelectronic systems, even if it can be hardly considered as "nano", is the optical fibre. Optical fibres are the optical equivalent of the electric cable, enabling transmission of optical signals over long distances with minimal losses also on curved paths. They consist in a transparent core surrounded by a transparent coating, with lower refraction index, which reflects the light toward the core. The big advantage of optical signal transmission is related to the high frequency of the light, which allows for very high data rates. Each fibre can carry many independent channels, each using a different wavelength of light. In combination with laser and solid-state detectors, they represent the backbone of Internet.

8.7 Nanoelectronics and Nanophotonics for Our Lives

Nanophotonics has profited from the semiconductor technology developed for nanoelectronics, and it is from the combination of nanophotonics and nanoelectronics that comes those applications that are revolutionizing our daily life. The fields in which the two technologies are deeply interconnected cover almost all aspects of modern world:

- *Transportation*: the impact of nanoelectronics in the car industry is quite evident to everybody. A survey of iSupply gave an average semiconductor content of 350$ per car, growing to more than 1,000$ for electrical cars. Possibly less evident, but not less important, has been the impact on other transport domains such as aviation, railways, and shipping. In the car industry, elements such as mechanical sensors combined with intelligent decision-making software, power electronics, and embedded multicore processors have allowed the introduction of assisted driving features (anti-lock braking system (ABS), Electronic Stability Control, rear radar, navigators), while power electronics has opened the way to hybrid and electrical cars to reduce pollution. More recent has been the introduction of nanophotonics: starting with flat screen displays, it has quickly grown to encompass LED head and taillights, and rear cameras. The trend toward automated driving is pushing the introduction of LIDAR[5], wide-spectrum cameras for obstacle recognition, and possibly photovoltaics as energy source. The future widespread diffusion of full electrical vehicles will go along with the addition of various degrees of driving automation, supported by car to car and car to infrastructure communications. Nanophotonics and nanoelectronics will both play a leading role (Figure 8.6): the one by providing environmental awareness and interface to users and the other the control and decisional power and the energy management system.
- *Communications and IT*: Digital data are increasingly generated and distributed by users' mobile devices that are combining communication and consumer services. The progress in miniaturization of nanoelectronics has allowed packing wireless communication and data processing services in handy mobile devices, but nanophotonics has provided the low-consumption LED or OLED (Organic LED) displays and the image capture facility. Moreover, while local links are based on radiofrequency transmission, a high-capacity fixed backbone is needed to transfer and store this amount of data. Today, the major highways of communication and information flow are optical, based on lasers, optical fibres, and optical detectors. While nanoelectronics is the key enabler for data storage and management, data compression, signal modulation and demodulation for digital communications, nanophotonics provides the communication backbone and the user interface to the Internet world,

5 LIDAR: acronym for acronym of Light Detection And Ranging or Light Imaging, Detection, And Ranging, is the equivalent of RADAR, using laser pulses in place or radio waves.

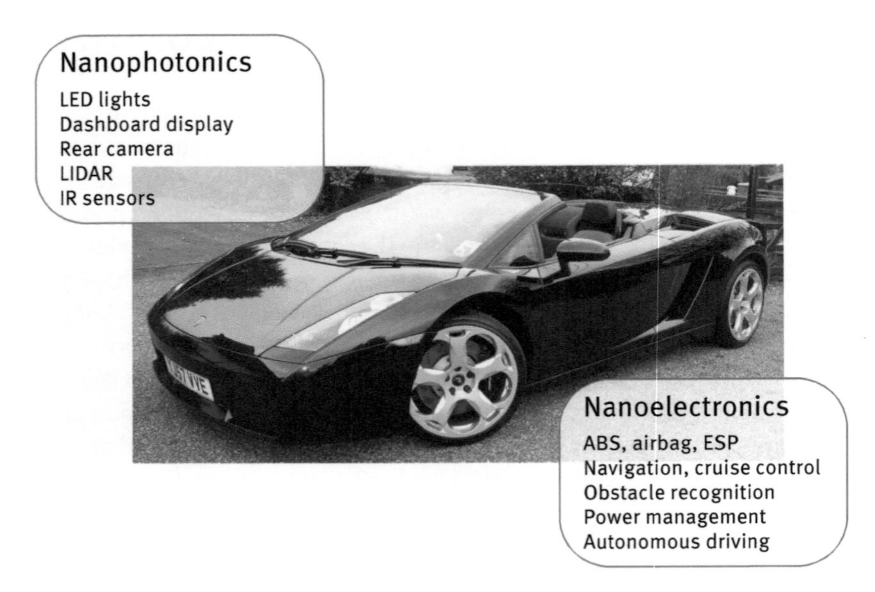

Figure 8.6: Nanophotonic and nanoelectronics in future cars.

- *Lighting and displays*: on a global scale, lighting generates 1,900 million tonnes of carbon dioxide emissions each year. In Europe, the replacement of incandescent lamps by solid-state light sources would have translated in 2015 into savings of 40,000 MW electrical peak power supply or an equivalent of 2 billion barrels of oil and 50 million tonnes CO_2. Nanophotonics has enabled solid-state light sources – LEDs and OLEDs – that outperform almost all other sources in terms of efficiency and would push potential savings up to 50% by 2025 [12]. However, it is nanoelectronics that provides the necessary drivers to adapt LED to the grid current and, coupled with sensors in intelligent light management systems, can regulate light output according to ambient lighting conditions or people's presence and activities and allows saving another 20%.
- *Energy*: The growing concerns for global warming are pushing to replace fossil sources with renewable ones, while the energy demand is increasing. The European Union (EU) has established a target of 20% of RE (Renewable Energy) as a proportion of gross final consumption by 2020, and EU leaders agreed to at least 27% of RE for 2030. Expanding the use of renewable energy sources and energy saving are the two main tools for implementing this policy. Progress in nanophotonics has enabled a dramatic decrease in the costs of photovoltaic energy generation, which is the fastest growing among renewable energy sources [13], also thanks to its almost unlimited potential, even if it has not yet reached the volume of hydroelectric, biomass, and eolic generation. On the other hand, nanoelectronics is the main enabler of energy saving, which is expected to contribute on the same level of new energy sources. Nanoelectronics is also a key element for

the exploitation of photovoltaics by providing the solid-state conversion units to interface solar cells with the grid, thanks also to the development of power semiconductor technology based on new materials,

- *Life science and health*: Optical technologies, starting from the microscope, have always played a decisive role in biology and medicine. The extensive image processing systems powered by nanoelectronics, like in CAT (Computerized Axial Tomography) and NMR (Nuclear Magnetic Resonance), have allowed to exploit a wide range of electromagnetic frequencies, from microwaves to X-rays in advanced diagnostic tools. Endoscopy, a combination of optical fibres and imagers, enables minimally invasive optical inspection inside the body, and it can be a useful support to surgery. Progress in nanoelectronics integration density is making possible the realization of smart pills that combine optical sensors, control logic, and wireless communication in a millimetre size package, thus reducing the need for invasive technologies. While nanoelectronics-driven ICT (Information and Communiation Technology) is making progress in diagnostics, with the introduction of machine learning techniques, nanophotonics has entered the field of surgery introducing laser sources for the cure of disturbs of the eye, in the treatment of dermal diseases, and in dentistry.
- *Industrial*: nanoelectronics has revolutionized modern manufacturing first with the introduction of numerical control machine tools, and then with the integration of fab equipment and logistic in an expert system. This trend has been recognized by Europe that has put Industry 4.0 among its top priorities for development. If nanoelectronics is providing the overall frame for the new industrial revolution, nanophotonics has provided some key components. Lasers are now one essential manufacturing tool, spanning from "heavy-duty" lasers, being able to focus up to hundreds of watt in a tight beam for industrial cutting and soldering, to metrological lasers used for precision measurements and holography, all integrated by computer-driven manufacturing. Lasers can also be used for material deposition: laser ablation is used to deposit also complex materials with little heating, because the electric field of the photons in high-peak power pulses is strong enough to literally rip atoms apart. Photonic sensors are another critical component, since they are able to provide chemical information without requiring a physical contact. Sensor arrays combined with the possibility of multispectral laser excitation will be able to provide real-time, three-dimensional (3D) measurement of key elements, allowing accurate control of production processes, but also helping in environmental monitoring.

8.8 A Glimpse into the Future

In Part 1, we have outlined a few of the fields in which nanoelectronics and nanophotonics are going to change and improve our everyday life. The list is by no way complete, not taking into account the progress enabled in the scientific fields and the contribution to other sectors, like security and mobility.

The growth will come both from the extension of the application of existing technologies to a variety of fields that have been only marginally impacted until now, and from the technological progress that is still proceeding at a very fast (somebody even says accelerating) pace, in spite of all prediction to the contrary.

Three areas are likely to be at the spearhead of the growth:

- *System integration*: the future will see an evolution from the solution of single issues to the integration of smart systems in a complete environment, able to give a global solution to our problems.
- *Extended control*: by incorporating sensors, data processing, and communication in many functions that now require human intervention, we could realize a true Ambient Intelligence.
- *Big data*: the integration of networks of sensors in a global communication network will make available plenty of data that can be processed with now available storage and computing power. It will therefore be possible to build comprehensive and validated models of all relevant processes and move from a problem solution approach to a problem prevention approach.

What are listed above is what could be obtained with already existing technology and with its expected evolution in the short term. Is there any factor that could stop technology evolution in the future? And which are the trends?

8.8.1 Nanoelectronics

Evolution of nanoelectronics has been closely associated with Moore's law, which has been valid for more than 50 years, and with the ITRS (International Technology Roadmap for Semiconductors) roadmap. At the moment, the 7 nm CMOS technology node has been already announced [14], and main companies are working on the 5 nm node (even if since more than a decade the definition of technology node has lost contact with actual geometrical size).

However, some clouds are appearing on the horizon of the continuation of the present rate of progress in computer performances. Already in the first years of 2000, it was realized that clock speed of computers could not go on increasing indefinitely, due to thermal dissipation problems. Parallel computing in multicore processors was introduced, but it cannot be applied to all problems. For conventional memories (DRAM (Dynamic Random Access Memory) and Flash), the reduction in the physical number of electrons that can be stored in a reduced size cell, and related worsening of noise immunity, has required the introduction of error correction strategies of increased complexity.

Anyhow, the main problem seems not to be related to the physics of the semiconductor device, but to the sharply increasing costs of the technology. As it was said before, scaling would bring huge economic advantages, by increasing the number of devices that could be produced on the same silicon substrate. The catch is that the cost of processing per substrate should increase only slowly with each new technology generation. Until now, this result was obtained by compensating the increasing cost of more advanced equipment with increases in productivity, and also through the increase of the size of silicon wafers. However, the lithographic process used to print the smaller geometry features of nanoelectronic devices seems to have reached a limit at around 40 nm, with the immersion lithography using a 193 nm wavelength. No material is available to build lenses for shorter wavelengths, and alternative equipment, based on direct e-beam writing or extreme UV light in vacuum, have until now proved to be very expensive, thus killing the economic convenience of scaling. Research is very active since several years in looking for alternative approaches to current CMOS technology. Investigations in the so-called beyond CMOS technologies range from information carriers different from electrical charges (electron spin, phonons, photons, magnetic dipole moment), to new computational approaches (neuromorphic computing, quantum computing, etc.) to innovative switches (magnetic, molecular, Josephson junctions, etc.).

None of the approaches seems to offer a comprehensive solution, but it is not easy to replace the result of 50 years of accelerated technology evolution. In the meantime, gap-stop solutions are being adopted, like moving to 3D structures for memories, using multiple lithographic steps for logic, and striving for a closer integration of nanoelectronics (for computing) and nanophotonics (for data transfer) at the chip level with silicon photonics.

8.8.2 Nanophotonics

Nanophotonics, even if already present in several important sectors, is still far from having fully deployed its potential, and no fundamental technological or economic limit is visible at the moment. However, a few critical sectors need special attention to extend the capability of the technology:

- *Integration of photonics with nanoelectronics*: as discussed above, nanoelectronic devices are used for data processing while nanophotonic devices for data transmission. One critical element for high-performance systems is the data transmission among high-performance processors and between processor and memory. Largely parallel electrical interfaces consume a lot of energy, while a photonic interconnection could offer a high-speed serial interface at low cost. Critical research and development issues for what is called silicon photonics are the integration of the optical drivers and detectors on the nanoelectronic device and the interconnection of the optical fibre to the device package.
- *Integrated photonics*: even when using optical interconnections, nanoelectronic devices are still needed when signals must be split, recombined, or anyhow

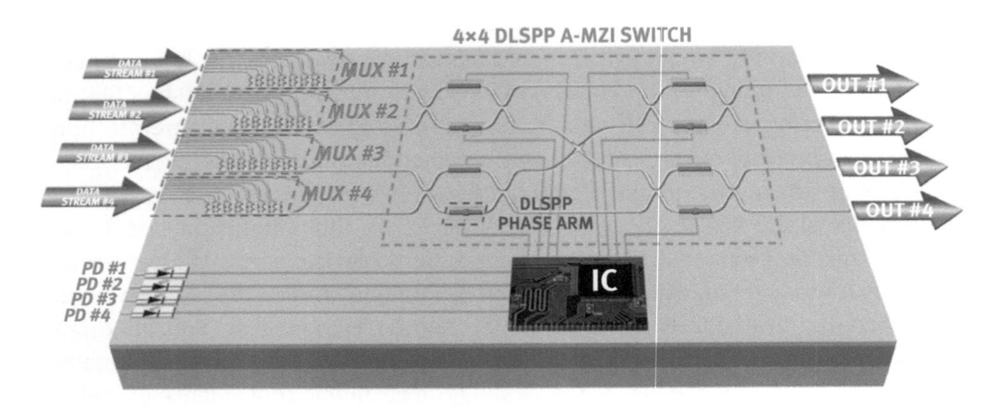

Figure 8.7: Example of photonic circuit based on plasmonics (source: Intechopen.com).

merged. The conversion from optical to electronic signals and back uses time and energy that could be saved if several optical components could be integrated on the same substrate, using the tools (lithography, deposition, and etching) taken from nanoelectronics.

Integrated photonic devices can be realized on different materials, according also to the desired performances. III–V materials are used when the integration of active devices is required. Integrated devices include low loss interconnect waveguides, power splitters, optical amplifiers, optical modulators, filters, lasers, and detectors (Figure 8.7). The primary application for photonic integrated circuits is in the area of optical fibre communication, where integrated multiplexers/de-multiplexers have replaced discrete elements.

- *Exploitation of new materials*, including new semiconductors and nanophotonic materials and technologies. For example, photonic metamaterials (artificial materials composed of subwavelength structured layers of different materials) can be engineered so as to achieve functional properties that may not be found in nature, like magnetism at optical frequencies, a negative refractive index, or enhanced optical nonlinearities. Another example is plasmonics that takes advantage of the unusual dispersion relation of light at the interface between a metal and a dielectric, both for ultrasensitive sensors and for improving light emission or light absorption properties, in LED and in photovoltaics, respectively.

Other more advanced applications are the use of photons in quantum computers. Since photons do not interact with one another under normal circumstances, photon qubits could be insensitive to decoherence by stray electromagnetic fields. Also "optical tweezers" are used to manipulate matter on the atomic or molecular scale.

8.9 Conclusions

Both electronics and photonics have profited from the rapid development of solid-state technology, from the middle of the twentieth century, even if the theoretical basis had been established already at the beginning of the century. Nanoelectronics has been the first one to develop and to enter into everyday use, while nanophotonics has been able to make use of the established technology basis of nanoelectronics to accelerate its development. Today both technologies are deeply influencing our everyday life in all its aspects and are at the basis of much larger application markets.

Exploitation potential is still far from being exhausted, and system integration will probably further extend their impact on our life and their contribution to the main societal problems.

Europe is well aware of the potential of nanoelectronics and nanophotonics for its social and economic growth, and both technologies have been classified among the six key enabling technologies (KETs). According to the Communication of the Commission: "A significant part of the goods and services that will be available in the market in 2020 are as yet unknown, but the main driving force behind their development will be the deployment of key enabling technologies (KETs). Those nations and regions mastering these technologies will be at the forefront of managing the shift to a low carbon, knowledge-based economy, which is a precondition for ensuring welfare, prosperity and security of its citizens" [15].

In accordance with this decision, important initiatives have been launched to support the development of nanoelectronics and nanophotonics in Europe under the form of two public–private partnerships, ECSEL (Electronic Components and Systems for European Leadership), and Photonics PPP (Public Private Partnership), respectively, which will mobilize public investments for more than 3 billion Euro by 2020.

References

1. Digital transformation of European Industry and Enterprises, A report of the Strategic Policy Forum on Digital Entrepreneurship, Brussels, European Commission, 2015. Available at: http://www.digitaleurope.org/ Press-Room/Publications/Digital-Transformation-of-EU-Industry-and-Enterprises-report. Accessed: 20 May 2017.
2. SIA 2016 Factbook. Washington, DC, USA: SIA, 2016. Available at: http://go.semiconductors.org/2016-sia-factbook-0-0. Accessed: 20 May 2017.
3. Photonics Industry Report: Update 2015. Dusseldorf DE 2015: Photonics21. Available at: http://www.photonics21.org. Accessed: 20 May 2017.
4. Class 977 Nanotechnology Cross-Reference Art Collection, The United States Patent and Trademark Office. Available at: https://www.uspto.gov/patents/resources/classification/class_977_nanotechnology_cross-ref_art_collection.jsp. Accessed: 20 May 2017.
5. Lilienfeld J. US Patent 1900018, 1928.

6. Dennard RH, Gaensslen FH, Rideout VL, et al. Design of ion implanted MOSFET's with very small dimensions. IEEE J Solid-State Circuits 1974;9(5).
7. Moore G. Cramming more components onto integrated circuits. Electronics 38(8):1965.
8. Moore G. Progress in digital integrated electronics. Proceedings of IEDM, 1975.
9. Zhang Q, Zhao K, Chang Z. Attosecond extreme ultraviolet supercontinuum. In The supercontinuum laser source. (edit. Robert R. Alfano) New York: Springer, 2016:337–370.
10. Einstein A. Über einen die Erzeugung und Verwandlung des Lichtes betreffenden heuristischen Gesichtspunkt. Ann Phys 1905;17(6):132–48.
11. Einstein A. Zur Quantentheorie der Strahlung. Phys Z 1917;18:121–8.
12. Consolidated European photonics research initiative. photonics for the 21st century. VDI – The Association of German Engineers. Available at: http://www.romnet.net/ro/flash news/flash news32/ visionpaperPh21.pdf. Accessed: 20 May 2017.
13. Energy policies of IEA countries – European Union – 2014 review. International energy agency. Available at: https://www.iea.org/publications/freepublications/publication/EuropeanUnion_2014.pdf. Accessed: 20 May 2017.
14. Chang J, Chen YH, Chan WM, et al. A 7 nm 256 Mb SRAM in high-K metal-gate FinFET technology with write-assist circuitry for low-VMIN applications. Solid-State Circuits Conference (ISSCC), 2017 IEEE International.
15. Preparing for our future: Developing a common strategy for key enabling technologies in the EU. COM(2009) 512 final, Brussels, BE, 2009. EUR-Lex – 52009DC0512 – EN – EUR-Lex – Europa.eu

Livio Baldi

9 Quantum computing

9.1 Introduction: A Short Story of Computing

We live in a digital world, and in a computation-intensive one. As all revolutions, it started slowly. The first conceptually true computer was the Babbage's analytical engine, which incorporated an arithmetic logic unit, control of the program flow through conditional branching and loops, and integrated memory. It could be programmed with punched cards (like the IBM 360 one century later). However, due to its mechanical complexity, it was not realized physically until 1991, and in practice, computation was limited to mechanical adding machines until the end of the Second World War. The invention of the triode by Forest in 1906 opened the way to the realization of computing machines that were not purely mechanical. However, it took a long time before the basic concept of Babbage was implemented with electronic components and the outcome was ENIAC,[1] a vacuum-tube-based monster of 30 tons, realized in 1946.

The basic concept behind the exponential expansion of computers in all fields was that a vast class of problems can be solved by recursive operations, following a procedure (algorithm) that can be implemented and made easier with the help of a computing machine. The formal basis of the computer were put down by Alan Turing [1], who described an abstract machine that manipulates symbols on an endless strip of tape according to a table of rules. Despite the model's simplicity, in principle a Turing machine (TM) can simulate any logic algorithm (Figure 9.1). Church further extended the concept, leading to the Church–Turing thesis that states that any function of natural numbers is computable by a human being following an algorithm, if and only if it is computable by a TM, if problems of resource limitations are not considered.

The TM is just a conceptual device to prove fundamental limits of computation and cannot be used for practical calculations. Most real computers are based on the von Neumann architecture that introduced the distinction between data and program storage.

The invention of the transistor in 1947 and of the integrated circuit in 1958 strongly reduced the cost of computers, and allowed them to get out of the military and scientific world and to enter civil applications. With computing cost becoming compatible with the consumer market, and personal computers making their appearance, the economic push behind the evolution of microelectronics has led to an exponential progress of the computing power, closely following the Moore's law of integrated circuit density.

1 ENIAC: acronym of Electronic Numerical Integrator and Computer. The first general purpose electronic computer of history, realized in 1946 at the Moore School of Electrical Engineering of the University of Pennsylvania

https://doi.org/10.1515/9783110547221-009

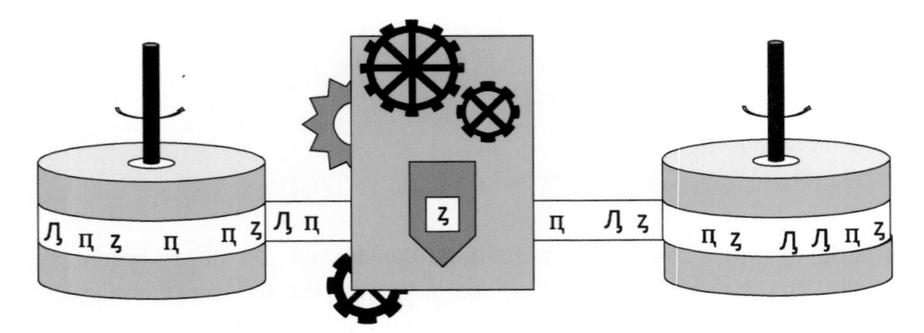

Figure 9.1: Idealized Turing machine.

Special applications supercomputers are the top runners of computing power. The TOP500 project [2] ranks and details the 500 most powerful nondistributed computer systems in the world, testing them on a standardized benchmark, based on the solution of a system of linear equations. Currently, the China-made Sunway TaihuLight is the world's most powerful and fastest supercomputer, reaching 93.015 petaflops, closely followed by the US machines. China and the United States are leading the pack, with both nations now claiming 171 systems apiece in the latest rankings, accounting for two-thirds of the list. Japan recently announced a new supercomputer for 2017, which should be capable of 130 petaflops.

However, plenty of problems still exist that are requiring resources/time not compatible with the foreseeable advances in computer technology, even if the problems themselves are theoretically computable. Some down-to-earth problems concern optimization issues and the factorization in prime numbers. The latter is a problem for which quantum computing is of special interest, and that justifies the large effort spent on it, because of its practical implications. It is a well-known theorem of arithmetic that every positive integer can be decomposed in a unique way in the product of prime numbers. This property is used in most of the cryptographic systems, like the asymmetric key cryptography, that are based on the practical difficulty of identifying the two large prime numbers that are used to generate the keys of the system. Even if theoretically solvable, key lengths of 128 or 256 bits would keep the most advanced computing systems busy for such a long time as to make computation practically impossible.

Moreover, the continuation of the present rate of progress in computer performances appears doubtful (Figure 9.2) [3]. Already in the first years of 2000, it was realized that clock speed of computers could not go on increasing indefinitely, due to thermal dissipation problems. Parallel computing in multi-core processors was introduced, but it cannot be applied to all problems. Cloud computing has been boosted by Internet and offers the advantage of spreading the computing task over a large number of processors, but again it is advantageous only if the algorithm lends itself to parallelization. A second roadblock has been the approach of the end of Moore's law, not so much related to physical problems, but to the sharply increasing costs of the technology. It

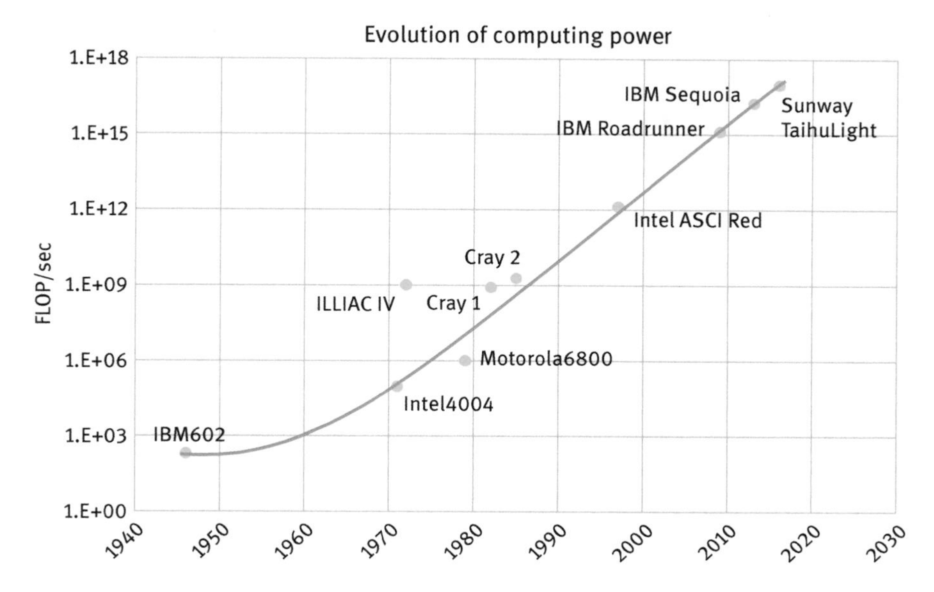

Figure 9.2: Evolution of computing power (data source: Wikipedia).

does not exclude the possibility that a few more advanced chips could be realized for special applications, for which cost is not an issue, but anyhow the foreseeable progress is not such to overcome the computing time limitations.

9.2 Quantum Theory to the Rescue

The promise of a possible solution is coming from quantum theory. In quantum theory, the state of a variable describing a system at a given time results from a complex wave function that gives the probability for the state having a specific value.[2] The act of measurement makes the variable fall into a defined state with a certain probability distribution, but before the measurement, the variable can exist in a superposition of possible states: an electron exists *both* in the spin-up *and* the spin-down states, before a measurement freezes it in one of the two states (Figure 9.3).[3]

2 Even if it does seem contrary to normal experience, the fact that a particle can be considered as "spread" over a statistic distribution is confirmed by experimental evidence; for example, by the fact that a single electron or photon can interfere with itself, or that an electron can cross a potential barrier, which is at the base of Flash memories.

3 This interpretation of quantum mechanics was not accepted by Einstein, who said that: "God does not play dice with the universe," maintaining that the variable was already in one of the two states, and the measurement was only revealing it. However, the reality of the coexistence of different states has been experimentally proved.

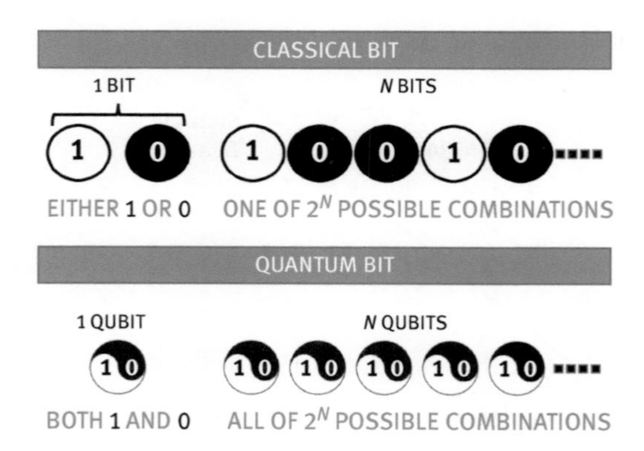

Figure 9.3: Bits versus qubits.

How can this principle be applied to computing? In normal computers, a variable exists only in one well-defined state, a "one" or a "zero." When a recursive operation is performed starting from a number of N bits, 2N combinations must be tested, and the time required increases exponentially with N. If however we are dealing with quantum variables, each variable can be in a superposition of two states, that is, in both "1" and "0" at the same time. In this case, performing the same operation on these N quantum variables, we can test all possible combinations in parallel, and the number of required iterations is increasing only slowly with N. It is therefore evident that a significant gain in performance can be achieved by operating on quantum variables [4].

Theory of quantum computation has started much before real quantum computing systems could be investigated. In a way similar to classical computing, a quantum algorithm is a step-by-step procedure, where each of the steps can be performed but on a quantum computer. Although all classical algorithms can also be performed on a quantum computer, quantum algorithms use some essential features of quantum computation such as quantum superposition or quantum entanglement. Problems that are undecidable using classical computers remain undecidable using quantum computers. What makes quantum algorithms interesting is that they might be able to solve some problems faster than classical algorithms.

As for the classical computer, the theoretical basis for the realization of quantum computers have been set much in advance of their practical realization. Already in 1985, mathematician David Deutsch gave a formal definition of a quantum Turing machine (QTM) [5]. He suggested that the classical TM could be generalized in a QTM, replacing the classical internal states with quantum variables and the transition function by proper operators. As its classical version, the QTM has no practical applications but can provide a very simple model, which captures all of the power of quantum computation. Any quantum algorithm can be expressed formally as a particular QTM.

Since then, several algorithms have been already defined for quantum computing. The most well-known algorithms are Shor's algorithm for factoring, and Grover's algorithm for searching an unstructured database or an unordered list. Shor's algorithm runs exponentially faster than the best-known classical algorithm for factoring, the general number field sieve [6]. Grover's algorithm runs much faster than the best possible classical algorithm for the same task.

Shor's algorithm has received special interest because it could be used to break public-key cryptography schemes such as the widely used RSA[4] cryptography scheme, which are based on the assumption that factoring large numbers requires unacceptable computing times by conventional computers when the numbers are sufficiently large.

9.3 How to Realize Quantum Computing?

In micro- and nanoelectronics the basic unit of information is the bit. The bit has two possible states typically called 0 and 1, which are physically represented by a charge on a capacitor (DRAM, Flash memory) or by a current flowing in a circuit.

In quantum computing, the unit of information is the qubit, which is a two-state quantum-mechanical system. A qubit is similar to a classical bit, in the sense that, if we measure it, we obtain one of the two states, like for a classical bit, but it is intrinsically very different because the qubit can exist also as a superposition of both states. Also, if we measure several times a bit, we are supposed to get always the same result, be it "1" or "0," while if we measure several times qubits that are in a superposition of states[5] we obtain an even distribution of "1 and "0."

Another important distinguishing feature between a qubit and a classical bit is that multiple qubits can exhibit quantum entanglement. Two entangled qubits have a higher correlation than is possible in classical systems, since the variables that describe their state are always coupled. For example, if a pair of particles are generated in such a way that their total spin is known to be zero, and one particle is found to have clockwise spin on a certain axis, the spin of the other particle, measured on the same axis, will be found to be counterclockwise. What is surprising is that, when measured, *both* particles collapse in their final states *at the same time*, irrespective of the distance. It is not a violation of relativity, since it cannot be used to transmit information, but it surely gets close to it. This property can be exploited in cryptography and it is one of the reasons of the strong interest in quantum computing.

Realizing qubits in practice is not such an easy task [7]. In principle, any system small enough to allow isolating its properties on the nanoscale can do the job. A large variety of alternatives exists, and each group of scientist is claiming to have the best solution. In general, what is needed is some physical support for a system that can

4 Rivest–Shamir–Adleman, from the name of the inventors, one of the first public-key cryptography systems

5 Or repeat several times the same measurement on the same qubit after restoring the initial condition.

be described by a simple, easily measurable, two-state quantum variable. Most used physical implementations of qubits are the polarization state of photons, the spin of electrons, atomic nuclei, or atoms, the sense of current or the phase in a superconducting Josephson junction, the position of charge in a quantum dot pair, but several more exotic versions are being explored (Figure 9.4). The large number of candidates demonstrates that the topic, in spite of rapid progress, is still in its infancy.

The next problem is to realize the quantum logic gates that should be used to implement the desired algorithm. The quantum logic gates can consist of laser beams and optical paths, microwaves, electric fields, or other probes, designed to let the system's wave function evolve in a well-defined way such that, upon measurement, there's a high probability that it will collapse to the classical state corresponding to the right answer for the problem. However, one catch of quantum computing is that quantum algorithms are often nondeterministic, and they provide the correct solution only with a certain known probability. Repetition of the computation is therefore needed to obtain a probability distribution. We should point out that also the well-known Shor's algorithm will return the correct answer *only* with high probability.

Aside from the practical realization of the quantum gates, the implementation of the algorithms in a quantum computer is not an easy task. Qubits must be manipulated, yet simultaneously protected from any external source of interference. Heat or electromagnetic radiation would react with the qubits in a way similar to what measurement is doing and destroy the superposition of states. This process is known as decoherence [8] and requires isolating the system from its environment. For this reason, most of the implementation of quantum computing operate at temperatures very close to absolute zero, down to a few tens of millikelvin. Decoherence is irreversible and

Figure 9.4: Qubit quantum processing circuit (source: D:wave).

is characterized by a typical decay time that can range between nanoseconds and seconds even at low temperature. The fight for longer decoherence times is one of the key issues of quantum computing: to have meaningful results, the decoherence time must be much longer than the computation time.

Error correction can be used to compensate for failing qubits [9], but usual error correction algorithms adopted for classic bits do not work for qubits. Other procedures exist that spread out the information over a larger number of qubits. However, also the extra qubits can also decohere, so adding them to the quantum system can actually increase the system's vulnerability to outside interference. All of this imposes a fault-tolerance threshold on a given error correction scheme – a maximum frequency of errors that the computer can encounter when operating on qubits, such that adding more qubits makes more likely for the algorithm to give the wrong answer.

Another practical problem is the fact that the very low temperatures required to reduce parasitic effects leading to decoherence are posing severe engineering problems to the interfacing of quantum devices to conventional electronics for the handling of the results of the calculation.

In general, however, all the approaches to quantum computing tested until now are far enough from classical binary computing which executes programs stored in a memory. Even if basic logic quantum gates and elementary quantum memory systems have been demonstrated and compatibility with Von Neumann computer architecture have been theoretically proved, most of implementation are more in the way of dedicated machines, able to perform a single well-defined operation, in this way similar to hard-wired logic or finite-state machines.

An alternative approach is the adiabatic quantum computation, which is used in the only commercial quantum computer built until now. The basic concept, based on the adiabatic theorem, is that a system can be made to evolve from one ground state (the initial conditions of the problem) to the final state (the solution of the problem) without losing coherence, if the transformation is slow enough, and at any moment the distance from other possible states of the system is large enough [10]. The critical issue here is to avoid falling in the trap of the Heisenberg's uncertainty principle $\Delta E \bullet \Delta t \geq h$.

9.4 Where We Stand Now

Theoretical work on quantum computing started already in the 1970s, but it is only in the 1980s that sound bases were established with the paper on the QTM. Work went on in the 1990s with the definition of several algorithms for quantum computing, and the first example of qubits and quantum gates were realized. From then, there has been a continuous progress in the number of qubits and in the practical demonstration of some algorithms on elementary quantum structures of a few qubits.

Probably the most advanced qubit technology is the one based on Josephson junctions. The Canadian company D-Wave claims a computing machine with 1,000 qubits [11] that are working at little more than 10 mK under extreme vacuum (Figure 9.5). The system works on the adiabatic quantum computation principle, looking for the lowest energy point of a system of interconnected qubits, under the assumptions set by the user. The processor considers all possibilities simultaneously to determine the lowest energy and the values that produce it and returns multiple solutions to the user. Because a quantum computer is probabilistic, the computer returns not only the best solution found but also other very good alternatives close to it. Machines have been sold to some key customers such as NASA, Google, and the US defense contractors. Contrasting opinions exist on this machine: scientists of ETH in 2014 tested a previous generation machine (503 qubits) on some problems and were not able to evidence any quantum-related speed improvement over conventional computers [12]. At the beginning of 2016, however, Google researchers published an internal paper claiming a speed increase by a factor of 100 million on a simulated annealing problem [13], which proves the difficulty in assessing the real performances of quantum computing. Skepticism concerns the number of qubits that really participate in the computation, and the decoherence time of the machine. It should be noted also that adiabatic quantum computing requires the system to be made to evolve very slowly (ideally in an infinite time) to avoid losing coherence, which rather conflicts with the speed of calculation.

Figure 9.5: The D-Wave 2000Q quantum computer (source: D:Wave).

Another popular approach involves photons. Photons do not interact with one another under normal circumstances, and therefore a superposition state, for example of spin, could be insensitive to decoherence by stray electromagnetic fields. However, in order to realize quantum gates, photons have to be made to react with each other, which could require interacting with matter. Some basic quantum computation with photons has been already achieved, thanks also to the development of sources and detectors for single photons. A potential advantage of the photon-based system is given by the advances in nanophotonics that are opening the possibility of realizing waveguides in millimeter-sized silicon chips, on which photon sources and detectors could be mounted (Figure 9.6).

Photons are also widely used to initialize or read solid-state qubits, for example, the spin of electrons that are trapped by the so-called nitrogen vacancy centers in diamond – point defects in the diamond's rigid lattice of carbon atoms where a nitrogen atom has been substituted.

Probably less advanced, but capable to open a wider spectrum of opportunities, is the use of silicon as the substrate for qubits. Physicists at the University of New South Wales in Sydney, Australia, are developing a semiconductor device involving spin qubits made from phosphorous atoms embedded in a silicon substrate, in which qubits have been reported to remain in a superposition state for up to 30 s. A recent paper published by the CEA-LETI [14] illustrates the qubit formation in a nanowire silicon MOS device. The qubit is encoded in the spin of a hole in the quantum dot defined by one of the gates of a compact two-gate p-channel MOS (Metal Oxide Semiconductor) transistor. Coherent spin manipulation was possible by means of a radiofrequency field applied to the gate itself. The spin quantization required operation at 100 mK, due to the limits of manufacturable device size, but higher temperatures could be achieved

Figure 9.6: Photonic quantum gate from the University of California, Santa Barbara (source: Futurism.com).

by scaling the nanodot size. The big advantage of this approach is the possibility of interfacing with conventional CMOS technology to realize hybrid systems.

9.5 Quantum Communications

While one of the claimed potential uses of quantum computing is to break the most common cryptography codes, quantum theory also offers a solution to secure communications through quantum cryptography. Quantum cryptography exploits some properties of quantum theory, like Heisenberg's uncertainty principle or state entanglement, to transmit the key for a symmetric key cryptography in such a way that any attempt to intercept the transmission would be detected [15]. The basic concept is that whoever tries to intercept the transmission of the key must make a measurement and therefore alters the distribution seen by the receiver. A comparison of the distribution of a subset of data can evidence the presence of an eavesdropping. If not, the remaining data are the key. Of course, the system requires that data be transmitted with a high degree of coherence: any decoherence due to thermal noise or other casual sources would spoil the statistics like an eavesdropping. This limits the distance over which the key can be transmitted. A large statistical basis is also required. However, several companies already entered the business, such as MagiQ Technologies (United States), ID Quantique (Switzerland), QuintessenceLabs (Australia), and SeQureNet (France), and secure quantum key distribution up to 100 km has been claimed.

Another side application of quantum computing in secure communications is the generation of random numbers, which are used in cryptographic systems (Figure 9.7). Commercially available generators make use of computational algorithms that can produce long sequences of apparently random results, starting from a seed value

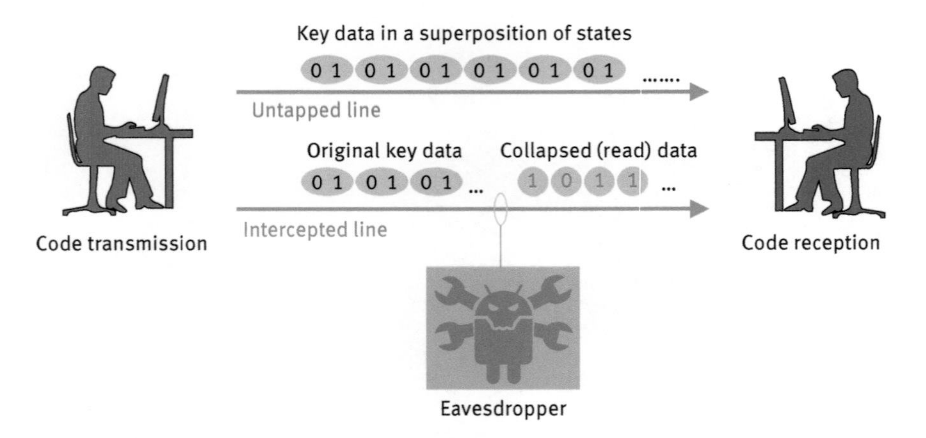

Figure 9.7: Principle of quantum cryptography.

or key. However, the entire sequence can be reproduced if the seed value is known. Quantum physics offers a perfect solution for the problem: measuring a system that exists in a superposition of states gives a perfectly random result. Due to the availability of high-quality optical components and the potential of chip-size integration, most of today's practical quantum random number generators are implemented in photonic systems.

9.6 Conclusions

In the last 15–20 years, impressive advances have been made in quantum computing, both on the theoretical and on the implementation side. Mathematical basis seem to have been firmly established and algorithms have been defined for a variety of problems. Several approaches to generate and handle qubits have been tried, with an increasing number of qubits and increasing decoherence times. Algorithms have been tested in elementary form with some of the available technologies. Even room temperature operation seems now to be within reach. However, the diversity of the approaches still being actively investigated clearly testify that no fully satisfying solution has been reached. Moreover, the probabilistic nature of quantum mechanics has the effect that the results of many experiments are based on statistics, and therefore open to criticism and different interpretations.

The strong push toward the development of a quantum computer, or something similar, comes also from its practical implications for security. It is not a chance that one of the first D-Wave machines ended up in the labs of a US defense contractor, and that documents by Edward Snowden affirm that the US National Security Agency is running a large research program to develop a quantum computer capable of breaking vulnerable encryption systems. In December 2015, IBM announced that it has been notified by the US Intelligence Advanced Research Projects Activity (ARPA) program of the intention to award a major multiyear research grant to advance the building blocks for a universal quantum computer [16].

If we take into account all the technical problems, including decoherence, error correction, and interfacing, we have to conclude that a universal quantum computer, capable of performing all the tasks of a classical supercomputer but at quantum speed, is still rather far in the future.

A useful alternative, which is closer at hand, is given by specialized devices known as quantum simulators, which are built to model a specific system, or to carry out a single algorithm. Quantum simulators that can outperform today's best supercomputers for specific tasks might need only 50–100 qubits, having little or no need for error correction, and could model, for example, the behavior of small molecules or perform multiple parameter optimization tasks, like those that are supposed to be solved by the D-Wave machine.

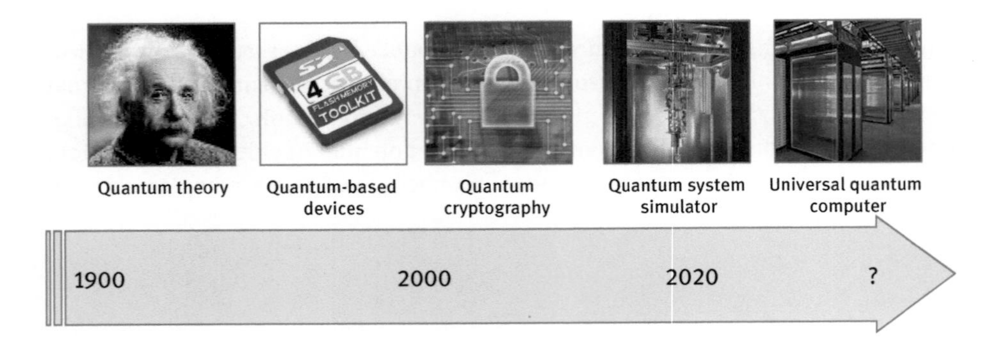

Figure 9.8: Evolution of quantum computing.

On the other side, when Einstein, Heisenberg, Bohr, and Schrödinger put the basis of quantum physics one century ago, they could not imagine that their abstract formulations would have found practical application in a variety of quantum effect-based devices to be found in everyday life, like tunnel diodes, quantum dot laser, and Flash memories (Figure 9.8). Even now, quantum cryptography is finding its way in secure communications, and first quantum simulators seem to be not far away. However, the realization of the universal quantum computer seems to be much more complex, and widely different estimations are given by scientists on the time needed.

In Europe, following a "Quantum Manifesto" [17] published with the support of more than 3,000 representatives from academia, industry, and governmental and funding institutions, the European Commission has decided to launch a €1 billion initiative to coordinate science, engineering, and application work on quantum computing. The initiative will be managed as part of the Future and Emerging Technologies (FET) program and has a time horizon of 10 years. In 10 years, we should have a much better view of the possible timing of the universal quantum computer.

References

1. Turing AM. On computable numbers, with an application to the Entscheidungs problem. Proc London Math Soc 1937;2(42):230–65.
2. TOP500 web page: https://www.top500.org/
3. Markov L. Limits on fundamental limits to computation. Nature 2014;512:147–54.
4. Nielsen MA, Chuang IL. Quantum computation and quantum information. Cambridge: Cambridge University Press, 2001.
5. Deutsch D. Quantum theory, the Church-Turing principle and the universal quantum computer. Proc R Soc A 1985;400(1818):97–117.
6. Hayward M. Quantum computing and Shor's algorithm. In Matthew Hayward's quantum algorithms page, imsa.edu. Available at: http://citeseerx.ist.psu.edu/viewdoc/download?doi=10.1.1.121.1509&rep=rep1&type=pdf. Accessed: 17 February 2005.

7. Cartlidge E. Quantum computing: how close are we? Opt Photonics News, Vol. 27, Issue 10, Oct. 2016:30–37.
8. Schlosshauer M. Decoherence, the measurement problem, and interpretations of quantum mechanics. Rev Mod Phys 2005;76:1267.
9. Gottesman D. Class of quantum error-correcting codes saturating the quantum Hamming bound. Phys Rev A 1996;54:1862.
10. Das A, Chakrabarti BK. Colloquium: quantum annealing and analog quantum computation. Rev Mod Phys 2008;80:1061.
11. D-Wave Systems Breaks the 1000 Qubit Quantum Computing Barrier. D-wave press release 22 June 2015.
12. Rønnow TF, Wang Z, Job J, et al. Defining and detecting quantum speedup. Science 2014;345:420.
13. Neven H. When can quantum annealing win? in Google Research Blog, Tuesday, 8 December 2015.
14. Hutin L, Maurand R, Kotekar-Patil D, et al. Si CMOS platform for quantum information processing. 2016 IEEE Symposium on VLSI Technology, Kyoto, April 2016.
15. Gisin N, Thew R. Quantum communication. Nature Photonics 1, 165–171 (2007).
16. Loeb L. Spy agencies fund IBM's quantum computing research information week 11 Dec. 2015.
17. Quantum Manifesto. Available at: http://qurope.eu/system/files/u7/93056_Quantum%20 Manifesto_WEB.pdf

Livio Baldi

10 Spintronics

10.1 Introduction

Spintronics (from spin and electronics) is a new branch of nanoelectronics that exploits the detection and manipulation of electron spin, while nanoelectronics, until now, has been mainly concerned with the exploitation and manipulations of electron charge. Even if the physics of electron spin (and of the spin of elementary particles in general) dates back to the beginning of the twentieth century, the interest in the manipulation of electron spin and the tools to enable it are relatively recent. A strong impulse to the research in this area has come from the investigation of an alternative to electrons as information carriers, in the quest for a solution to the perceived limits of Moore's law.

10.2 The Nature of Spin

In quantum mechanics, spin is a parameter linked to the angular momentum of elementary particles such as electrons, hadrons (e.g. protons and neutrons) and atomic nuclei. It is associated with a rotation of the particle on its axis, and it differs from the orbital angular momentum, associated with the trajectory of the particle (e.g. the electron rotating around the nucleus).

As a quantum variable, spin was introduced by Wolfgang Pauli in 1924 to differentiate electrons that occupy the same orbit. It was a consequence of the exclusion principle by Pauli that states that two electrons cannot have the same quantum number, and which was introduced to explain the structure of atom, as it was emerging at that time from chemical and spectroscopic studies. It was afterwards generalized in the Fermi–Dirac statistics that applies to fermions, particles endowed with mass, which are characterized by half-integer value of spin. The spin of electrons can have the value $\pm 1/2$.

The elementary explanation of the spin as the result of the rotation of a particle around its axis is not correct, but it can be a useful simplified model. For particles with electrical charge, like electrons, the angular momentum is associated with a magnetic dipole moment, as shown in Figure 10.1.

The quantization of the magnetic dipole moment associated with the spin was experimentally evidenced by the Stern–Gerlach experiment, where electrically neutral atoms were thrown through a non-homogeneous magnetic field. According to the spin, atoms were deflected in well-defined directions, indicating the presence of discrete momentum values.

https://doi.org/10.1515/9783110547221-010

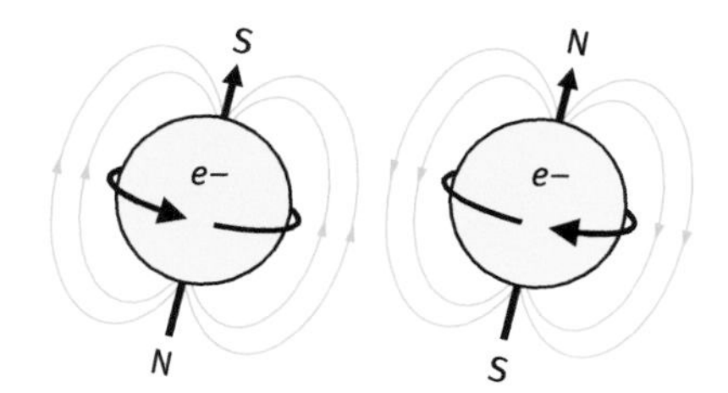

Figure 10.1: Schematic representation of electron spin.

However, aside from its importance in the quantum theory, and some application in spectroscopy (Zeeman effect[1]), the electron spin did not receive any special attention for a long time.

It was only towards the middle of the twentieth century that the progress made in microelectronics made possible practical applications based on the spin properties of electrons and nuclei.

10.3 Making Spin Resonate

The first practical application of spin was related to the spin of atomic nuclei. Nuclear magnetic resonance (NMR) is a physical phenomenon in which nuclei absorb and re-emit electromagnetic radiation due to the interaction of their spin with an external magnetic field. The absorption/emission takes place at a specific resonance frequency that depends on the strength of the magnetic field and the magnetic properties of the atoms. The principle of NMR usually involves two sequential steps:

- The application of a strong magnetic field to align the magnetic nuclear spins.
- The perturbation of this alignment with a radiofrequency (RF) pulse perpendicular to the polarizing field to maximize the signal. The RF pulse resonates at a frequency that depends on the strength of the polarizing field and on the type of nucleus.

NMR was first described and measured in molecular beams by Isidor Rabi in 1938, and in 1944, Rabi was awarded the Nobel Prize in Physics for this work. In 1946, Felix Bloch and Edward Mills Purcell applied the technique to the analysis of liquids and solids, which brought them the Nobel Prize in Physics in 1952. NMR spectroscopy is currently used to study molecular physics, crystals and non-crystalline materials. Perhaps the best-known application of NMR is as a technique for high-resolution, non-invasive, medical imaging. The basis of this technique is in the fact that the resonance frequency for a given substance is proportional also to the strength of the

1 When atoms are in strong magnetic fields, electrons on the same orbit have different energy, according to the orientation of their spin with respect to the external field. Therefore, the absorption or emission spectral lines of the atoms split into more lines, in relation to the different energy values.

applied magnetic field. Therefore, if a strong non-uniform magnetic field is used, the resonance frequency varies with the position and gives the possibility of mapping the spatial distribution of the substance.

10.4 Giant Magnetoresistance

The fact that the resistivity of some materials changes in the presence of an applied magnetic field was known since the nineteenth century as "magnetoresistance", even if it could not be explained until much later, when the electronic structure of the atom was completely understood. Several types of magnetoresistance exist, depending on the atomic structure, but in the most common case for iron and ferromagnetic materials the resistance increases when the magnetic field is parallel to the current flow, decreases when it is perpendicular, even if the effect is generally limited to a few per cent. The explanation is related to the interaction of the magnetic field with the intrinsic and orbital spin of the electrons. This effect has been widely used in magnetic sensors, especially for the reading heads of magnetic hard disk drives.

It was the progress in thin-layer technology, triggered by the evolution of microelectronics, that opened the way to the discovery of the giant magnetoresistance (GMR) effect in the second half of the twentieth century. In the late 1980s of last century, both Peter Grünberg of the Jülich Research Centre and Albert Fert of the University Paris-Sud discovered the GMR effect working on thin layers of ferromagnetic and non-magnetic materials. The theory was developed in the following years, and Grünberg and Fert obtained the Nobel Prize in 2007 for their discovery. Essentially, GMR is related to the (relatively) large change of resistivity in thin multilayer stacks in the presence of a magnetic field. Fert used the "Giant" term, because the change was relevant at cryogenic temperatures, even if much less at room temperature.

The qualitative model of the effect (Figure 10.2) is based on the fact that the conduction is due to an equal mixture of electrons with spin parallel or antiparallel to

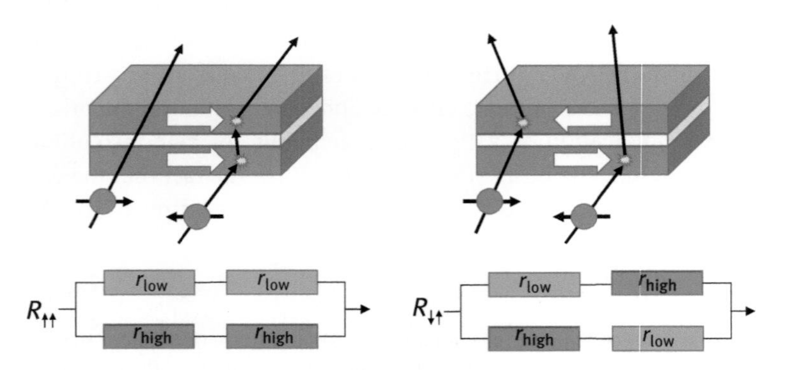

Figure 10.2: Qualitative model of giant magnetoresistance.

the magnetic field [1]. In metals, the spin of electrons has a longer lifetime than the scattering process with the reticle; therefore, spin-up and spin-down electrons can be considered as two independent currents. In ferromagnetic materials, the scattering of electrons is different for electrons with different spin: strong for electrons with spin antiparallel to the magnetization of the ferromagnetic material, and weak for electrons with spin parallel to it. Therefore, if two layers of ferromagnetic material have a parallel magnetization, one half of the electrons crossing them, having spin parallel, will find a low scattering, that is, a low resistance path, while if the magnetization is antiparallel, all electrons will find high resistance paths.

This effect is valid for currents perpendicular to the layers, as in Figure 10.2, but also for currents parallel to it, if the layers are thin enough to have surface scattering as the dominant effect.

The GMR effect takes place when two thin layers of ferromagnetic materials (Fe, Co, are) are separated by a thin layer of paramagnetic material (Cr, Cu, etc.). For a proper thickness of the non-magnetic material (in the order of nanometres), the coupling between the two magnetic layers becomes antiferromagnetic, that is, the magnetization in the two layers is oriented in opposite directions. In this case, the resistance is always high. However, if an external magnetic field is applied, the two layers can be forced into parallel magnetization, and the total resistance drops.

If one of the two ferromagnetic layers has a high coercitive field (fixed or pinned layer), while the other has a low coercitive field (soft layer), we have a spin valve [2]. In this case, the parallelism or anti-parallelism of the layers, and therefore the overall resistance, is controlled only by the external magnetic field. The fixed layer can be fixed to its magnetization direction by an adjacent antiferromagnetic layer (spin valve) or be simply made by a material with very high coercitive strength, like NiFe (pseudospin valve). In both cases, it is important that the separation layer is thick enough to prevent the interaction between the two ferromagnetic layers in order to enhance the control of the magnetization by the external field. On the other side, it must be such as to preserve the spin orientation of the crossing electrons, such as a conducting paramagnetic material. Replacing the conductive layer with a dielectric gives us a magnetic tunnel junction (MTJ). In this case, the mechanism that controls the conductivity is different and related to different density of states for parallel or antiparallel oriented electrons. If the two layers are parallel, majority electrons on one side will find a large density of states on the other side of the junction, and tunnel probability will be high. If they are antiparallel, majority electrons on one side will face a low density of states on the other side and tunnel probability will be low. Minority electrons will still face a large density of states, but their number is much lower.

The largest technological application of GMR is in the read head sensor of hard disk drives. As the magnetic domain in the hard disk drive passes under the read head, the soft magnetic layer of the spin valve aligns parallel to it. If the direction is also parallel to one of the fixed layer, the resistance to the motion of the electron with the correct spin will be minimal and a “1” will be read. If the alignment is antiparallel, all the electrons will encounter a high resistance path and a “0” will be read. Since the change between parallel and antiparallel orientation takes place around $H = 0$, at this point the change in resistance will be maximal. Thanks to the introduction of GMR reading heads, compact hard disk drives have become the most popular storage medium in portable computers, and have entered, even if for a short time, the consumer market (first-generation i-pod). Magnetic disks are still today the most cost-effective storage medium, at the price of about 20 GB/US$ (2017).

Due to its sensitivity and relative simplicity, the GMR effect is used also in other magnetic sensors, like solid-state compass. One special type of spin valve is also getting increased attention for non-volatile solid-state memories.

10.5 Solid-State Memories

Magnetic storage has been at the beginning of computer industry. Much before the introduction of semiconductor memories, the random-access memory (RAM) of computers was based on arrays of tiny ferrite rings, through which wires were threaded. The information was stored as the direction of magnetization in the ring, writing was achieved via the Ørsted effect by sending electric pulses of proper polarization through the two wires (*X* and *Y* addresses) crossing inside the ring, and reading by bringing the magnetization to a reference value, which induced a current pulse in a third wire. Magnetic core memories had also the advantage of being non-volatile, that is, of keeping the information even in the absence of an applied voltage. They were the main RAM of computers until replaced by dynamic RAM (DRAM) in the middle of the 1970s.

In the 1990s of the same century however, the discovery of the GMR effect and of the spin valve opened the way to a new version of the magnetic memory, this time based on the electron spin. The basic cell of the magnetoresistive RAM (MRAM) is formed by a tunnel barrier spin valve or Magnetic Tunnel Junction (MTJ), where two ferromagnetic layers, one soft and one fixed, are separated by a thin metal oxide layer (Figure 10.3 (A)) [3]. The information is stored in the orientation of the soft magnetic layer. If the two magnetic layers have parallel orientation, the tunnelling probability is larger than when the two layers have antiparallel orientation. Reading is performed by measuring the tunnelling current and it is non-destructive, even if it requires measuring very low currents. In the first implementation of the memory, the orientation of the soft layer was changed in a way similar to the old magnetic core memory, by sending current pulses in two wires (*X* and *Y* coordinates) that cross at the location of the cell (at 45 degrees with respect to the magnetic field orientation in the cell, to minimize interferences).

The investigations in this kind of memory had a marked acceleration at the beginning of the new millennium, culminating in several presentations at International

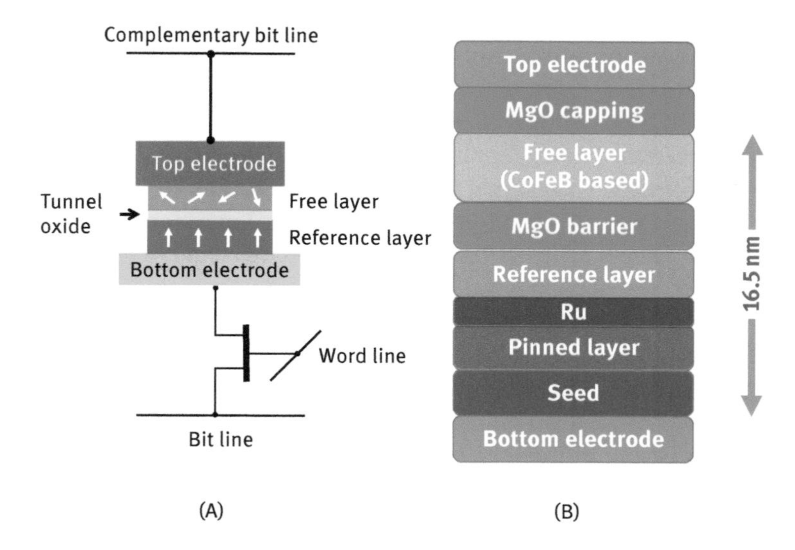

Figure 10.3: (A) Schematic section and (B) stack composition of an STT MRAM.

Solid-State Circuits Conference 2003, and the announcement of a 4 Mb memory by Motorola at International Electron Devices Meeting 2003 [4]. The interest in this approach was related to the fact that the memory promised to combine the high switching speed of DRAM with the non-volatility of Flash, and to offer good radiation resistance compared to SRAM and DRAM. It was this last characteristic that gave impulse to the development of the memory for military and space applications. However, for less demanding applications the memory had several drawbacks, like the large currents required for programming via the Ørsted effect, the small I_{on}/I_{off} ratio and the poor cell scalability, also due to the programming crosstalk between neighbouring cells, since it was not possible to contain the magnetic field in the programming wires.

The main breakthrough that has revamped interest in magnetic RAM has been the introduction of the spin-transfer torque (STT) effect that exploits the spin of electrons for the polarization of the soft layer. To achieve a parallel polarization, an electron current is forced from the fixed layer through the tunnel junction towards the soft layer. Passing through the thick fixed layer, the spin of the electrons is forced to be parallel to the dominant magnetic field. When the electrons enter the soft layer, they transfer their magnetization to it, if the current is large enough. Reversing polarization is achieved by forcing an electron current in the opposite direction. The mechanism is more complex, but it could be simplified in this way: electrons coming from the soft layer and going to the fixed layer are randomly oriented. The ones with orientation parallel to the fixed layer go through; the others are reflected and force the soft layer to an antiparallel spin orientation. This effect is minimal in programming because the STT effect of the electrons is not strong enough to change the orientation of the thick fixed layer.

STT memories have the advantage of being programmed through an electric current, and therefore are immune to the parasitic magnetic fields that were limiting the scaling of previous generation GMR RAM. Further progress has been achieved by introducing MgO as dielectric, which increased the I_{on}/I_{off} ratio to more than 300% [5], and by using fixed and soft layer with magnetic polarization perpendicular to the layers and parallel to the electron flow. The scalability of the STT MRAM cell has been demonstrated by the announcement of a 4 Gbit memory by Hynix-Toshiba in 2017 [6], even if the largest commercially available device at the moment is the 256 Mbit memory by Everspin.

STT MRAM is considered a viable alternative especially for the replacement of DRAM, due to high programming speed, endurance and non-volatility. However, some issues are still to be solved, like the low I_{on}/I_{off} ratio, the small voltage window between programming and breakdown voltages, problems of stability of magnetic polarization and the complexity of the metal stack (Figure 10.3 (B)). Several additional layers are needed for contact, layer adhesion, matching of crystalline structure and compensation of stray magnetic fields, bringing the total to more than 10 layers, with thickness in the range of a few nanometres.

Several alternative approaches are being considered to reduce the programming current and to improve the reliability of the memory, like

- the introduction of local heating, possibly induced by the same programming current, to decrease the switching energy (thermally assisted MRAM) and
- the use of voltage control of anisotropy (VCMA) or of more exotic effects, like Spin-Hall effect or topological insulators, to achieve the switching of the free layer with a current parallel to the tunnel junction.

A completely different approach is the racetrack memory, originally proposed in 2003 by IBM [7, 8]. It is a sort of shift-register memory, in which domains of different polarity are moved along a magnetic wire by the application of a current. In principle, high densities could be achieved, especially if vertical wires could be used, as in the original patent, but scalability, controlled domain motion and current densities are still an issue.

10.6 Spin Oscillators

Another potentially interesting application of the spin valve is the spin-transfer nano-oscillator (STNO) [9]. The basic structure is the spin valve, with a fixed magnetic layer, a thin non-magnetic separation layer and a second thin “soft” magnetic layer. The non-magnetic material can be a conductor, as in the GMR spin valve, or an insulator as in the tunnel spin valve of STT MRAM. As discussed above, when the electrons, spin polarized by their passage through the fixed magnetic layer, are injected into the soft layer, the angular moment they are carrying exerts a torque on the magnetization vector of the soft layer leading to either reversal (the effect exploited in the memory) or to a persistent precession, with the generation of microwaves. An external magnetic

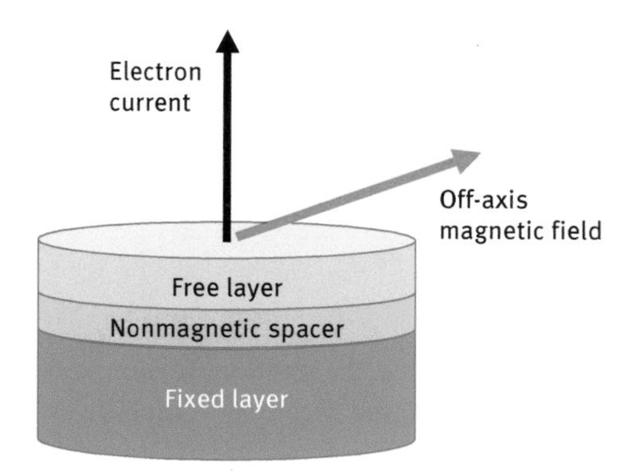

Figure 10.4: Schematic description of a spin transfer nano-oscillator.

field, applied off-axis with respect to the easy axis of the free layer, ensures the continuation of the precession and avoids the stabilization of the layer (Figure 10.4). The frequency of the oscillation is related to the applied field strength and the current density.

To produce a significant oscillation effect, the current *density* must be high, in the order of 10^7 A/cm^2, which means that the device geometry must force the current through a very small cross section. Therefore, STNOs are built either as thin pillar-like structures or as point-contact structures on extended multilayers.

The use of a conductive or of an insulating non-magnetic layer has an influence on the characteristics of the oscillator: in the first case, the series resistance is low and therefore the required large current density can be achieved at low voltages. However, the output signal is also small. In the second case, a larger output power can be obtained, because of the larger output resistance, but the breakdown voltage of the tunnel junction, which is around 1–1.5 V, limits the current.

STNOs can be realized with the magnetic field in the fixed and soft layer either parallel or perpendicular to the plane of the junction. Another option is to have the two polarizations perpendicular to each other, which can lead to permanent oscillations even in the absence of the external magnetic field, but the structure can be less stable.

The interest in the STNOs lays in the several potential advantages over standard LC (Inductance-Capacitance) voltage-controlled oscillator (VCO):

- They are tuneable by bias current and magnetic field over a range of several GHz, which is much larger than what is achieved with VCOs.
- They work over a broad range of temperatures and can be biased at low voltages (<1.0 V).
- The structure of STNO is quite simple and compact, and in principle compatible with standard CMOS (Complementary MOS) making it especially suitable for on-chip integration.

The main problems to be solved are the very low output power, which is anyhow improved by using the tunnel spin valve approach, the large external magnetic field, which would be an obstacle to on-chip integration and the large phase noise, which is mostly due to the impact of thermal fluctuations on the dynamics of magnetic nanostructures. Research is very active, and more complex device structures, including additional magnetic layers and the use of more exotic spin effects, like topological insulators, are being investigated for this purpose.

10.7 The Quest for Spin Logic

The discovery in the early 2000 that limits imposed by power dissipation to microcontroller scaling were much closer than the end of Moore's law triggered a large investigation in alternative devices, architectures and information carriers. There were several contributors to the problem:

- While the geometrical size of the transistors had been scaled according to the Dennard's scaling rules, the supply voltage was not, leading to a significant increase in switching power for the elementary gate.
- The progress of the technology allowed an increase in chip size, which did mean longer interconnections and larger parasitic capacitances to be charged for signal transmission.
- The clock frequency, that is the computer speed, has been constantly increasing, paralleling Moore's law.

The increasing awareness of the problem is evident from the extrapolated maximum power dissipation of high-performance MCU (Micro Controller Unit)s in the subsequent versions of ITRS (International Technology Roadmap for Semiconductors) from 2001 to 2011 (Figure 10.5), which at the end had to comply with the package thermal dissipation limit of around 150 W.

Waiting for a better technology to become available, the problem was bypassed with the introduction of multicore processors, using parallel processing techniques and limiting the clock frequency to around 3 GHz, even if this solution is not ideal for all classes of problems.

Already in the 2001 release of ITRS (International Technology Roadmap for Semiconductors) several alternatives were considered, such as spin devices and spin-based logic. The interest in spin as information carrier was based on two facts:

- If information could be transferred by changing the spin orientation of electrons with spin waves, instead of physically moving the electrons, the power dissipation due to capacitance charging could be strongly reduced.
- The discovery of the GMR and the recent results with magnetic memories based on tunnel junction spin valves were showing that spin manipulation was possible and exploitable even outside the research labs.

An all-spin logic, where the transfer of spin orientation replaces the electron propagation in interconnections, could be the ideal solution to the power dissipation problem. To this purpose, it is required to inject, propagate and detect spin signals.

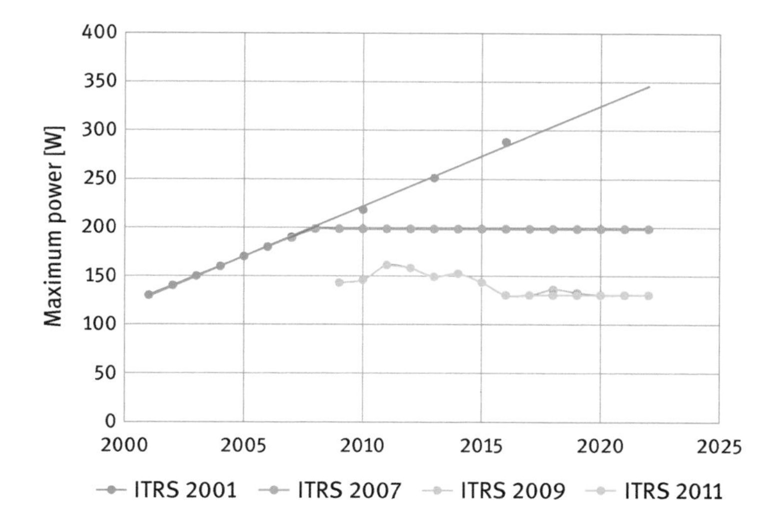

Figure 10.5: Allowable maximum power forecast – high-performance MCU with heat sink (source data: ITRS).

- **Spin transistor**

  A theoretical model for a spin transistor already exists since 1989 and it is the Datta-Das model [10]. In its essence, it looks like a metal oxide semiconductor (MOS) device, in which a spin-polarized current is injected through a ferromagnetic material at the source and is collected by another ferromagnetic contact at the drain (Figure 10.6).

  The spin direction of the electrons in the channel can be manipulated by influencing the spin precession angle with the voltage applied to the gate, exploiting what is known as Rashba effect [11]. Assuming that source and drain ferromagnetic contacts have antiparallel orientation, when the gate voltage is such as to invert the

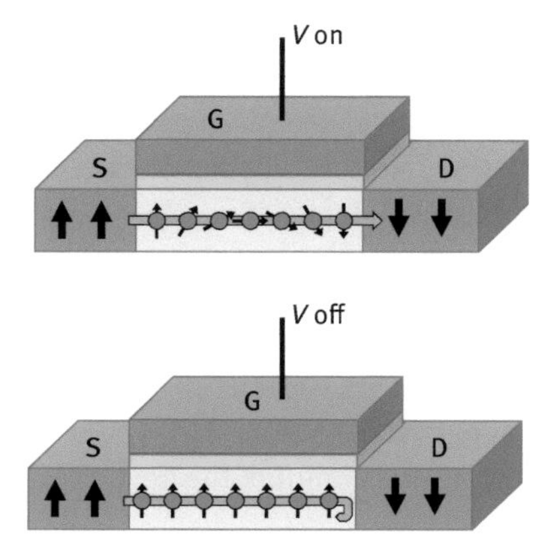

Figure 10.6: Schematic model of a spin transistor.

spin orientation of the electrons in the channel, they become parallel to the drain, all the current goes through and the transistor is on. On the contrary, if the voltage is such as not to invert the spin or to invert it an even number of times, electrons are antiparallel to the drain contact, the drain current drops and the transistor is off. However, injecting spin-polarized electrons into the semiconductor channel can be hard to achieve, due to the impedance mismatch between semiconductor and ferromagnetic source and drain. The electrical injection of spin-polarized electrons in a semiconductor can be improved by inserting a thin insulating tunnel barrier, as in STT MRAM, but, in this way, it is not easy to achieve large currents.

- **Spin propagation**
  Signal propagation is another problem since the excess spin injected into a semiconductor (or a metal) tends to relax to zero, which is the equilibrium value of non-magnetic semiconductors. It has been demonstrated that spin can propagate up to 350 μm through a silicon wafer at 77 K. Similar values have been reported for graphene at low temperatures. At room temperature the spin diffusion length is reduced to approximately 200 nm, while for other non-magnetic materials, values of 100–1,000 nm have been achieved. More recently, spin transport lengths up to 20micron have been reported in Silicon at room temperature (T. Sasaki et al., 2014). Confining the electrons with thin multilayers further reduces the spin lifetime, because of the scattering events at the interfaces. These values appear to be still far away from the target needed to give an effective contribution to the reduction of power dissipation in microprocessors. Power dissipation analysis on microprocessors has shown that the largest part of interconnection-related power dissipation takes place over lengths larger than 10 μm [12].
- **Spin detection**
  Spin detection is probably the least difficult of the problems related to full spin logic, since it can be performed with existing spin valves. One problem is however the low signal-to-noise ratio that in other existing implementations (read head sensors, STT MRAM) is overcome by amplification with conventional nanoelectronic circuits.

In summary, even if simple logic gates, involving short-distance spin transfer, and elementary logic operations based on spin manipulation have been demonstrated by several authors, the realization of an all-spin logic seems still far away.

Among the stumbling blocks on the way to make spintronics a replacement for conventional nanoelectronics in complex circuits, there is:

- the fact that most of the experiments to realize logic operation have been performed at cryogenic temperatures, while performances are severely degraded at room temperature where normal electronics is expected to operate;
- devices realized until now have a low output current thus limiting the device fan out, and making impossible to drive long interconnections;

– the combination of ferromagnetic materials with semiconductors gives rise to many different compatibility problems. Many different combinations of materials are being investigated, including ferromagnetic semiconductors, but no consolidated solution has emerged yet.

10.8 Mixed Logic

STT MRAMs have been proposed also as basic components of non-volatile logic gates. As shown in Figure 10.7, STT MRAM cells could be combined to execute basic logic operation; in this case, an AND function. If both A and B cells have parallel orientation (both "1"), the sum of the two currents is enough to switch the cell C to a parallel orientation. If at least one of the two cells A and B is in antiparallel orientation ("0"), the resulting current does not suffice to switch C, and the result is "0". More complex basic devices, like flip-flops, buffers and other logic gates, can be obtained by properly combining more cells and can perform basic combinatorial logic operations [13].

However, this kind of logic suffers from a few disadvantages, like the proper balancing of currents to achieve or avoid switching, the immunity to disturb, the very low fan out and the possible interactions among connected cells. Intermediate CMOS circuitry can be used to decouple cells, performing extra read/write operations to read out the information stored in the output (target) and to write it to an input (source) MTJ, but at the cost of increased complexity, increased delay and energy consumption.

Even if full spin-based logic is still far away, more immediate results could be obtained by merging spin-based memories, like STT MRAM, with conventional CMOS memories. One of the advantages of STT MRAM over the most diffused memory types like Flash and DRAM is that it can be integrated on top of conventional CMOS logic, with just the addition of a multilayer stack and some interconnections. With respect to SRAM, which is the other commonly embedded memory, it has an area advantage, being ideally a 1T solution instead of a 6T or 8T one, and does not present the same disturb sensitivity and stability problems with scaling. Moreover, STT MRAM

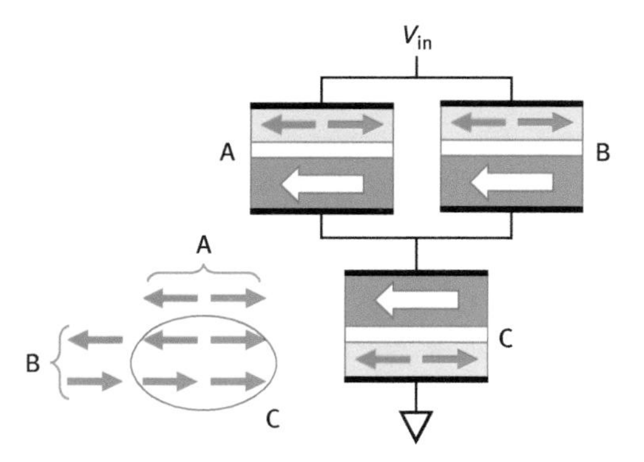

Figure 10.7: Schematic of STT MRAM-based AND gate.

is non-volatile and therefore can act as buffer to store intermediate results, without power dissipation. Since one of the techniques used to reduce overall power dissipation in logic circuits is to turn down power supply to the circuit part that are not used at the moment, the insertion of non-volatile STT MRAM elements can allow a quick restart, without the need to download the circuit past status from an external memory.

STT MRAM can also be used in FPGA (Field-programmable Gate Array) wcircuits, to store internally circuit configurations, avoiding the use of external memories, and allowing for fast reconfiguration. A more advanced possibility would be to use STT-programmed embedded FPGA blocks in processors as hardware accelerators to execute specific tasks, thus reducing overall power dissipation.

10.9 Spintronics: An Outlook

Spintronics is a quite new science. It really started with the discovery of the GMR effect in the 1980s of last century, and became quickly a front-end technology with its successful application to the read head of hard disk drives. The invention of the spin valve has led to two other applications that are on the verge of industrial exploitation: the STT MRAM, already commercialized for special applications, and the spin-torque oscillator, which still requires some research effort. The increased sophistication of analytic tools, the availability of new materials and the possibility to create multilayer metamaterials and well-controlled surfaces with advanced deposition tools like MBE (Molecular Beam Epitaxy) or ALD (Atomic Layer Deposition) are increasing the interest in the study of more exotic spin-related effects. The fact that electrons behave as magnetic dipoles gives rise to a variety of effects that tends to separate electrons with different spin orientation. The most interesting ones for a possible application in spintronics are the ones related to interaction with electric fields:

- *The spin–orbit interaction* is the interaction of a particle's spin with a magnetic field generated by its motion. The first and best known example of spin-orbit interaction, whence it comes the name, is the interaction of the spin of an electron with the magnetic field related to its orbital motion around the nucleus, which results in the splitting of the electron energetic levels associated with that orbit.
- *The Spin Hall effect* consists in the accumulation of electrons with opposite spin directions on the lateral surfaces of a sample in which an electric current is flowing, and it is related to the interaction between the spin of the electron and the magnetic field generated by the current flow.
- *The Rashba effect* is a more complex form of interaction that can take place in a bidimensional layer in the presence of a vertical electric field. When the electron spin is not aligned with the magnetic field induced by the electric field, spin precession takes place with a frequency that depends on the magnitude of the field.

- *The magnetoelectric effect*: in recent years, it has been published that an applied electric field can modify in a reversible way the magnetocrystalline anisotropy of thin ferromagnetic intermetallic compounds. This effect, which seems to be related to the spin–orbit interaction, could be of primary importance for magnetic memories and transistors.
- Other effects concern the interaction of electron spin with thermoelectrical effect (spin Seebeck effect) and with plasmons at the surface of conducting layers (spin plasmonics).

Most of these effects have been primarily detected at very low temperatures, but can become relevant in quantum wells, realized with very thin, or bidimensional (as for graphene) layers. At the moment, they are mainly being investigated at the research laboratory level.

Elementary magnetic gates have been demonstrated with a variety of technologies and even more complex serial logic chains have been realized with magnetic cellular automata, in which magnetic nanodots interact via magnetic coupling, on the same principle of quantum cellular automata based on single electron transistors. The advantage of magnetic quantum automata is in the capability to work at room temperature, even if external magnetic fields are requested to inject data or to perform the logic operations [14].

10.10 Conclusions

In spite of its quite recent origins, spintronics has managed the transformation from basic science to industrial technology, taking profit also from the progress in microelectronic technology, which on one side has made possible the mass production of spintronic devices, and on the other side has allowed the detection and amplification of the relatively weak spin-related signals. The breakthroughs like the mass introduction of GMR-based reading heads in hard disk drives, and the commercialization of the STT MRAM, have given a further push to both basic and industrial research.

Early enthusiasm in this field appears to be premature because no viable alternatives to the MOS transistor have been realized. Proposed circuit solutions are characterized by low fan out and lack of capability for driving long-range interconnections, which are the ones that are responsible for the greater part of power dissipation in complex logic circuits. However, on the short and middle term STT MRAM can replace SRAM and DRAM in the cache levels of processor architectures. Their capability to be integrated on top of advanced CMOS logic allows introducing non-volatility very close to logic blocks to reduce static power consumption, and to make use of innovative architectures with memory blocks finely distributed among logic to minimize the distance that data have to travel. In the longer term, the number of spin-related effects that are being investigated may make new logic approaches possible. Material research and better physical understanding of the phenomena that take place in thin layers and at their interfaces, coupled to the availability of tools to reliably control layer deposition at the molecular layer, could open the way to the realization of innovative devices and open new fields of technology.

References

1. Chang L, Wang M, Liu L, et al. A brief introduction to giant magnetoresistance. Ithaca: Cornell University Library, 2014. arXiv:1412.7691 [cond-mat.mtrl-sci]
2. Dieny B, Speriosu VS, Metin S, et al. Magnetoresistive properties of magnetically soft spin-valve structures. J Appl Phys 1991;69:4774.
3. Daughton J. Magnetoresistive memory technology. Thin Solid Films 1992;216(1):162–8.
4. Durlam M, Addie D, Akerman J, et al. A 0.18 μm 4 Mb toggling MRAM. IEDM Technical Digest, Washington, 2003.
5. Parkin SS, Kaiser Ch, Panchula A, et al. Giant tunnelling magnetoresistance at room temperature with MgO (100) tunnel barriers. Nature Mater 2004;3:862–7.
6. Rho K, Tsuchida K, Kim D, et al. A 4Gb LPDDR2 STT-MRAM with compact 9F2 1T1MTJ cell and hierarchical bitline architecture. Solid-State Circuits (ISSCC), IEEE International Conference, San Francisco, CA, 2017.
7. Parkin S. US Patent US 6834005 B1, 2003.
8. Parkin S. Spintronic materials and devices: past, present and future! Electron devices meeting, 2004. IEDM Technical Digest. IEEE International, San Francisco, CA, 2004.
9. Zeng Z, Finocchio G, Jiang H. Spin transfer nano-oscillators. Nanoscale 2013;5:2219–31.
10. Modarresi H. The spin field-effect transistor: can it be realized?. University of Groeningen 2009. Available at: https://www.researchgate.net/publication/255641305; https://www.rug.nl/research/zernike/education/topmasternanoscience/ns190modarresi.pdf
11. Manchon A, Koo HC, Nitta J, et al. New perspectives for Rashba spin-orbit coupling. Nat Mater 2015;14:871–82.
12. Magen N, Kolodny A, Weiser U, et al. Interconnect-power dissipation in a microprocessor. Proceedings of the 2004 international workshop on system level interconnect prediction. Paris, 2004.
13. Makarov A, Windbacher T, Sverdlov V, et al. CMOS-compatible spintronic devices: a review. Semicond Sci Technol 2016;31:11.
14. Welland ME, Cowburn RP. Room temperature magnetic quantum cellular automata. Science 2000:1466–8. Feb 25;287(5457):1466–8.

Part III: Nanotechnology for Industrial Applications

Urszula Narkiewicz

11 Nano-catalysts: research – technology – industrial applications

11.1 Introduction

The "nanoboom" at the turn of twentieth/ to twenty-first centuries was less crucial for catalysis than for other fields of research and technology. The first reaction of the catalytic scientific milieu at the appearance of nano was rather cool. Nano has not been perceived as an important breakthrough, but rather as an evolution, not a revolution, because catalytic people were always being familiar with nanoscale, or even smaller. Before the nano era, catalysis was used to apply nanoclusters of noble metals dispersed on a carrier, or nanocrystalline metals and alloys, or zeolites with nanochannels, and so on. François G. Gault in 1957 demonstrated the influence of platinum dispersion on the support on the selectivity of the hydrogenolysis of cyclopentanic hydrocarbons.

It has always been obvious that catalytic performance strongly depended on the size and shape of catalysis particles at nanoscale (or on the size and shape of nanochannels in catalysts or catalytic carriers), as with increasing surface-to-volume ratio of particles, the number of exposed active sites for reactants adsorption, and activation increased, too. The catalytic performance is extremely sensitive to particle size, as at nano- and subnanoscale surface structure and electronic properties change dramatically (for better, taking into account the catalytic phenomena). But, even if scientists in catalysis were (comparing with other fields of science) less excited by the dramatic increase of nanotechnology importance, the appearance of nano brought a lot of new insight in catalysis development.

The huge market for nanomaterials opened in many areas, among them for catalysts, sorbents, and sensors. Nanotechnology Products Database [1] created in 2016 reports on 7,819 nanoproducts produced by 1,253 companies in 52 countries. According to the National Science Foundation, the worldwide nanotechnology market surpassed $1 trillion in 2015.

What is the impact of nanorevolution on the development of such a well-established field of science and industry as catalysis is? What are the links between these two domains, nanotechnology and catalysis?

Catalysis always used to couple this nano (or sub-nano) world, where chemical reaction occurs, with macroscopic world of reaction engineering, where in industry sometimes hundreds of tonnes of materials are applied to fill the reactor.

The significance of catalysis for chemical industry is crucial. More than 90% of processes in chemical industry run with the use of catalysts. The worldwide

https://doi.org/10.1515/9783110547221-011

market of catalysts is higher than $5 bn. Catalysts enable to accelerate the rate of chemical reactions, to minimize the volume of side products, and to save energy (offering the possibility to perform the processes at lower temperatures or under lower pressures).

Some processes could not be performed without catalysts. For example, synthesis of ammonia from elements, more than hundred-year-old industrial process, is performed in industry using an iron catalyst, with efficiency of 500–1,000 tonnes or Mg of ammonia per day. To perform the same process without a catalyst, an appropriate reactor should have the dimensions of our galaxy, and despite that, its efficiency would be 0.1 g of ammonia per day only, which is a miserable result when comparing with the performance of the simple and cheap iron catalyst, over hundred year old and having a nanocrystalline structure.

Recently, there is a growing demand for fine chemical industries and a need to fulfil increasingly stringent ecological standards. To satisfy these needs, new technologies are required as well as new, more effective, and more selective catalysts.

The most important issues to address in heterogeneous catalysis are:
1) stability of the active phases in time,
2) reversibility of poisons' adsorption on the catalyst's surface,
3) real catalytic reaction, which means that the reaction is exclusively controlled by kinetics.

How can nanotechnology contribute to the development of the efficient catalysts?

Catalysis did not contribute to the creation of nanotechnologies and nanosciences, for example, through important breakthrough catalytic discoveries, but the movement was in the opposite direction – development of nanotechnology brought a lot of benefits for catalysis.

The influence of nanotechnology started roughly 50 years ago, but unfortunately at that time scientists were not able to give clear scientific explanations of that phenomenon. There were no experimental tools and quantum chemistry was not so developed as it is now. Then we have had to wait until the end of the twentieth century to explain the influence of particles size on catalytic reaction and only at this time the "material" aspect of catalysis became as important as the kinetic one.

In the first half of twentieth century and before, catalytic reactor has been perceived as a magic black box, in which reactants entered and products were obtained, but nobody knew what really happened at the catalyst surface. The appearance of surface science techniques in the second half of twentieth century and a development of modelling and simulation techniques improved the situation, enabling to explain the mechanism of many catalytic reactions or to confirm the mechanism proposed before as a hypothesis.

11.2 Size and Shape of Catalyst Particles and Structure Sensitivity

At the end of twentieth century, the explosion of nano brought a considerable added value in material science and in catalysis (both in science and in industrial practice).

Industrial catalysts are often in the form of nanoparticles. The examples of heterogeneous nanocatalysts can be dispersed metal nanoparticles (Au, Pt, Pd, Rh, etc.), oxide nanoclusters (FeO_x, VO_x, MoO_x, CrO_x, CoO_x, etc.), and mesoporous molecular sieves (silicates, zeolites, and nanotubes). In homogeneous catalysis, various metal complexes, colloidal metals, or enzymes can be applied.

The size and shape of catalyst nanoparticles have been hardly tunable before the nano era. Recently, the *structure sensitivity* and the effect of catalyst particle *size* in nano and subnano level were successfully addressed in many cases.

Before the nano era, the interesting surface science studies of catalysts were carried out on monocrystals at ultrahigh vacuum (UHV) conditions, so far away from real conditions of catalytic processes, carried out at ambient or elevated temperatures, at higher pressures, and using complex catalytic systems. Then, it was a huge "pressure gap" and "material gap" between surface science studies and real catalytic processes.

Nevertheless, there was some research bridging the gap, for example the paper of Dauscher et al. [2], in which the isomerization and hydrocracking reactions of 2-methylpentane and *n*-hexane and the hydrogenolysis of methylcyclopentane were investigated on a clean Pt(311) single crystal at 350°C under atmospheric pressure in an isolation cell housed within an UHV chamber. The Auger electron spectroscopy was applied to monitor the surface composition before and after the catalytic reactions. The results were compared with those obtained in the same experimental conditions on Pt(111), Pt(557), and Pt(119) surfaces, a polycrystalline foil, and two Pt/Al_2O_3 catalysts of both low and high dispersion. According to the obtained results, the reactions were structure sensitive taking into account catalytic activity, selectivity towards isomers produced, and selectivity as regards mechanisms. On the contrary, the distribution of cracked products presented no significant structure sensitivity on "massive" metal surfaces although the highly dispersed alumina-supported platinum catalysts behave differently. The so-called B_5 site configuration was responsible for the bond-shift plus cracking reactions and that the atoms located at the corners and edges were involved in the cyclic mechanism.

Surface science studies enabled to discover structure sensitivity of some catalytic reaction – as in the case of the ammonia synthesis, which occurs much more quickly on Fe(111) face than on other crystalline planes [3–5]. Recently, Wang et al. [6] applied density functional theory (DFT) calculations in combination with atomistic thermodynamics to study N_2 dissociative adsorption on various iron surfaces at different coverages and conditions. Since the Fe(100) surface has 25 times higher activity in ammonia synthesis than Fe(110) surface, this surface reconstruction should be of great significance in improving activity of ammonia synthesis reaction. It can be done under N_2 pretreatment, when Fe(100) is the most exposed surface. Separate attention should be paid for iron catalyst with promoters, mainly potassium [7, 8], which significantly alternates the surface state.

The important discoveries achieved at the end of twentieth century and being a real breakthrough in catalytic knowledge were done on monocrystals under UHV conditions only and have never been reflected in industrial practice. Together with the arrival of the nano era, new techniques were developed, enabling to prepare materials with specific crystalline facets exposed and to study the real catalysts under real work conditions.

In the nanoworld, according to the "bottom-up" approach, it is possible to create nanomaterials by assembling them from building nanoblocks, so it is possible to expose a preferred crystalline plane. Then, thanks to nanotechnology, the dream about catalytic performance improvement using a morphology-induced activity, through an exposition of the preferred (the most active) crystalline facet, has become a reality. It became finally possible to change the traditional catalytic approach "smaller is better" to a "specific morphology is better" one.

One of the crucial goals in catalytic studies nowadays should be to design and fabricate nanostructures with defined morphology, to obtain high density of active sites on the surface, and to reach an optimum activity and/or selectivity.

It seems to be difficult to design and fabricate nanocatalysts through atom-by-atom synthesis and to obtain a volume needed at least for bench-scale testing. Fortunately, there are also much simple ways to fabricate nanocatalysts without the use of building nanoblocks. There are more and more studies demonstrating that the external catalytic morphology is more crucial for catalytic activity than specific surface area.

Xie and Shen [9] described the importance of the morphology and crystal plane of Co_3O_4 nanomaterials in CH_4 combustion and CO oxidation reactions.

Co_3O_4 nanorods [10] were much more active in low-temperature oxidation of CO than conventional Co_3O_4 nanoparticles. Hu et al. [11] investigated the effect of the morphology and crystal plane effects of Co_3O_4 nanosheets, nanobelts, and nanocubes on methane combustion. The reaction proceeded about two times faster over the nanosheets than on the nanocubes.

Tao et al. [12] thoroughly studied nanocatalysis of two shapes of individual palladium nanoparticles – cube with (100) facet and octahedron with (111) facet. The studies were carried out with single-turnover resolution, and the facet-dependent activities and dynamics were observed. The winner was Pd octahedron, achieving higher intrinsic catalytic activity per site than Pd nanocube. Studies of this single-molecule catalysis could deepen the knowledge about catalytic kinetics and dynamics on different facets and to encourage researchers to design and perform controllable syntheses of nanocatalysts with high activity and stability.

Crampton et al. [13] proved that a procedure enabling to classify a reaction as structure sensitive or insensitive must be reexamined and reevaluated for particle sizes in the nonscalable subnanometre regime. In this regime, surface chemistry becomes strongly dependent on the precise number of atoms in the catalysing particle. The authors showed that ethylene hydrogenation is indeed a structure-sensitive reaction in the subnanometre particle size regime of Pt_7–Pt_{40}, with a maximum reactivity observed for Pt_{13}.

Pt catalyst used for benzene hydrogenation produces both cyclohexene and cyclohexane on the Pt(111) face, while only cyclohexane is formed if Pt(100) face is exposed [14, 15].

Truncated triangular, cubic, and near-spherical silver nanoparticles were produced and their activity in the process of the oxidation of styrene in colloidal solution was investigated by Xu et al. [15].

Chimentao et al. [16] applied poly(vinylpyrrolidone) as the template to obtain two different Ag nanoparticle morphologies: silver nanowires and nanopolyhedra, supported on α-Al_2O_3. The authors stated that most exposed crystal face of the silver nanowires and nanopolyhedra was (111), which may have a beneficial effect on the selective oxidation of styrene. Silver nanopolyhedra are more reactive because they have more edge and corner atoms with a higher coordination number than terrace atoms with lower coordination number.

Then, the synthesis of shape-controlled silver nanoparticles can have potential applications for the selective oxidation of olefins.

The process of styrene oxidation was also investigated on very small platinum clusters [17] supported on titania. Platinum nanoclusters have been synthesized by reduction of $H_2PtCl_6{\cdot}6H_2O$ salt with $NaBH_4$ in the presence and absence of water-soluble l-cysteine ligand. The applied method allowed producing a very narrow size distribution of platinum clusters ~1 nm, supported over anatase titanium dioxide (TiO_2) by 1% wt/wt. The authors applied the obtained catalyst in the process of styrene oxidation. The monodisperse clusters showed an extremely high catalytic activity in oxidation of styrene using H_2O_2 as oxidizing agent with 100% selectivity to benzaldehyde selectivity. The conversion and selectivity were better than using oxygen as an oxidizing agent.

Another example of structure sensitivity of nanoparticles is ceria, which is very important in exhaust car catalysts because of its well-known oxygen storage capacity. In this case, even if ceria (111) plane gives larger surface area than (100) and (110) facets, the latter ones are more active in CO oxidation, so it is better to apply ceria nanorods instead of the traditional ceria carrier [18].

The size of ceria nanoparticles is of crucial importance in the case of CO oxidation on gold catalyst supported on ceria. Carrettin et al. [19] showed that the activity of ceria with particle size of 3.3 nm (specific surface area of 180 m^2/g) was two orders of magnitude greater compared with a ceria material having average particle size of 15.9 nm (specific surface area (SSA) of 70 m^2/g). Such a result was explained by a synergistic effect of nano-Au and nano-ceria, enhancing the formation of highly reactive oxygen species at one-electron defect sites of the support.

As it has been emphasized here above, the most important and specific for nanocatalysts is their ability to create nanoparticles that present a specific crystal face to the reaction medium. As a result, significant changes in yield balances between competing end-products are achieved. Then, the key issue in practical catalysis nowadays should be to find an appropriate synthesis procedure in order to obtain a preferred nanocatalyst morphology, leading to an increase in the density of highly active sites at the surface and inhibition of the exposition of less reactive sites, and/or the optimum porous structure. All the "wet" preparation techniques seem to be the most flexible and facile in this end. Both old, traditional solution-based methods (as precipitation or sol–gel techniques) and new ones (as microwave-aided sonochemical or solvothermal techniques, or deposition of colloidal systems on supports) can be used. However, for some purposes the solid-state or gas-phase methods (as atomic layer deposition) can be recommended as well.

Various bio- and biomimetic techniques are recently more often applied, for example, synthesis of highly active Pd nanocatalysts using biological buffers (HEPES and Tris), described by Janairo et al. [20]. The size of the obtained palladium nanoparticles was below 10 nm and they were produced in a one-pot synthesis under ambient conditions. The catalyst was successfully tested in the reduction of nitroaminophenol isomers.

Fei et al. [21] used a mussel-inspired approach to design and fabricate a recycling-free nanocatalyst system based on the stabilization of in situ-reduced noble metal nanoparticles on silicone nanofilaments. The authors prepared the nanocatalyst based on Pd nanoparticles and demonstrated its high activity and high selectivity in single and successive Heck coupling reactions, but the approach can be extended to other supported noble metals and other reactions.

The fabrication of nanocatalysts can occur in various environments – liquid, solid, gas, or mixed phases. The interplay of homogeneous and heterogeneous catalysis can be observed, as well as the growing implementation of enzymatic catalysis to perform complex synthetic transformations.

There are some common bottlenecks in the nanocatalysts fabrication, mainly lack of reproducibility, agglomeration, sophisticated equipment, high cost, large volume of additives/solvents, and lack of stability of obtained nanostructures. These key barriers are not specific for nanocatalysts, but common for all nanomaterials and will be overcome step by step, as a result of combined research in heterogeneous catalysis, material sciences, and surface sciences studies.

Various capping agents (organic ligands, polymers, surfactants, dendrimers, cyclodextrins, and polysaccharides) are crucial in nanocatalysis, which was thoroughly discussed in the review paper of Campisi et al. [22]. The role of capping agents is to protect or to functionalize catalytic nanoparticles (e.g. to facilitate the anchoring onto the support resulting in a high metal dispersion) and, as in the standard catalysis, they can act in a promoting or poisoning way. The issue of tuning the activity and accessibility of gold clusters through ligand engineering was addressed by Li et al. [23].

Despite the significant importance of capping agents in nanocatalysis, there are also reports [24] about one-pot wet chemical synthesis of nanocatalysts without the use of capping agents, just by combining nucleation/growth with the dissolution of crystals in a reaction–diffusion system.

The issue of role of surface oxides in the catalytic oxidation reactions is often addressed, as in the paper of Lee et al. [25], dealing with surface oxidic motifs of Cu on Au(111), depending on the oxygen atmosphere and the type of surface defects introduced in the oxidic layer.

There is a crucial need to develop reproducible synthesis methods, resulting in controllable and tunable properties of nanocatalysts in a reproducible way, giving always the same size (and size distribution), shape, composition (bulk and surface), dispersivity, and so on. Despite the catalyst surface morphology, the fabrication of the appropriate porous nanostructures is an important challenge. The meso- and macroporous materials offer much lower resistance to molecular diffusion than microporous ones. Then, it is of cardinal importance as to how to better control the shape and size of the channels and cavities of the final catalyst. However, the microporous materials have also negative features – severe diffusion resistance and blocking of the entrance to the pores by side-products. A solution can be the application of hybrid meso-micro materials, microporous materials with interconnected array of mesopores, serving as highways for the reactants, and facilitating their access to micropores.

11.3 Nanostructured Noble Metals in Catalysis

Nanocatalysis is based mainly on noble metals, in particular gold. A spectacular career of gold in catalysis commenced together with development of nanotechnology. Bulk gold has been perceived before as unreactive until Haruta [26] demonstrated an exceptional highly selective catalytic activity of gold nanoparticles in the range of 3–5 nm in CO oxidation. The discovery opened a new approach to the understanding of fundamental mechanisms controlling the behaviour of atomic clusters and nanoparticles, which form a bridge between atoms and their bulk counterparts. Then, even catalytically inactive and inert materials can become active catalysts thanks to the size, structure, morphology, and support effects at nanoscale [27]. Another important discovery about gold catalytic properties in the 1980s concerned the prediction that Au would be best catalyst for ethyne hydrochlorination [28]. Recently, gold has been shown to be a very versatile redox catalyst [29].

Examples of application of nanoparticulate noble metals in catalysis are shown in Table 11.1.

Table 11.1: Noble nanometals in catalysis.

Noble metal	Catalyst	Reaction	Results	Ref.
Au	Uniformly sized Au nanoparticles as the core of the rattle nanoarchitecture system, composed from silica sphere and silica shell	Direct synthesis of hydrogen peroxide under ambient conditions	Formation rate of H_2O_2 on the nanorattle of Au@SiO_2 was 2–3 orders of magnitude higher than that of other Au catalysts	[30]
Pd–Au	Pd–Au/titania	H_2O_2 synthesis	Pd_2Au catalyst exhibited the best performance with the H_2O_2 selectivity of 48.1%	[31]
Pd–Au	Pd–Au colloid catalyst in the presence of NaBr and H_3PO_4	H_2O_2 synthesis	H_2O_2 yield (57%) and H_2 conversion of 85%)	[32]
Au	Yolk–shell nanocomposites of Au nanocore encapsulated in electroactive polyaniline shell	Catalytic aerobic oxidation of alcohol in aqueous solution	High catalytic efficiency	[33]
Au	Gold-based unsupported nanocomposites – nanoreactors	Pyrrole polymerization	Au nanoclusters (2–3 nm) and atomically thin gold nanosheets (30–50 nm) incorporated as active centres in self-assembled polymers. The activity of the obtained Au-nanoreactors was better than that of similar Pt-based nanoreactor.	[34]
Au	Self-assembled biomimetic nanoreactors with gold active centres	Nanoreactors produced using poly(styrene-alt-maleic acid)	For the first time pure atomically thin gold sheets were obtained.	[35]
Au	Gold nanoparticles supported on mesoporous cerium–tin mixed oxide	Aerobic oxidation of benzyl, cinnamyl, and 4-methylbenzyl alcohol, 2-octanol, and geraniol	Synergistic effect of Au and Sn on the CeO_2 support was evidenced by catalytic activity results. Au-Cs/Sn (95/5) catalyst showed the best catalytic performance.	[36]
Au	Naked mesoporous Au–γ-Fe_2O_3 nanocrystal clusters (5–8 nm)	Reduction of 4-nitrophenol by $NaBH_4$	Activity of Au–γ-Fe_2O_3 – about 22 times higher than that of self-assembled Au–Fe_3O_4 clusters	[37]

(continued)

Table 11.1: (continued)

Noble metal	Catalyst	Reaction	Results	Ref.
Au	Gold nanowire on MgO(100)	Low-temperature water-gas shift reaction	DFT calculations, microkinetic modelling, and kinetic experiments combined to investigate the mechanism of the reaction. Carboxyl mechanism – the most energetically favourable; and the COOH formation – the rate-limiting step. A very high barrier (2 eV) to activate water on the clean Au(111) surface drops to zero at the Au/MgO interface.	[38]
Au	Au (3–5 nm) on the mesoporous ZrO_2–TiO_2	Reduction of 4-nitrophenol to 4-aminophenol by $NaBH_4$	The catalyst exhibited higher activity compared with TiO_2/Au and ZrO_2/Au due to mixed oxide synergistic effect	[39]
Au–Pd	Noble metal catalytic nanoalloys-multifaceted Au–Pd nanorods	Catalytic hydrogenation of 4-nitrophenol by ammonia borane	The reaction kinetics depended on the surface atomic coordinations and the compositional stoichiometries	[40]
Au, Ag, Cu	Au, Ag, and Cu nanocrystals on titania nanosheets	Catalytic reduction of 4-nitrophenol to 4-aminophenol	Potential catalytic application of the materials in the investigated reaction. The Au cocatalyst was also the most active in photocatalytic H_2 evolution.	[41]
Pd	Narrow distributed (2–3 nm) and highly dispersed Pd nanoparticles encapsulated into mesoporous ionic copolymer	Atmospheric pressure oxidation of benzyl alcohol to benzaldehyde	The special ionic framework with –COOH group – a major role in the formation and stabilization of ultrafine Pd nanoparticles. Efficient and stable catalyst.	[42]
Pd	Fine Pd nanoparticles supported on fluorite-structured redox $CeZrO_{4-\delta}$	Heterogeneous Suzuki coupling in water	100% conversion	[43]
Au–Pt	Core-shell nanostructures, spherical gold core of 8 ± 2 nm, and a 3 Å atomically thin platinum shell (24 mol% of Pt and 76 mol% of Au)	Electrooxidation of sustainable fuels (i.e. formic acid, methanol, and ethanol), and selective hydrogenation of benzene derivatives	Au core increased the activity of the Pt shell by up to 55% and improved catalytic selectivity compared to pure Pt. Especially high activity in formic acid oxidation, 3.5-fold higher than a commercial Pt nanoparticle catalyst.	[44]

(continued)

Table 11.1: (continued)

Noble metal	Catalyst	Reaction	Results	Ref.
Pt–Co	Pt–CoO_x nanoparticles supported on microporous titanosilicate ETS-10	CO conversion	A complete CO conversion in the temperature range 120–150°C was achieved on the catalyst containing 1.4 wt% of nanoparticles.	[45]
Pt–Co	Bimetallic platinum–cobalt catalysts supported on ceria–zirconia mixed oxide	Aqueous-phase reforming reaction of ethylene glycol	Enhanced activity for an aqueous-phase reforming as well as water-gas shift reaction observed on the PtCo/CeO_2-ZrO_2 at a Co/Pt molar ratio of 0.5.	[46]
Pt–Co	Core-shell Pt–Co bimetallic nanoparticles encapsulated within metal-organic frameworks (MOFs) (UiO-66)	Hydrogenation of nitrobenzene under atmospheric pressure of H_2 and room temperature	Superior synergy and catalytic activity in the reaction. Thanks to the confinement effect from MOF shell, the catalysts were much more active, selective, and stable than the supported PtCo/UiO-66 material.	[47]
Pt	Pt nanostructures encapsulated in a polymer shell	Organic transformations	Role of the protective and stimuli-responsive shell – to promote efficient mass transfer to encapsulated Pt nanoparticles, to prevent the coalescence of Pt nanocores, to provide a void space for organic transformation on the surface of the ligand-free Pt nanocluster in a controlled manner.	[48]
Pt	Pt/large pore zeolites (acid forms of highly siliceous zeolite Y (HUSY) and zeolite beta (HBEA)) with different metal/acid ratios	Toluene hydrogenation	Pt/HBEA catalysts were more active per total adsorbing site than Pt/HUSY. The catalytic activity depended on the accessible metal, but also on the total acidity for a given zeolite.	[49]
Pt–Ni	Pt–Ni alloy nanoparticles supported on carbon	3-Pentanone hydrogenation	Excellent performance and good stability owing to high dispersion of Pt–Ni alloy nanoparticles and electron synergistic effect between Pt and Ni species.	[50]
Rh	Rh nanoparticles	Asymmetric transfer hydrogenation of carbonyl compounds	Activity in the reaction.	[51]
Rh	Rh in the form of colloid systems with optically active stabilizers	Enantioselective hydrogen transfer hydrogenation	Activity in the reaction.	[52]

(continued)

Table 11.1: (continued)

Noble metal	Catalyst	Reaction	Results	Ref.
Rh	Rh subnanoclusters supported on TiO_2	Oxidation of carbon oxide at cryogenic temperatures	A complete CO conversion at 223 K. At least three orders of magnitude higher TOF than the best Rh-based catalysts and comparable to Au/TiO_2. The size range of 0.4–0.8 nm Rh clusters – critical to the facile activation of O_2 over the Rh–TiO_2 interface.	[53]

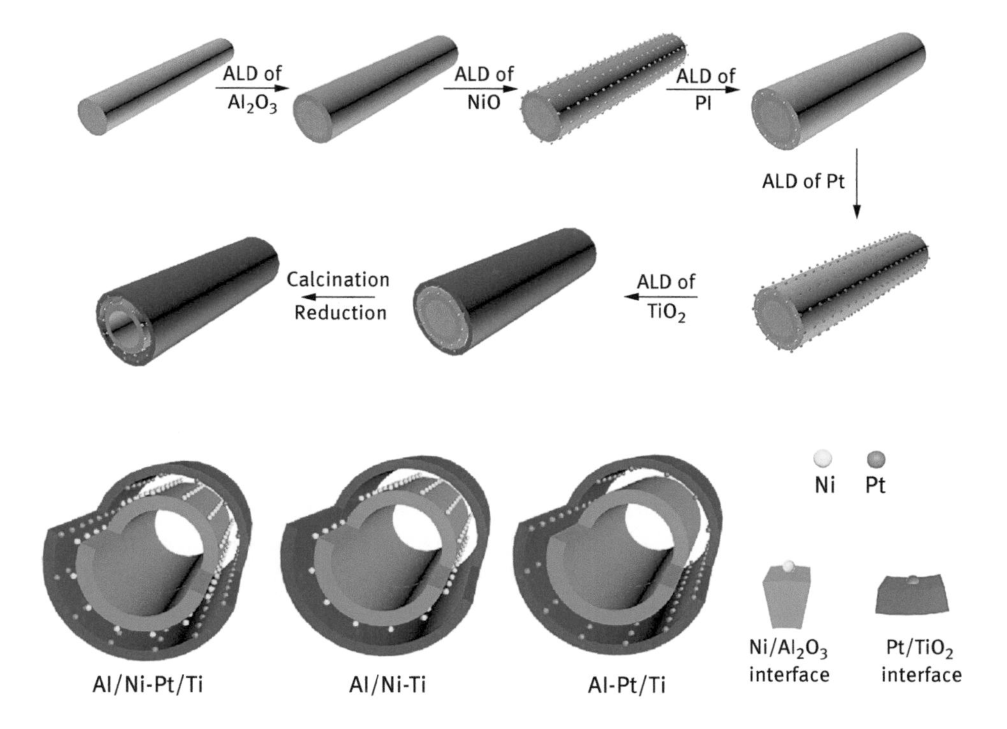

Figure 11.1: Schematic illustration of the procedure for the synthesis of the tandem catalyst with both Ni/Al_2O_3 and Pt/TiO_2 interfaces and semi-sectional views of different catalysts for comparison [54].

A very interesting Pt-based tandem nanocatalyst was recently described by Ge et al. [54] Pt nanoparticles were attached to the inner surface of the outer TiO_2 nanotube (Pt/TiO_2 interface) and Ni nanoparticles were supported on the outer surface of the inner Al_2O_2 nanotube (Ni/Al_2O_3 interface). A schema of the preparation procedure of such a system is shown in Figure 11.1.

The catalyst showed exceptionally high catalytic efficiency in nitrobenzene hydrogenation over Pt/TiO_2 interface with H_2 formed in situ by the decomposition of hydrazine hydrate over Ni/Al_2O_3 interface. The authors ascribed the remarkable activity to the synergy effect of the two interfaces and the confined nanospace favouring the instant transfer of intermediates.

11.4 Nanocatalysis as an Answer to Global Societal Challenges

Nanocatalysis plays a crucial role in the project that can be called "Chemistry for better life" [55] and can help to answering the needs of the increasing world population, foreseen to reach 6.3 billion people in 2050. Among main challenges here are *clean water and air, climate changes, and safe energy* issues.

In the field of energy, nanocatalysis can be applied for both energy production and conversion, for example, new combustion catalysts or nanocatalysts for fuel cells. In the latter case, the highest theoretical efficiency has been found for quantum dots, but there are also other convenient nanomaterials, for example quantum wells, carbon nanotubes, fullerenes and graphene, nanowires and dendrimers. CO_2 activation for the preparation of fuels and bulk chemicals from it is the other challenge for nanocatalysis, which is very important to inhibit climate changes. The concentration of CO_2 in the earth atmosphere is continuously increasing, contributing to the greenhouse effect and resulting in severe climate changes. On the other hand, CO_2 can be treated as a source of carbon and a raw material for production of useful chemicals, but new catalysts and new processes are necessary to be implemented to achieve this goal. A solution to decrease concentration of two greenhouse gases, carbon dioxide and methane, would be a successful implementation of carbon dioxide reforming, leading to syngas (H_2 and CO) formation. Recently, the process of carbon reforming is still at the laboratory scale due to the unfavourable reaction thermodynamics, complexity of the reaction system, and coking effect. Overcoming of these bottlenecks is one of the most important challenges for nanocatalysis.

Nanocatalysts can be used not only for improvement of selectivity or reactivity of already known processes but also for new applications. Thanks to nanocatalysts the use of raw materials and energy will decrease and the environment will be less contaminated with gas, solid, and liquid wastes. The latter aim can be reached through better selectivity of nanocatalysts, which is being more important nowadays than productivity. Nanocatalysts can be applied mainly in petroleum, in refinery industry, in fine chemistry (production of pharmaceuticals, cosmetics), and in protection of environment (removal of volatile organic compounds and nitrogen and carbon oxides), production and purification of new energy sources (mainly hydrogen).

The examples of applications of nanocatalysis for the solution of global challenges are presented in Table 11.2.

Table 11.2: Applications of nanocatalysis to environment protection and energy issues.

Area	Process	Details	Ref.
Water treatment	Photocatalysis	TiO_2 incorporated into MCM-41 nanostructure for the photocatalytic degradationof methylene blue.	[56]
		Controlled thermal treatment, influencing the phase composition of titania.	[57]
		Modification of titania with carbon for the enhancement of the photocatalytic performance.	[58]
Inhibiting the greenhouse effect	Methane dry reforming	NiO nanoparticles supported on mesoporous silica.	[59]
	Carbon dioxide hydrogenation	Core-shell nanocatalyst composed of iron and carbon. Influence of amount of Fe_3O_4, Fe_5C_2 in the core and graphitized carbon in the shell on catalytic activity with higher selectivity to C_2–C_4 olefins.	[60]
		Nano-Ru/Ni catalyst with oxide passivation layer on the surface. A 100% conversion and a high TOF of 940/h at 200°C.	[61]
Energy	Methane oxidation	Lean methane oxidation on core–shell-structured $NiO@PdO/Al_2O_3$ catalysts.	[62]
		More than 99% of CH_4 conversion at 400 °C.	
		Partial oxidation of CH_4 to synthesize gas over Ni/CeO_2 catalysts (6 wt% of Ni). All the samples reached 98% conversion with CO selectivity values >95% at 700–800 °C, under atmospheric pressure and feed gas in molecular ratio $CH_4/O_2 = 2$.	[63]
		New technology of CH_4 catalytic combustion in small turbines. A three-stage catalyst package, each step with different chemical composition, shape, and size of granules. Excellent results over 99.97% CH_4 combustion at 470–580°C.	[64]
	Biomass gasification	Catalytic steam reforming of bio-tar using toluene as a model compound of biomass tar. Drastic promoting effect of Ba addition on the catalytic performance of $Ni/LaAlO_3$.	[65]
	Biodiesel production	The process of biodiesel production carried out over bifunctional solid nanocatalysts delivered excellent results in the biodiesel production by catalysing simultaneously both esterification and transesterification reactions.	[66]
	Hydrogen production	Transition metal carbides (Co_2C) synthesized using a facile one-pot bromide-induced wet chemistry exhibited high electrocatalytic activity and long-term stability in the H_2 evolution reaction.	[67]

(continued)

Table 11.2: (continued)

Area	Process	Details	Ref.
		Cu catalysts supported on well-shaped nanorod and nanocube CeO_2 were applied for photocatalytic water splitting. Surface "in situ" functionalization of Cu^0 nanoparticles resulted in 753 μmol/h/g of hydrogen.	[68]
		The specific activity of hematite facets in the photocatalytic splitting of water was found to be in the order {110} > {012} >> {001} and it was improved dramatically on sensitization with Cd quantum dots (CdS QDs).	[69]
		Highly dispersed Ru nanoparticles supported on graphene, applied in NH_3 to produce clean hydrogen, free from CO_x. Despite the excellent catalytic activity, a gradual decrease of performance was observed at 500°C within 20 h, due to the sintering of Ru nanoparticles and to the methanation of the graphene nanosheets under H_2.	[70]
		Effect of various additives on the activity of Co catalysts applied in NH_3 decomposition was investigated. The highest activity was demonstrated for Co(0) catalyst, promoted by oxides of Al, Ca, and K.	[71]

11.5 Characterization and Simulation Tools for Nanocatalysis

The development of nanotechnology brought for catalysis an important input in the field of characterization and simulation tools. It is the question of crucial importance to deepen a knowledge of how once prepared materials evolve under reaction conditions, and then, the issue of the characterization of chemical, structural, morphological, and textural features of catalysts conducted in situ (in the working state under real reaction conditions) should be still addressed more efficiently.

Computer-aided simulation and modelling complete the experimental techniques, accelerating the research and enabling to save time and money. Modelling of catalyst synthesis is still an open challenge, because the process often involves the self-assembly of metastable structures, governed by weak interactions of molecules in the system. The system is a non-equilibrium one, as a reconstruction of the as-prepared structures can next occur under reaction conditions. Both weak interactions and non-equilibrium systems are hardly treated using computational methods.

Recent spectroscopy techniques provide information on chemical and physical properties of materials at the atomic scale, not only in two-dimensional, surface-oriented systems, but also in three-dimensional (3D). The modern 3D tomography enables to observe the objects with spatial resolution of 1 nm or less.

Current challenge is to determine a structural determination of chemical composition in 3D. Thanks to important improvement of the performance of characterization devices, it is not only possible to better characterize the catalysts before and after the reaction, but also in real time, during the catalytic process. New characterization tools enable researchers to monitor and characterize phenomena occurring in the catalytic reactor, which are of crucial importance both from scientific and practical point of view. It has been already possible to apply the synchrotron radiation source as well as differential pumping to perform the x-ray photoelectron spectroscopy measurements of the catalyst surface not under UHV, but under mbar reaction pressure [72], then much closer to the real reaction conditions.

There are also sample cells for transmission electron microscopy studies [73], especially dedicated for observations of catalyst nanoparticles under reaction environment (elevated temperatures and mbar pressure).

A novel environmental scanning transmission electron microscope with 0.1 nm resolution was applied by LaGrow [74] in studies of complex dynamic oxidation and reduction mechanisms of Cu nanoparticles. It is a very important issue, because in nanocatalysis the nanoparticles can undergo oxidation or reduction in situ, and thus the redox species are not what are observed before and after reactions.

Single-molecule fluorescence microscopy [75] was applied to study size-dependent kinetics and dynamics of Pd nanocubes at the single nanoparticle level. It has been possible to monitor the catalysis of individual nanoparticles in real time with single-turnover resolution. The same group of researchers investigated Au single-molecule nanocatalysis [76], and in situ deactivation of Pt/C electrocatalysts during the hydrogen–oxidation reaction [77], and finally summarized the results of their studies in a recent feature paper [78].

Nanotechnology can also contribute to the combinatorial catalysis or high-throughput catalyst testing, improving the miniaturization and accuracy of analysis methods.

Another field of application of nanotechnology is the catalytic reaction engineering for testing. Recently, the attention in this area is focused on microchannel reactors, but one can imagine nanochannel reactors, enabling to achieve a maximum reaction selectivity and adapted for specific needs.

11.6 Outlook

The new nanocatalytic processes in the future will be carried out under much *softer conditions* and will *mimic the nature*, for example, nitrogen fixation from ambient air or CO_2 transformation into useful products. Another important challenge for nanocatalysis is water splitting to obtain hydrogen. This process, if successful and efficient, will finally solve the problem of the clean energy, climate changes, and decreasing stocks of fossil fuels.

Nanotechnology allows designing and producing catalysts with tuned properties not only from the quantitative point of view (small particles, high surface-to-volume ratio) but also qualitative – optimum facets and surface atom coordination. Then, thanks to the nanotechnology, a breakthrough in catalysis is performed – from traditional size-dependent catalysis to the concept of morphology/plane-dependent nanocatalysis.

Nanocatalysts enable to efficiently and precisely control reaction pathways.

Thanks to nanocatalysis, industrial processes will be carried out with high selectivity at high yield. The environment will be still better protected using nanocatalysts applied for the removal or neutralization of contaminants.

The key point of the European strategy is the sustainable development; it means an intensive economic growth respecting environmental needs. Then, the development of nanocatalysis is crucial for European industry and science development and could become an European *competitive strength*, as according to the report on catalysis of the World Technology Evaluation Center delivered in 2009, European investment in catalysis research is higher than that in the USA.

References

1. http://product.statnano.com/, visited on 24 May 2017.
2. Dauscher A, Garin A, Maire A. Correlations between the surface structure of platinum single crystals and hydrocarbon skeletal rearrangement mechanisms: approach to the nature of the active sites. J Catal 1987;105(1):233–44.
3. Schloegl R. Preparation and activation of the technical ammonia synthesis catalyst. In: Jennings JR, editor. Catalytic ammonia synthesis, fundamentals and practice, fundamental and applied catalysis, 1st ed. New York: Plenum Press, 1991:19–108.
4. Schuetze J, Mahdi W, Herzog B, Schloegl R. On the structure of the activated iron catalyst for ammonia synthesis. Topics Catal 1994;1:195–214.
5. Alstrup I, Chorkendorff I, Ullmann S. The Interaction of nitrogen with the (111) surface of iron at low and at elevated pressures. J Catal 1997;168:217–34.
6. Wang T, Tian X, Yang Y, Li Y-W, Wang J, Beller M, Jiao H. Coverage-dependent N_2 adsorption and its modification of iron surfaces structures. J Phys Chem C 2016;120:2846–54.
7. Ertl G, Lee SB, Weiss M. Adsorption of nitrogen on potassium promoted Fe(111) and (100) surfaces. Surf Sci 1982;114:527–45.
8. Tsai MC, Ship U, Bassignana IC, Küppers J, Ertl G. A vibrational spectroscopy study on the interaction of N_2 with clean and K-promoted Fe(111) surfaces: π-bonded dinitrogen as precursor for dissociation. Surf Sci 1985;155:387–99.
9. Xie X, Shen W. Morphology control of cobalt oxide nanocrystals for promoting their catalytic performance. Nanoscale 2009;1:50–60.
10. Xie XW, Li Y, Liu ZQ, Haruta M, Shen WJ. Low-temperature oxidation of CO catalysed by Co(3)O(4) nanorods. Nature 2009;458:746–9.
11. Hu LH, Peng Q, Li Y D. Selective synthesis of Co_3O_4 nanocrystal with different shape and crystal plane effect on catalytic property for methane combustion. J Am Chem Soc 2008;130:16136–7.

12. Chen T, Chen S, Song P, Zhang Y, Su H, Xu W, Zeng J. Single-molecule nanocatalysis reveals facet-dependent catalytic kinetics and dynamics of palladium nanoparticles. ACS Catal 2017;7:2967–72.
13. Crampton AS, Rötzer MD, Ridge CJ, Yoon B, Schweinberger FF, Landman U, Heiz U. Assessing the concept of structure sensitivity or insensitivity for sub-nanometer catalyst materials. Surf Sci 2016;652:7–19.
14. Bratlie KM, Kliewer CJ, Somorjai GA. Structure effects of benzene hydrogenation studied with sum frequency generation vibrational spectroscopy and kinetics on Pt(111) and Pt(100) single-crystal surfaces. J Phys Chem B 2006;110(36),17925–30.
15. Somorjai G, Li Y, Surface Structure and Selectivity. In: Introduction to Surface Chemistry and Catalysis, 2nd ed. Hoboken: Wiley, 2010:627–30.
16. Chimentao RJ, Kirm I, Medina F, Rodríguez X, Cesteros Y, Salagreb P, Sueirasa JE. Sensitivity of styrene oxidation reaction to the catalyst structure of silver nanoparticles. Chem Comm 2004:846–7.
17. Farrag M. Monodisperse and polydisperse platinum nanoclusters supported over TiO_2 anatase as catalysts for catalytic oxidation of styrene, J Mol Catal A: Chem 2016;413:67–76.
18. Yi G, Xu Z, Guo G, Tanaka K, Yuan Y. Morphology effects of nanocrystalline CeO_2 on the preferential CO oxidation in H_2-rich gas over Au/CeO_2 catalyst. Chem Phys Lett 2009;479:128–32.
19. Carrettin S, Concepcion P, Corma A, Nieto JML, Puentes VF. Nanocrystalline CeO_2 increases the activity of Au for CO oxidation by two orders of magnitude. Angew Chem Int Ed 2004;43:2538–40.
20. Janairo JIB, Carandang JS, Amalin DM. Facile synthesis of highly active Pd nanocatalysts using biological buffers. Chiang Mai J Sci 2017;44(1) 243–7.
21. Fei X, Kong W, Chen X, Jiang X, Shao Z, Lee JY. A Recycling-free nanocatalyst system: the stabilization of in situ- reduced noble metal nanoparticles on silicone nanofilaments via a mussel-inspired approach. ACS Catal 2017;7:2412–18.
22. Campisi S, Schiavoni M, Chan-Thaw CE, Villa A. Untangling the role of the capping agent in nanocatalysis: recent advances and perspectives. Catal 2016;6:185–206.
23. Li J, Nasaruddin RR, Feng Y, Yang J, Yan N, Xie J. Tuning the accessibility and activity of $Au_{25}(SR)_{18}$ Nanocluster catalysts through ligand engineering. Chem Eur J 2016;22:14816–20.
24. Liu LJ, Lai YD, Li HH, Kang LT, Liu JJ, Cao ZM, Yao JN. The role of dissolution in the synthesis of high-activity organic nanocatalysts in a wet chemical reaction. J Mater Chem A 2017;5:8029–36.
25. Lee T, Lee Y, Kang K, Soon A. In search of non-conventional surface oxidic motifs of Cu on Au(111). Phys Chem Chem Phys 2016;18:7349–58.
26. Haruta M. Size and support dependency in the catalysis of gold. Catal Today 1997 36(1) 153–60.
27. Lyalin A, Gao M, Taketsugu T. When inert becomes active: a fascinating route for catalyst design. Chem Rec 2016;16(5), 2324–37.
28. Hutchings GJ. Vapor phase hydrochlorination of acetylene: correlation of catalytic activity of supported metal chloride catalysts. J Catal 1985;96:292–5.
29. Hutchings GJ. Catalysis by gold. Catal Today 2005;100:55–61.
30. Ouyang L, Tan L, Xu J, Tian P-T, Da G-J, Yang X-J, Chen D, Tang F, Han Y-F. Functionalized silica nanorattles hosting Au nanocatalyst for direct synthesis of H_2O_2. Catal Today 2015;248:28–34.
31. Ouyang L, Da G, Tian P, Chen T, Liang G, Xu J, Han Y-F. Insight into active sites of Pd–Au/TiO_2 catalysts in hydrogen peroxide synthesis directly from H_2 and O_2. J Catal 2014;311:129–36.
32. Ishikara T, Nakashima R, Ooishi Y, Hagiwara H, Matsuka M, Ida S. H_2O_2 synthesis by selective oxidation of H_2 over Pd–Au bimetallic nano colloid catalyst under addition of NaBr and H_3PO_4. Catal Today 2015;248:35–9.
33. Li X, Cai T, Kang E-T. Yolk–Shell Nanocomposites of a gold nanocore encapsulated in an electroactive polyaniline shell for catalytic aerobic oxidation. ACS Omega 2016;1:160–7.

34. Shah V, Malardier-Jugroot C, Jugroot M. Mediating gold nanoparticle growth in nanoreactors: role of template-metal interactions and external energy. Mater Chem Phys 2017;196:92–102.
35. McTaggart M, Malardier-Jugroot C, Jugroot M. Self-assembled biomimetic nanoreactors II: noble metal active centers. Chem Phys Lett 2015;636:221–7.
36. Santra C, Pramanik M, Bando KK, Maity S, Chowdhury B. Gold nanoparticles on mesoporous Cerium-Tin mixed oxide for aerobic oxidation of benzyl alcohol. J Mol Catal A: Chem 2016;418–19: 41–53.
37. Shang L, Liang Y, Li M, Waterhouse GIN, Tang P, Ma D, Wu L-Z, Tung C-H, Zhang T. "Naked" magnetically recyclable mesoporous Au–Fe_2O_3 nanocrystal clusters: a highly integrated catalyst system. Adv Funct Mater 2017;27:1606215.
38. Zhao Z-J, Li Z, Cui Y, Zhu H, Schneider WF, Delgass WN, Ribeiro F, Greeley J. Importance of metal-oxide interfaces in heterogeneous catalysis: a combined DFT, microkinetic, and experimental study of water-gas shift on Au/MgO. J Catal 2017;345:157–69.
39. Huang M, Zhang Y, Zhou Y, Zhang C, Zhao S, Fang J, Gao Y, Sheng X. Synthesis and characterization of hollow ZrO_2–TiO_2/Au spheres as a highly thermal stability nanocatalyst. J Coll Interf Sci 2017;497:23–32.
40. Sun L, Zhang Q, Li GG, Villarreal E, Fu X, Wang H. Multifaceted gold–palladium bimetallic nanorods and their geometric, compositional, and catalytic tunabilities. ACS Nano 2017; 11:3213–28.
41. Shoaib A, Ji M, Qian H, Liu J, Xu M, Zhang J. Noble metal nanoclusters and their in situ calcination to nanocrystals: Precise control of their size and interface with TiO_2 nanosheets and their versatile catalysis applications. Nano Res 2016;9(6):1763–74.
42. Wang Q, Cai X, Liu Y, Xie J, Zhou Y, Wang J. Pd nanoparticles encapsulated into mesoporous ionic copolymer: Efficient and recyclable catalyst for the oxidation of benzyl alcohol with O_2 balloon in water. Appl Catal B: Environ 2016;189:242–51.
43. Burange AS, Shukla R, Tyagi AK, Gopinath CS. Palladium supported on fluorite structured redox $CeZrO_{4-\delta}$ for heterogeneous Suzuki coupling in water: a green protocol. Chem Sel 2016; 1:2673–81.
44. Engelbrekt C, Seselj N, Poreddy R, Riisager A, Ulstrup J, Zhang J. Atomically thin Pt shells on Au nanoparticle cores: facile synthesis and efficient synergetic catalysis. J Mater Chem A 2016; 4:3278–86.
45. López A, Navascues N, Mallada R, Irusta S. Pt-CoO_x nanoparticles supported on ETS-10 for preferential oxidation of CO reaction. Appl Catal A: Gen 2016;528:86–92.
46. Jeon S, Park YM, Saravanan K, Han GY, Kim B-W, Lee J-B, Bae JW. Aqueous phase reforming of ethylene glycol over bimetallic platinum-cobalt on ceria-zirconia mixed oxide. Int J Hydr Energy 2017;42:9892–902.
47. Chang L, Li Y. One-step encapsulation of Pt-Co bimetallic nanoparticles within MOFs for advanced room temperature nanocatalysis. Mol Catal 2017;433:77–83.
48. Li X, Cai T, Kang E-T. Hairy hybrid nanorattles of platinum nanoclusters with dual-responsive polymer shells for confined nanocatalysis. Macromol 2016;49:5649–59.
49. Mendes PS, Lapisardi G, Bouchy C, Rivallan M, Silva JM, Ribeiro MF. Hydrogenating activity of Pt/zeolite catalysts focusing acid support and metal dispersion influence. Appl Catal A: Gen 2015;504:17–28.
50. Zhu L, Zheng T, Yu C, Zheng J, Tang Z, Zhang N, Shu Q, Chen BH. Platinum-nickel alloy nanoparticles supported on carbon for 3-pentanone hydrogenation. Appl Surf Sci 2017;409:29–34.
51. Nindakova LO, Badyrova NM, Smirnov VV, Kolesnikov SS. Asymmetric transfer hydrogenation of carbonyl compounds catalyzed by rhodium nanoparticles. J Mol Catal A: Chem 2016;420:149–58.

52. Nindakova LO, Badyrova NM, Smirnov VV, Strakhov VO, Kolesnikov SS. Enantioselective Hydrogen Transfer Hydrogenation on Rhodium Colloid Systems with Optically Active Stabilizers. Rus J Gen Chem 2016;86(6):1240–9.
53. Guan H, Lin J, Qiao B, Yang X, Li L, Miao S, Liu J, Wang A, Wang X, Zhang T. Catalytically active Rh sub-nanoclusters on TiO_2 for CO oxidation at cryogenic temperatures. Angew Chem Int Ed 2016;55:2820 –2824.
54. Ge H, Zhang B, Gu X, Liang H, Yang H, Gao Z, Wang J, Qin Y. A tandem catalyst with multiple metal oxide interfaces produced by atomic layer deposition. Angew Chem Int Ed 2016;55:7081–5.
55. Bornscheuer UT, Hashmi ASK, García H, Rowan MA. Catalysis at the heart of success! Chem Cat Chem 2017;9:6–9.
56. Hassan HM, Mohamed SK, Ibrahim AA, Betiha MA, El-Sharkawy EA, Mousa AA. A comparative study of the incorporation of TiO_2 into MCM-41 nanostructure via different approaches and its effect on the photocatalytic degradationof methylene blue and CO oxidation. React Kinet Mech Cat 2017;120:791–807.
57. Grzechulska-Damszel J, Morawski AW, Grzmil B. Thermally modified titania photocatalysts for phenol removal from water. Int J Photoenergy 2006;ID 96398:1–7.
58. Janus M, Kusiak E. Morawski AW. Carbon modified TiO_2 photocatalysts with enhanced adsorptivity for dyes from water. Catal Lett 2009;131:506–11.
59. Baktash E, Littlewood P, Pfrommer J, Schomäcker R, Driess M, Thomas A. Controlled formation of nickel oxide nanoparticles on mesoporous silica using molecular Ni_4O_4 clusters as precursors: enhanced catalytic performance for dry reforming of methane. Chem Cat Chem 2015;7:1280–4.
60. Gupta S, Jain VK, Jagadeesan D. Fine tuning the composition and nanostructure of Fe-based core–shell nanocatalyst for efficient CO_2 hydrogenation. Chem Nano Mater 2016;2:989–96.
61. Polanski J, Siudyga T, Bartczak P, Kapkowski M, Ambrozkiewicz W, Nobis A, Sitko R, Klimontko J, Szade J, Lelatko J. Oxide passivated Ni-supported Ru nanoparticles in silica: A new catalyst for low-temperature carbon dioxide methanation. Appl Catal B: Environ 2017;206:16–23.
62. Zou X, Rui Z, Ji H. Core–Shell NiO@PdO nanoparticles supported on alumina as an advanced catalyst for methane oxidation. ACS Catal 2017;7:1615–25.
63. Pantaleo G. La Parola V, Deganello F, Singha RK, Bal R, Venezia AM. Ni/CeO_2 catalysts for methane partial oxidation: Synthesis driven structural and catalytic effects. Appl Catal B Env 2016;189:233–41.
64. Ismagilova ZR, Shikina NV, Yashnik SA, Zagoruiko AN, Kerzhentsev, Ushakov VA, Sazonov VA, Parmon VN, Zakharov VM, Braynin BI, Favorski ON. Technology of methane combustion on granulated catalysts for environmentally friendly gas turbine power plants. Catal Today 2010;155(1–2):35–44.
65. Higo T, Saito H, Ogo S, Sugiura Y, Sekine Y. Promotive effect of Ba addition on the catalytic performance of Ni/$LaAlO_3$ catalysts for steam reforming of toluene. Appl Catal A: Gen 2017;530:125–31.
66. Baskar G, Aiswarya R. Trends in catalytic production of biodiesel from various feedstocks. Renewable Sustainable Energy Rev 2016;57:496–504.
67. Li S, Yang C, Yin Z, Yang H, Chen Y, Lin L, Li M, Li W, Hu G, Ma D. Wet-chemistry synthesis of cobalt carbide nanoparticles as highly active and stable electrocatalyst for hydrogen evolution reaction. Nano Res 2017;10(4):1322–8.
68. Clavijo-Chaparro SL, Hernández-Gordillo A, Camposeco-Solis R, Rodríguez-González V. Water splitting behavior of copper-cerium oxide nanorods and nanocubes using hydrazine as a scavenging agent. J Mol Catal A: Chem 2016;423:143–50.

69. Yuan D, Zhang L, Lai J, Xie L, Mao B, Zhan D. SECM evaluations of the crystal-facet-correlated photocatalytic activity of hematites for water splitting. Electrochem Comm 2016;73:29–32.
70. Li G, Kanezashi M, Tsuru T. Catalytic ammonia decomposition over high-performance Ru/graphene nanocomposites for efficient CO_x-free hydrogen production. Catalyst 2017;7(1):23–35.
71. Lendzion-Bielun Z, Narkiewicz U, Arabczyk W. Cobalt-based catalysts for ammonia decomposition. Materials 2013;6:2400–9.
72. Salmeron M, Schloegl R. Ambient pressure photoelectron spectroscopy. Surf Sci Rep 2008;63(4):169–99.
73. Hansen P, Wagner J, Helveg S, Rostrop-Nielsen J, Clausen B, Topsoe H. Atom-resolved imaging of dynamic shape changes in supported copper nanocrystals. Science 2002;15:2053–5.
74. LaGrow AP, Ward MR, Lloyd DC, Gai PL, Boyes ED. Visualizing the Cu/Cu_2O interface transition in nanoparticles with environmental scanning transmission electron microscopy. J Am Chem Soc 2017;139:179–85.
75. Chen T, Zhang Y, Xu W. Size-dependent catalytic kinetics and dynamics of Pd nanocubes: a single-particle study. Phys Chem Chem Phys 2016;18:22494–502.
76. Chen T, Zhang Y, Xu W. Single-molecule nanocatalysis reveals catalytic activation energy of single nanocatalysts. J Am Chem Soc 2016;138:12414–21.
77. Zhang Y, Chen T, Alia S, Pivovar BS, Xu W. Single-molecule nanocatalysis shows in situ deactivation of Pt/C electrocatalysts during the hydrogen-oxidation reaction. Angew Chem Int Ed 2016;55:3086–90.
78. Zhang Y, Chen T, Song P, Xu W. Recent progress on single-molecule nanocatalysis based on single-molecule fluorescence microscopy. Sci Bull 2017;62:290–301.

Ioana Fechete and Jacques C. Vedrine

12 Nanotechnology and catalysis: Supramolecular templated mesoporous materials for catalysis

12.1 Introduction

Nanoscience is concerned with the size of matter and is the science of matter that occurs in systems with at least one dimension on the 1–100 nm scale. Nanostructure science and technology is a broad and interdisciplinary area of research and development activity that has been growing explosively worldwide in the past few years. Nanoscale materials have much larger surface areas than similar masses of larger-scale materials. As the surface area per unit mass of a material increases, a greater amount of the material can come into contact with surrounding materials, thus influencing reactivity [1–8].

Scientists and engineers typically have approached the synthesis and fabrication of high surface area nanostructures from one of two directions:

- The "bottom-up" approach in which the nanostructures are built up from individual atoms or molecules. This is the basis of most "cluster science" as well as crystal materials synthesis, usually via chemical means. Both high surface area particles and micro- and mesoporous crystalline materials with high void volume (pore volume) are included in this "bottom-up" approach.
- The "top-down" approach in which nanostructures are generated from breaking up bulk materials. This is the basis for techniques such as mechanical milling, lithography, post-synthesis procedure such as synthesis of hierarchical mesoporous materials, precision engineering, and similar techniques that are commonly used to fabricate nanoscale materials [1–8].

> The noblest pleasure is the joy of understanding – Leonardo da Vinci

Mesoporous materials are cornerstones of nanoscience and nanotechnology [1–8]. The rapid development of nanoscience with mesoporous materials has gained strong support from chemistry and supramolecular chemistry. Supramolecular chemistry and chemical self-assembly provide many ways of forming nanostructures and mesoporous materials. The field of supramolecular chemistry has become important, with Lehn, Cram, and Pedersen winning the Nobel Prize in 1987. The concept of supramolecular chemistry is that molecules can self-organize into definite structures, without forming covalent bonds, but rather through weaker interactions such as hydrogen bonding or hydrophobic interactions. The supramolecular chemistry helped to highlight such systems, and chemists from Mobil were the first to realize

https://doi.org/10.1515/9783110547221-012

that supramolecular chemistry could be applied to catalyst design. Whereas initial approaches to mesoporous materials relied on larger and larger template molecules, Mobil researchers found that they could use supramolecular assemblies of molecules as templates [9]. These supramolecular-templated mesoporous materials, known as MCMs (from Mobil composition of matter), can be prepared with a range of pore sizes. These materials are highly unusual in their textural characteristics: uniform pore sizes, surface areas in excess of 1,000 m^2/g, and long-range ordering of the packing of pores. The mesoporous materials are derived with supramolecular assemblies of surfactants, which template the inorganic components during synthesis. Many researchers have since exploited this technique of supramolecular templating to produce materials with different compositions, new pore systems, and novel properties. The discovery of the M41S family [9] of nanostructured mesoporous molecular sieves, well-defined mesostructures, has attracted a pervasive interest from academic and industrial researchers due to the huge surface area, uniform pore size distribution, large pore size, large pore volume, adjustable framework, and surface properties and large number of current and foreseen applications of these materials. Applications of these materials include catalysis, environmental chemistry, organic synthesis, adsorption, medical imaging, molecular collection, and storage [10–23]. The field of mesoporous materials offers several advantages that cannot be found elsewhere. Materials chemistry of responsive mesoporous systems became a mature discipline with high standards and challenging goals. The successful synergy between sol–gel techniques, self-assembly, polymer synthesis, surface engineering, and physical chemistry in confinement represents a unique and versatile toolbox to achieve the building up of complex nanosystems with well-defined physical and chemical properties at several length scales: molecular, supramolecular, and mesoscopic.

One of the newest areas in the realm of catalysis is that of tailored mesoporous materials, which find many uses as highly selective catalysts in a range of applications. A mesoporous material is one that has pores with diameter in the range of 2–50 nm (a typical chemical bond is of the order of 0.1 nm). In catalysis the key goal is to promote reactions that have high selectivity with high yield. Catalysts with one, two, or three spatial dimensions in the nanometer size range exhibit unique (compared to the bulk) catalytic or chemical activity. Here we discuss recent development of heterogeneous catalysis in ring opening of methylcyclopentane (MCP) and dry reforming of methane (DRM) on mesoporous catalysts.

12.2 Conversion of MCP

The worldwide depletion of oil reserves is expected to increase the future demand for diesel fuel. Concomitantly, environmental constraints necessitate the development of environmental-friendly refining processes. With increasingly stringent environmental and economic regulations, the foremost challenge facing in all industries is to find greener processes with better atom efficiency [24–28]. In particular, the petrochemical industry has placed a strong research emphasis on the ring opening of naphthenic molecules over noble metal catalysts [29–34] to improve the cetane number of diesel fuels and to minimize harmful emissions. Diesel fuels can be improved by selective ring opening of naphthenic compounds following hydrogenation of aromatic compounds [35]. Taking into account the fact that each dearomatization reaction followed by a hydrogenation step leads to a cyclopentane, the understanding of MCP chemistry is very important. In this case, although MCP is not a component of diesel fuel, the ring opening of MCP must first be understood because it serves as a model molecule and provides the basis for understanding the behavior of all other molecules. The ring opening of MCP is one of the most used model reactions for exploring the structural sensitivity of hydrocarbon conversion catalyzed by noble metals. Excellent works have been reported concerning the use of noble metals [29–34]. These catalysts have generated widespread interest, because at low temperatures they exhibit the ability to promote the ring-opening reactions with atom efficiency and without unwanted side reactions such as cracking and enlargement. Following a careful analysis of previous studies, we ascertained that the distribution of these desired ring-opening products is governed by several factors, including the intrinsic nature of the metal [36], the platinum particle size [37–39], the interface length between the metal and support [40–42], the presence of carbonaceous residues [42–44], sulfiding [43], and the reaction conditions [41–42]. The ring-opening reaction of MCP can proceed by different mechanisms, which include associative [45] and dissociative [46] pathways. For the ring opening and cyclization of C5, a common adsorbed surface intermediate has been defined [37], which has sparked myriad discussions. Several structures have been proposed for this intermediate, including edgewise or flat geometries with associative or dissociative structures. The dissociative structure is a function of the degree of dehydrogenation of the surface intermediate. On Pt catalysts, a flat-lying intermediate with a low degree of dissociation has been suggested for selective ring opening [41, 47]. This intermediate suggests that the rupture of the C–C bond in the proximity of the tertiary carbon atom was hindered [48]. A plausible explanation is based on the assumption that two metal atoms form the active ensemble. In this case, the chemisorption takes place by the rupture of the weakest C–H bond on the tertiary carbon atom, followed by C–C rupture on the neighboring metal atom. The presence of chemisorbed hydrogen is also necessary to prevent excessively deep dissociation. A flat-lying intermediate and an associative mechanism have also been suggested by Liberman [45], with hydrogen–metal ensembles attacking the chemisorbed ring without the dehydrogenation of the ensembles' surface. Gault et al. [38, 39] identified three different paths for the ring opening of MCP: dicarbene, pi-adsorbed olefin, and metallocyclobutane intermediates. The dicarbene path occurs only at the unsubstituted secondary–secondary C–C bond [46]. In this case, the formation of 2-methylpentane (2-MP) and 3-methylpentane (3-MP) prevails in a nonstatistical distribution (selective mechanism), and these products were observed on Pt particles larger than 1.8 nm with high coordination numbers [37]. This mechanism also prevails on Ir. The pi-adsorbed olefin and metallocyclobutane paths occur at substituted C–C bonds, and the formation of 2-MP:3-MP:*n*-hexane (*n*-H) occurs with a statistical distribution of 2:1:2 (nonselective mechanism). The pi-adsorbed olefins require a flat adsorption of three neighboring carbon atoms, which interact with a single metal site on the catalyst surface. The formation of metallocyclobutane competes with the dicarbene path but exhibits a higher activation energy than that exhibited by the dicarbene mechanism. Gault found that the relative importance of the selective and nonselective mechanisms was a function of the size of the Pt particles used. Other researchers, however, attempted to attribute the nonselective mechanism to the presence of adlineation sites on the metal–support boundary, whereas the selective mechanism was assigned to pure metallic sites [40, 49]. The influence of particle size has been found to be different for reactions that involve Ir catalysts, which confirms that the nature of the ring-opening mechanism also depends on the catalytic metal [43]. Irrespective of the effects of particle size on Ir catalysts, the selective mechanism (dicarbene path) was found to prevail. Ir/SiO_2 catalysts open the ring via a selective mechanism, whereas the Ir/Al_2O_3 catalyst opens the ring via a nonselective mechanism. In the latter case, the selectivity was found to be a support effect rather than a particle effect. However, the conversion of MCP has been studied on Pt/TiO_2 [50–52]. A widespread view is that nonstoichiometric TiO_2 supports provoke a surface migration of reduced TiO_x species from the support to the metallic phase [53–55]. Thereafter, the catalytic behavior changes as a result of the decreased adsorption capacity of metals, the reduced catalytic activity and chemisorption capacity, and the changes in the selectivity of the MCP reaction [56]. Unique metal–support interactions between titanium and noble metals were observed for other reactions [57–58]. When Pt/TiO_2 was reduced at 500°C, the complete elimination of hydrogen and carbon monoxide chemisorption occurred; this result was associated with strong metal–support interactions (SMSI). Otherwise, the extent of chemisorption elimination was strongly correlated with the reducibility of the support. In this case, the mechanism proposed is explained by increased *d*-orbital occupancy of the support, followed by covalent bond formation between support metal ions and supported metal atoms. Moreover, theoretical studies [59] on surface clusters formed between Pt and the $(TiO_6)^{8-}$ octahedron has indicated that covalent sharing between Pt and the support surface is indeed likely if one of the oxygen atoms is removed from the octahedron. The modification of chemisorptive and catalytic properties of dispersed metals by substrates, observed first by Schwab [60] and Solymosi [61], has found renewed interest since the discovery by Tauster et al. of so-called SMSI between group VIII noble metals and TiO_2. Thereafter, the SMSI phenomenon was reported for other support–metal catalysts, including partially reducible oxides other than TiO_2.

However, the conversion of MCP has been studied on Pt–Ir/TiO_2 bimetallic catalysts [62], because the coexistence of two metal sites is expected to modulate the peculiar behavior of such catalysts for catalytic reactions in a manner different from that of their two components. These catalysts exhibit high selectivity, activity and stability compared to their component pure-metal particles. For the sake of argument, the bimetallic Pt–Ir particles supported on γ-Al_2O_3 exhibited higher selectivity and a lower tendency for coke formation [63] compared to a Pt catalyst. In the case of the catalysts that contain two metals, the specific behavior necessary for a desired catalytic reaction is tailored as a function of the distribution and topology of the two types of metal particles: (i) the two particles can be made bimetallic with a homogeneous composition by forming alloys; (ii) the two metals may also exist as separate phases but in intimate contact via the surface segregation of one metal; in particular, the metal with the lower melting point, the smaller atomic radius, and the lower heat of vaporization is typically segregated on the surface [64–66]; and (iii) metals may segregate into separate islands with each crystallite containing one of the respective metals.

The conversion of MCP on Pt/TiO_2, Ir/TiO_2 and Pt–Ir/TiO_2 catalysts, each prepared with low amounts of noble metals (0.5 wt%), was studied for the first time over the 180–400°C temperature range under hydrogen at atmospheric pressure [62]. The order of reactivity as a function of the temperature and total conversion rates were Ir/TiO_2 at 180°C >Pt–Ir/TiO_2 at 220°C >Pt/TiO_2 at 260°C. All catalysts exhibited the ability to open the ring of MCP with an atom efficiency. For the catalysts, a change in the reaction temperature provokes an alteration in the selectivity to ring opening of MCP in favor of selectivity to cracking reactions. The cracking products were formed in large amounts at high temperatures, with C_1 compounds as the major products. Ir/TiO_2 operates via the *selective* mechanism at low temperatures and via the *nonselective* mechanism at 220°C due to the SMSI phenomenon. Pt/TiO_2 operates by the nonselective mechanism, whereas the Pt–Ir catalyst operates by the selective mechanism and shows Ir-like character. The synergy between Pt–Ir as bimetallic particles was assessed by total conversion of MCP. The MCP ring-opening results indicated that the reaction takes place on Ir sites, which suggests that the bimetallic catalyst contains separate entities of the two metals. Ring elargement (RE) was absent on the Ir/TiO_2 and Pt–Ir/TiO_2 catalysts and present on Pt/TiO_2 only at high temperatures. Taken together, the results [62] revealed that the Ir/TiO_2 catalyst was more active and was the most promising catalyst for the ring opening of MCP at low temperatures.

Despite the effectiveness of supported noble metals, which efficiently catalyze the conversion of MCP with atom efficiency, the high cost and limited availability of precious metals restrict their wide application. A practical solution is to find a clean and efficient heterogeneous catalyst to replace noble metal catalysts. It is cheaper to employ base-metal oxide catalysts because of the environmental and economic advantages, good catalytic performance, and the competitiveness with the noble metals. In this context, mesoporous materials appear to be good candidates for the conversion of hydrocarbons because of a large internal surface area, high thermal stability, and surface properties that create new opportunities for heterogeneous catalysis [9, 67, 68]. However, it is noteworthy that few reports are available on the catalytic properties of mesoporous materials for the conversion of MCP. Although the discovery of mesoporous materials has brought about the beginning of a new age in the field of material synthesis, these materials are known to be of limited use as catalysts because the catalytic functionalities are missing. These mesoporous materials can be functionalized by depositing heteroatoms onto the silica framework [69–71]. A promising approach for the creation of active sites is that of substituting heteroatoms for Si by one-pot synthesis, which creates active sites that enhance the activities of mesoporous solids and allow catalysis [69–71]. The nonnoble metal oxides of molybdenum, tungsten, and iron appear to be efficient and environmental alternatives to the expensive noble metals. The results showed [71] that the efficiency of the catalyst is dependent on the density of active sites. Mo, Fe, and Mo–Fe catalysts have attracted significant interest because of their high efficiency as catalysts in environmentally important processes [72–78]. The efficiency of these catalysts is governed by the size of their oxide nanoparticles, which can be controlled by the dispersion on the support. A high dispersion leads to an increased number of catalytically active sites; however, the presence of strong particle–support interactions often reduces their activity [72–78]. In contrast, weak interactions with the support favor oxide particle agglomeration at elevated temperatures. Therefore, the strategy for optimizing the size of supported oxide nanoparticles to create uniform and well-dispersed supported transition metal oxides (Mo, Fe, Mo–Fe) for the conversion of MCP is to use the mesoporous support. The goal is to understand the basic surface chemistry of these oxides in a reductive environment greatly lags behind that of noble metals. Iron is of high interest due to its wide range of applications in catalysis [74, 79–81]. However, most studies of redox mesoporous catalysts published thus far dealt with oxidative environments. Despite the successful development of catalytically active Fe-containing mesoporous materials, little is known about their behavior in reductive environment and the influence of isomorphous substitution on their properties. It is known [82] that the isomorphous substitution of Fe^{3+} in the silicate framework leads to acid catalysts by creating Si-O-Fe bonds and being the genesis of Brønsted acidity. To obtain a gradation of the acidity of the Fe-mesoporous samples, a challenging controlled-synthesis method was used to substitute iron for silicon atoms to obtain several Si/Fe ratios. Fe-technische universitat delft (Fe-TUD-1) catalysts with Si/Fe ratios of 85, 65, and 45 were prepared [83] via hydrothermal one-pot synthesis. The samples were characterized by X-ray diffraction (XRD), N_2 physisorption, UV–vis, and X-ray photoelectron spectroscopy (XPS), which show the incorporation of Fe^{3+} in tetrahedral coordination. The ability of these catalysts was tested by examining the conversion of the reaction of MCP with hydrogen at atmospheric pressure and temperatures between 200 and 400°C. The active sites, tetrahedrally coordinated Fe and isolated atomic Fe sites, were responsible for the endocyclic C–C bond rupture between substituted secondary-tertiary carbon atoms, while the small clusters serve as active sites for the successive C–C bond rupture. The Fe content in the calcined samples is almost the same as the Fe content in the wet gel. This correlation

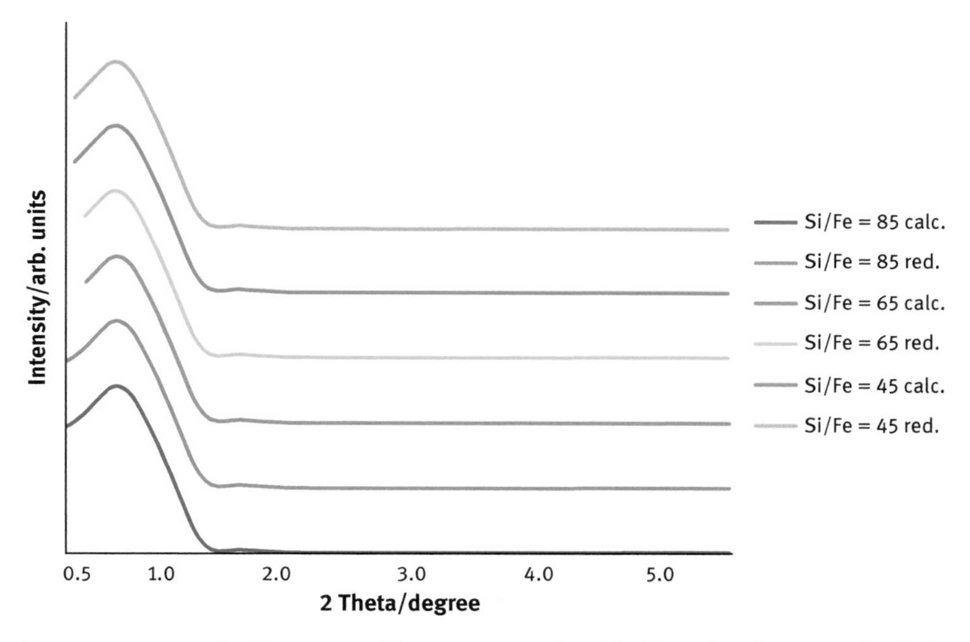

Figure 12.1: Low-angle XRD patterns of the Fe-TUD-1 samples with Si/Fe ratios of 85, 65, and 45 [83].

indicates that most of the Fe was incorporated into the mesoporous TUD-1 by hydrothermal one-pot synthesis at 180°C.

The XRD patterns at low angles for the calcined and reduced Fe-TUD-1 samples with different Si/Fe ratios are shown in Figure 12.1, which clearly shows a single diffraction peak at low angles of 0.7–0.8°. A significant change was not observed between the Fe-TUD-1 samples after metal loading and after calcination and reduction treatment. To identify any nonframework Fe oxide phases, XRD patterns for all samples were recorded in the wide-angle region (not shown).

The diffractograms do not show any patterns in this region. A diffraction peak corresponding to the crystalline phase of Fe was not detected. It is deduced that the added Fe may have been highly dispersed on the TUD-1 samples. These results were confirmed by the color of the calcined and reduced samples, which were completely white, indicating that iron was part of the framework of the mesoporous solid and that no iron oxides are formed [83]. It must also be noted that the absence of a diffraction peak can be attributed to the low content of Fe when the particles are small and thus might not be visible in the powder XRD pattern due to the detection limit of the XRD apparatus. The highly dispersed Fe species in the silica framework of TUD-1 and the absence of the crystalline phase of Fe were also confirmed by the transmission electron microscopy (TEM) analysis (not shown) [83].

The N_2-physisorption isotherms [83] of calcined/reduced Fe-TUD-1 samples (not shown) were all type IV isotherms, according to the Brunauer, Deming, Deming and

Teller (BDDT) classification [84], and showed a hysteresis loop characteristic of wormhole-type disordered mesoporous materials [85]. All samples showed steep increases in the volume of adsorbed nitrogen at relative pressures of $P/P_0 = 0.6–0.8$ due to the onset of capillary condensation within mesopores [83]. The surface area decreased gradually with increased metal loading, from 640 m^2/g for Si/Fe = 85 to 612 m^2/g and 570 m^2/g for Si/Fe = 65 and Si/Fe = 45, respectively. The incorporation of more metal decreases the pore volume from 0.93 cm^3/g for Si/Fe = 85 to 0.91 cm^3/g for Si/Fe = 45. Meanwhile, all Fe-TUD-1 samples exhibit a pore diameter of 6.6 nm, suggesting that pore shrinkage does not occur with increasing Fe content and that pore diameter is virtually independent of Fe loading.

For the sake of gaining insight into the Fe-TUD-1 structure and clarifying the nature/coordination environment of the active sites, UV–vis spectroscopy was employed [83]. The UV–vis spectra of calcined and reduced Fe-TUD-1 samples with different Fe contents are shown in Figure 12.2.

The UV–vis spectra of Fe-TUD-1 with Si/Fe ratios of 85 and 65 mainly show one absorption band between 200 and 280 nm with a maximum centered at ca. 235 nm [83]. This maxima value should be ascribed to Laporte-allowed ligand-to-metal charge transfer that involves isolated tetrahedrally coordinated Fe^{3+}. This means a $t_1 \rightarrow t_2$ and $t_1 \rightarrow e$ transition with Fe^{3+} in an isolated tetrahedral geometry ($Fe^{3+}O_4$) [83, 86, 87]. For low iron content, Si/Fe ratios of 85 and 65, the Fe-O-Si bond of the isolated tetrahedral iron ions in the silica framework of TUD-1 is dominant, indicating that Fe was successfully incorporated into the silica TUD-1 network.

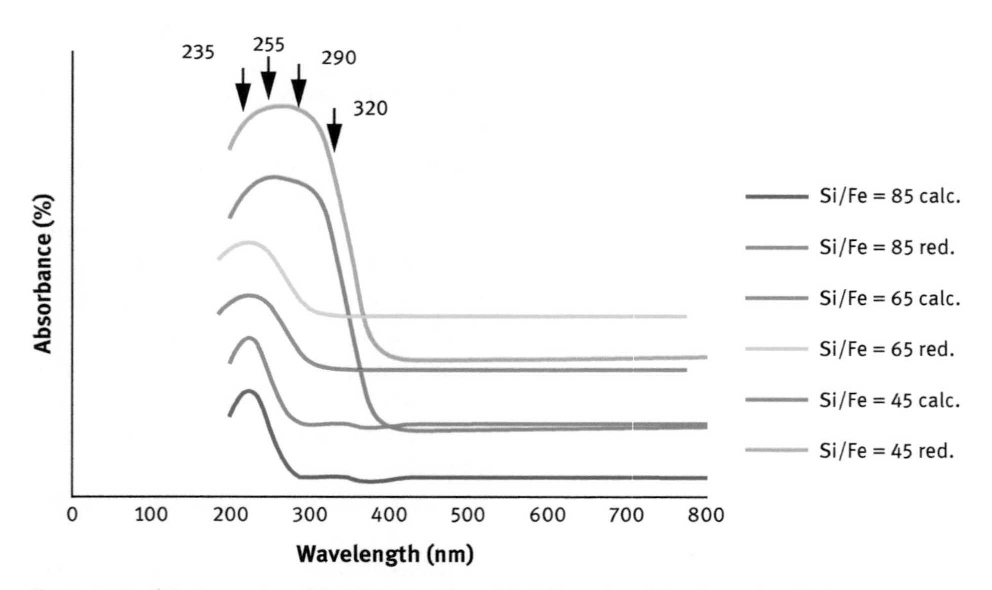

Figure 12.2: UV–vis spectra of Fe-TUD-1 samples with Si/Fe ratios of 85, 65, and 45 [83].

Moreover, the concentration of tetrahedral species increases with the crystallographic incorporation of Fe in TUD-1, as observed by the increased intensity of the absorption band for the sample with a Si/Fe ratio of 65. It is known that the ligand-to-metal charge transfer transition for isolated Fe^{3+} sites gives rise to bands below 300 nm and becomes red-shifted with increasing numbers of coordinating oxygen ligands [83, 86], reflecting the fact that our samples with Si/Fe ratios of 85 and 65 contained only the isolated sites. Absorption at higher wavelengths is essentially absent for the samples with Si/Fe ratios of 85 and 65, indicating that these samples contain predominantly isolated tetrahedral framework Fe^{3+} species and that the formation of the Si-O-Fe bond was not prevented from forming in our experimental conditions. A characteristic band above 320 nm, typical of octahedral coordination ($Fe^{3+}O_6$), was absent in the samples with Si/Fe ratios of 85 and 65, indicating that these samples were free of the ferric oxide species. Compared with the samples with Si/Fe ratios of 85 and 65, a red-shift is observed as the iron content increased for the sample with a Si/Fe ratio of 45. Nevertheless, the Fe-TUD-1 sample with a Si/Fe ratio of 45 shows a broad absorption band in the UV region between 260 and 400 nm [83]. The UV–vis spectra of this sample shows three absorption bands at 255, 290, and 320 nm. The band at 255 nm is typical of dπ–pπ charge transfer between the Fe and O atoms in the framework of TUD-1 and demonstrated the formation of Fe-O-Si bonds and the presence of tetrahedrally coordinated iron atoms. The *d–d* transition band at 290 nm indicates the presence of highly dispersed, isolated, octahedrally coordinated Fe entities, which are the prevalent species [83, 88]. The absorption band at 320 nm observed for the Fe-TUD-1 with a Si/Fe ratio of 45 stresses that the Fe-TUD-1 samples contain oligomers of low nuclearity 2*d*-FeO_x species in low amounts. In Fe-TUD-1 with Si/Fe ratio of 45, some Fe-O-Fe clusters seem to coexist with isolated Fe sites of Si-O-Fe bonds. However, it is worth noting that the charge transfer transition bands between 300 and 400 nm are attributed to octahedral Fe^{3+} in small oligomeric Fe_xO_y clusters [86, 88]. No absorption band was observed between 400 and 600 nm, revealing the absence of bulk iron oxide. The UV–vis spectra in the wavelength range of 200–800 nm for the samples reduced for 4 h under H_2 are also displayed in Figure 12.2 [83]. It was observed that the reductive treatment does not cause a noticeable change in the mode of iron-ion stabilization and that the charge transfer bands are not essentially shifted. This observation is in good agreement with the high stability of isolated iron atoms in a reductive environment, which suggests that no partial or total breaking of the Si-O-Fe framework linkages take place. These results agree with those already reported [89], which show that the isolated tetrahedral sites incorporated into the framework is not reduced up to 500°C.

Figure 12.3 illustrates the parts of the XPS spectra associated with the Fe $2p_{1/2}$ and Fe $2p_{3/2}$ binding energy regions for calcined and reduced Fe-TUD-1 samples, with Si/Fe ratios of 85, 65, and 45. All the spectra are centered at 724.1 and 710.8 eV for Fe $2p_{1/2}$ and Fe $2p_{3/2}$, respectively, which are typical values for Fe^{3+} [90].

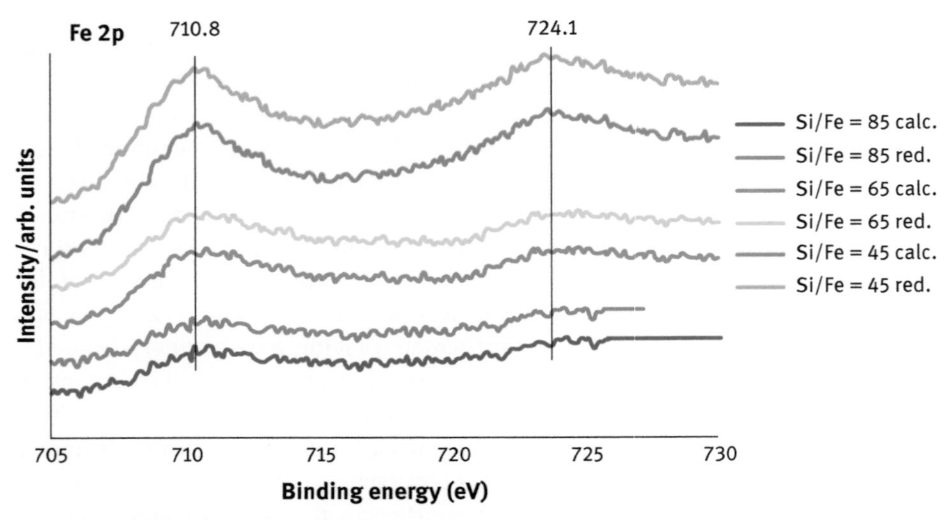

Figure 12.3: Fe 2*p* XPS spectra of Fe-TUD-1 samples with Si/Fe ratios of 85, 65, and 45 [83].

The spin orbit separation between the peaks is 13.4 eV and is identical for all the samples. The Fe $2p_{3/2}$ peaks (Fe $2p_{3/2}$ has a degeneracy of four states) are narrower and stronger than those of Fe $2p_{1/2}$, which only has two states, because of the spin–orbit (j–j) coupling. The intensity of the Fe 2*p* peaks increases with the Fe content [83]. It is clearly seen from this result that the surface iron species of the Fe-loaded samples mainly existed in the form of Fe^{3+} species. As shown in Figure 12.3, the XPS signal is unaffected by the reduction treatment, suggesting that the isolated Fe ions in the TUD-1 situated in tetrahedral sites are stable under reductive conditions up to 500°C. The Fe in Fe-TUD-1 cannot be reduced to Fe^{0} up to 500°C, or even to Fe^{2+} by one electron transfer. However, it was observed [91] that isolated Fe atoms in Fe-ZSM-5 mobile five (MFI) cannot be reduced to Fe^{0}, even up to 800°C. This means that the Si-O-Fe framework containing the isolated Fe atoms is stable in the reductive environment. When introduced in the reductive environment at a high temperature, the presence of Fe^{3+} in the catalysts can be due to the nonmobility of Fe ions which are located in Si-O-Fe framework. It must be noted that for the Fe-TUD-1 sample with Si/Fe = 45, the UV–vis results stressed the presence of octahedral coordination ions situated on their surface, but the XPS analysis did not detect Fe^{2+}. Fe^{3+} is present in the Fe-TUD-1 with a Si/Fe ratio of 45, but the nature of this oxidation state is unclear [83]. One may expect the presence of Fe^{2+} or Fe^{0}, but it is not detected in the XPS spectra. Thus, the amount of these species either is below the detection limit or is absent. A slight but systematic increase in the Si/Fe ratio was observed compared with the chemical analysis. The spectrum of O 1 s from the XPS spectra of each sample is composed of a single peak centered at 532.4 eV [83]. Other oxygen peaks attributable to iron oxide (at approximately 530 eV) and to adsorbed –OH groups (at approximately 531.1 eV) are very low (Figure 12.3).

12.2.1 Catalytic Activity in the Conversion of MCP

Isolated tetrahedral Fe^{3+} entities appear to be selective sites for *n*-H, while the small aggregates serve as active sites for the C_1 cracking products, as previously proposed for Fe/KIT-6 prepared by a solid–solid method [76, 83]. For completeness, the reactivity of isolated tetrahedrally/octahedrally coordinated Fe cations and small aggregates in the Fe-TUD-1 catalysts were evaluated in the conversion of MCP. The Fe-TUD-1 catalysts with different Si/Fe ratios were investigated for the first time [83], in this study, with respect to MCP conversion as a function of reaction temperature and metal loading. The effect of the temperature on the activity was investigated in the temperature range of 200–400°C for Si/Fe ratios between 85 and 45. Before the catalytic tests, all samples were reduced for 4 h in hydrogen at 500°C. The potential products in the conversion of MCP are *n*-H, 2-MP, and 3-MP, which are formed from a ring-opening reaction of MCP. Cyclohexane (Ch) and benzene (Bz) can be formed via ring-enlargement reactions, while methane (C1), ethane (C2), propane (C3), butane (C4)/i-butane (i-C4), and pentane (C5)/i-pentane (i-C5) can be formed from cracking reactions. It must be noted that olefins were observed for the Fe-TUD-1 with a Si/Fe of 45 over the entire range of studied temperatures, but the olefins were present in small concentrations (<0.02 %).

All catalysts displayed an activity in the conversion of MCP [83]. Increasing the iron content of TUD-1 significantly enhanced the MCP conversion, suggesting that iron can function as the active sites in this reaction. For all catalysts, the activity increased with temperature. If evaluated at the same reaction temperature, the order of conversion (catalytic performance) of the samples was as follows: Fe-TUD-1, Si/Fe = 45>Fe-TUD-1, Si/Fe = 65>Fe-TUD-1, Si/Fe = 85. This order is in good agreement with the sequence of the density of active sites of these catalysts.

12.2.2 Ring-Opening Selectivity

n-H was the only species among the ring-opening products that was formed irrespective of the reaction temperature or the Fe loading [83]. The ring-opening reaction was dominant in the samples with Si/Fe ratios of 85 and 65, irrespective of the reaction temperature. These results suggest that the highly isolated, tetrahedrally coordinated Fe entities on the samples with Si/Fe ratios of 85 and 65 were responsible for the rupture of endocyclic C–C bonds in MCP between the substituted secondary–tertiary carbon atoms [83]. The results were very interesting because a different behavior was observed when compared with noble metal catalysts. For example, Pt/Al_2O_3 catalysts can open the MCP ring by breaking the substituted or the unsubstituted C–C bonds in MCP [46, 48] to form 2-MP, 3-MP, and *n*-H. Ir/Al_2O_3 catalysts exhibit the tendency to break the endocyclic C–C bonds in MCP between unsubstituted secondary–secondary carbon atoms, which only form 2-MP and 3-MP [46, 48, 92]. However, the ability of

Ni/Al_2O_3 catalysts for the conversion of MCP must be noted [93]. A recent work [94] reported the selective formation of *n*-H on noble metal catalysts, but the experimental conditions were different. The conditions in that work were 10–50 torr under H_2. Logically, the 1–6 cyclization is favored under these conditions, as already reported [95]. For high Fe loading on the sample with a Si/Fe ratio of 45, the ring-opening reaction with the selective formation of *n*-H prevailed at the low temperature of 200°C. Only the highly isolated, tetrahedrally coordinated Fe entities are favorable sites for the endocyclic rupture of C–C bonds at the substituted position. For the other reaction temperature, the cracking C_1 product was dominant. Taking into account the UV–vis results, isolated tetrahedrally/octahedrally coordinated Fe species and small Fe clusters seem to coexist in this sample [83].

This result is consistent with the suggestion that the tetrahedrally coordinated and atomically isolated sites are responsible for the rupture of endocyclic C–C bonds between substituted secondary–tertiary carbon atoms. It must be noted that the ring-opening reaction is dominant but not exclusive, suggesting that only a part of the Fe ions are involved in the ring-opening reaction and that the mononuclear species of Fe formed in these cases can be responsible for the successive rupture of C–C bonds (which will be discussed in the next section). A decrease in the *n*-H selectivity with the temperature is a common feature observed in all samples [83]. However, these results were significantly better than what was observed for W mesoporous catalysts [71], for which no products derived from ring-opening reactions were observed under the same conditions or for Fe/KIT-6, Mo/KIT-6, and FeMo/KIT-6 prepared by solid–solid methods in which the cracking products prevailed [76]. This reaction of ring opening of MCP, also called "methylcyclopentane hydrogenolysis," was extensively studied in the 1970s and was exclusively performed on noble metals, mainly on alumina-supported platinum catalysts with various mean metallic particle sizes. As a consequence of such studies, it was noted that in the skeletal rearrangements of hydrocarbons, two types of isomerization mechanisms occur: the bond shift and the cyclic mechanism. To distinguish between them, the ^{13}C tracer technique was used (Scheme 1) [46].

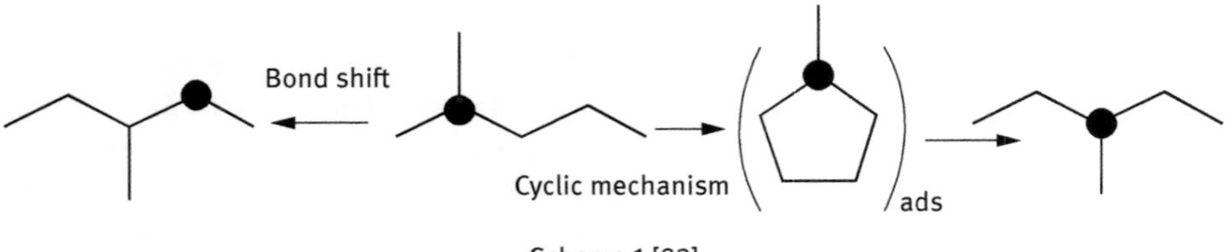

Scheme 1 [83]

It was noticed that the CM was strongly favored on small Pt aggregates with diameters <2 nm and that three types of hydrogenolysis take place: a nonselective one, occurring on highly dispersed catalysts and corresponding to an equal chance of breaking any C–C bond of the ring; a selective one, allowing only the rupture of bisecondary C–C bonds; and a “partially selective” mechanism competing with the selective one on catalysts of low dispersion. For the results that we have only n-H formed, we have to assume the reaction of a metallocyclobutane intermediate with the exocyclic methyl, as already suggested [96]. After the metallocyclobutane exocyclic intermediate is created, the C–C bond rupture of the ring leads to the formation of a carbene–olefin, which gives a π-adsorbed vinyl group via a 1,2-hydride shift, and n-H is formed. In addition, using deuterium exchange reactions on Fe films, it was observed that the vinylic hydrogen is very mobile and is very easily exchanged [97]. Such observations on Fe catalysts reinforce the proposed mechanism, which may take place on the Fe-TUD-1, as noticed on Scheme 2. Such a mechanism can explain the predominant n-H formation, the aromatization, and the extensive cracking reactions when the conversion or the temperature is increased.

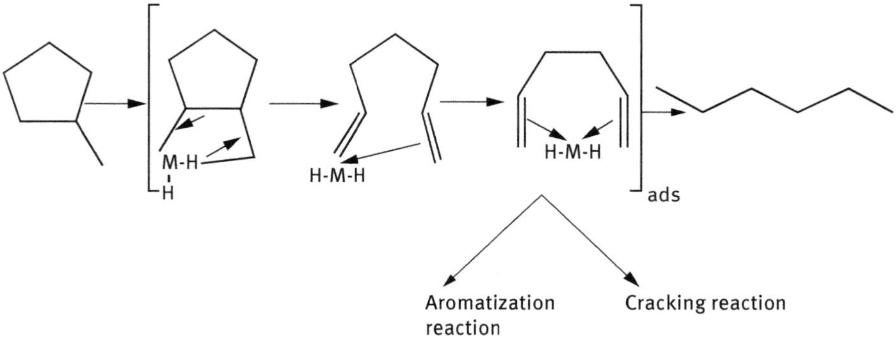

Scheme 2 [83]

12.2.3 Cracking Selectivity

Regardless of the Fe loading in the TUD-1 and the reaction temperature, the only major product formed was determined to be C_1, indicating that a deep cracking process occurred with the Fe-TUD-1 catalysts. The high selectivity toward C_1 is attributed, most likely, to the hindered desorption of products that are formed, followed by the successive C–C bond rupture on the Fe active sites present on the TUD-1. Compared with the samples with Si/Fe ratios of 85 and 65, impressive values of selectivity toward cracking reactions were observed for the sample with a Si/Fe ratio of 45 in the temperature range of 250–400°C, which suggests that octahedral Fe entities/small clusters and high temperatures increase the C_1 selectivity. The color of this sample with a Si/Fe ratio of 45 was gray, while the color of the samples with Si/Fe ratios of 65 and 85 remained white after the catalytic tests [83]. It must be recalled that the color of all reduced samples was white. In other words, a slight deactivation was observed on the sample with a Si/Fe ratio of 45. A plausible explanation could be the agglomeration of small clusters [83]. It seems that these small clusters are susceptible to easily forming the iron carbide, taking into account the fact that in the Fe-O-Fe clusters, the binding energy of O–Fe is small compared with that of C–Fe. When the Fe is isolated and tetrahedrally coordinated in the Si-O-Fe framework, the binding energy is strong compared with C–Fe and the Fe carbide cannot be formed, which is supported in this case by the white color after catalytic tests [83]. Noticeably, for the sample with a Si/Fe ratio of 45 (the point of the framework partial damage) and a high temperature, the convertibility between Fe clusters and Fe carbide was observed. These latter moieties are responsible for the cracking reaction [83].

12.2.4 Ring-Enlargement Selectivity

Benzene was only detected for the sample with a Si/Fe ratio of 45. Enlargement of the MCP ring was only observed at temperatures of 400°C. This result can be attributed to thermodynamic phenomena because aromatization is favored at high temperatures [83]. Keeping in the mind that ring-enlargement product formation was hindered at low temperatures and at low Fe contents, this result represents an advantage of Fe catalysts due to their ability to hinder aromatization. This also holds true for the samples with high Fe loadings but at low temperatures [83].

12.3 Dry Reforming of Methane

Methane and carbon dioxide are two of the cheapest and most abundant carbon-containing molecules. As methane and carbon dioxide are the two major greenhouse gases, their conversion into value-added chemicals is of great interest to reduce global warming. One of these processes is the conversion of carbon dioxide by reforming with methane, an environmentally friendly process since it utilizes two major greenhouse gases, in which carbon dioxide and methane produce valuable syngas. In the past few years, it has been proposed to utilize CO_2 an oxygen-transfer agent or a nontraditional oxidant for several reactions [98], but the utilization of CO_2 as a source of carbon as not been proposed. In this context, the reforming of methane with carbon dioxide to produce synthesis gas has received significant attention as an alternative process to steam reforming of methane. The increasing interest in this process is based on the lower energy requirements compared to steam reforming. CO_2 reforming also allows the production of syngas with a low H_2/CO ratio (theoretically 1/1; however, the presence of side reactions such as reverse water gas shift [RWGS] slightly reduces this ratio), which is suitable for the production of numerous chemicals, including methanol, dimethyl ether, Fischer–Tropsch chemicals, ammonia, acetic acid, and formic acid [99]. From the perspective of catalytic chemistry, the reforming of methane with carbon dioxide to syngas at low temperatures has been a challenge. Finding an adequate method to activate methane is essential for the DRM reaction. The reforming of methane with carbon dioxide is a thermodynamically unfavorable reaction. High reaction temperatures are required to shift the equilibrium to a state favorable for the formation of syngas. The high reaction temperatures cause the equilibrium conversion of carbon dioxide reforming (highly endothermic reaction) to be more thermodynamically favorable than side reactions, such as methane decomposition ($CH_4 \rightarrow C + 2H_2$), the Boudouard reaction ($2CO \rightarrow C + CO_2$) and the reverse carbon gasification reaction ($CO + H_2 \leftrightarrow C + H_2O$) [100, 101]. Coke formation at lower temperatures is mainly due to the Boudouard reaction. However, endothermic decomposition reactions become more significant at higher temperatures. These side reactions could contribute to the deactivation of the catalyst. DRM is an endothermic reaction, which, even at high temperatures, needs a suitable catalyst to achieve sufficient conversion. In the presence of a supported catalyst, such as metal nanoparticles on an oxide, the DRM process proceeds via a complex mechanism involving multiple reactions with the metal phases responsible for C–H bond cleavage and the oxide carriers contributing to C–O bond cleavage [102]. The stable C–H bonds in CH_4 (425 kJ/mol) and the stable C–O bonds in CO_2 (396 kJ/mol), along with the highly endothermic nature of the CRM process, require harsh operating conditions for practical reactant conversion. It is desirable to develop an efficient catalyst that is catalytically reactive at low temperatures and stable at elevated temperatures.

Noble metal catalysts have drawn attention for their superior coking resistance in the dry reforming of methane, higher stability and activity, especially for high-temperature applications [103–105]. However, noble metals cannot be applied on an industrial scale due to their high cost. Ni has been the most common metal utilized as a catalyst in dry reforming chemistries for technical and industrial standpoints due to its low cost, but the drawbacks to using methane in DRM are the intense formation of carbonaceous deposits that lead to rapid catalyst deactivation through coke deposition and the Boudouard reaction, the sintering effect that is caused by high reaction temperatures and metal oxidation. These drawbacks restrain the practical application of Ni in industry. For Ni-based catalysts in DRM, two major sources of deactivation have been identified, namely, carbon deposition at low temperatures and sintering at high temperatures [106, 107]. Consequently, research efforts related to DRM have been focused on the development of commercial catalysts able to achieve high and stable conversions while maintaining resistant to deactivation. To date, only two processes based on DRM have been industrially implemented: the SPARG process [108, 109] and the CALCOR process for CO production [110].

Different types of supported metal catalysts have been extensively studied to minimize the coke deposition problem. Assembling metal nanoparticles into mesostructured materials is an example where the interplay of the metal/carrier can be notably reinforced by the nanoconfinement effect [111]. The surroundings of the metal nanoparticles with concave internal surfaces of nanosized channels can optimize the metal/carrier interfaces.

It was reported in the literature that mesoporous catalyst supports, like MCM-41, were less susceptible to catalyst deactivation due to coke formation than the conventional microporous materials [112]. Ni-MCM-41 samples with various Si/Ni prepared directly by the conventional hydrothermal [113, 114] were tested in DRM. It have been observed that the conversion of CH_4 strongly depends on both reaction temperature and nickel content [114]. A high initiation temperature of 600°C is required for the samples with low Ni-loading samples (Ni–MCM-41 Si/Ni = 50) and Ni–MCM-41 (Si/Ni = 25) but 700°C for Ni–MCM-41 (Si/Ni = 100). An increase in the catalytic conversions with increasing temperature is observed for the lowest nickel content catalyst Ni–MCM-41(Si/Ni = 100). Ni–MCM-41(Si/Ni = 25) catalyst is found to be the most active within the temperature range of 650–800°C which show that the catalytic activities do not constantly increase with increasing Ni content [114]. It is possible that aggregated Ni species may predominate when the Ni content exceeds the dispersion-limit loading. Although H_2 and CO are formed simultaneously according to the stoichiometry of dry reforming, there is excessive CO with respect to H_2 at all temperatures investigated. The CO/H_2 ratio approaches the stoichiometric value of unity gradually as the temperature increases regardless of catalysts, which agrees well with the thermodynamics tendency caused by RWGS side reactions [114]. Slightly higher selectivities of CO are observed over Ni–MCM-41(Si/Ni = 50) and Ni–MCM-41(Si/Ni = 25) catalysts. At the optimal reaction temperature of 750°C temperature with high catalytic activity, moderate CO/H2 ratio, and low energy consumption, the initial catalytic activity of Ni–MCM-41 catalysts prepared by direct synthesis has the following sequence: Ni–MCM-41(Si/Ni = 25) > Ni–MCM-41(Si/Ni = 50) > Ni–MCM-41(Si/Ni = 100). The stability of the catalysts at 750°C shows the following sequence: Ni–MCM-41(25) > Ni–MCM-41(50), which coincides with the increase of Ni content. This indicates that the presence of sufficient surface active centers is crucial to maintain both high activity and long-term stability [113, 114].

When one correlates catalytic stability and carbon deposition, it has been observed that the Ni–MCM-41(Si/Ni = 50) and Ni–MCM-41(Si/Ni = 25) catalysts show low catalytic stability but low carbon deposition. For the catalysts with low Ni loading, the density of active sites on the pore wall surface is relatively deficient and therefore the activation ability is low and limited [113, 114]. It has been observed that the conversion is dependent on both the reaction temperature and the method of introducing Ni and the Ni loading [115]. 4% Ni-TUD-1 catalysts were prepared by post-synthesis grafting (graf), direct hydrothermal synthesis (DHT), and impregnation (imp) and tested in the carbon dioxide reforming of methane [114, 115]. The size of Ni particles is different as a function of method preparation: 8 nm has been observed on Ni-TUD-1-graf, 18 nm on Ni-DHT, and 24 nm on Ni-TUD-1-imp. The small nickel particle size on both Ni-graf and Ni-DHT samples has been attributed to the strong anchoring effect of TUD-1 support which restricts the migration of nickel clusters [114, 115]. Grafted nickel species may provide a better catalyst synthesis method compared to framework incorporated nickel species as higher dispersion and smaller nickel particle size is observed on Ni-graf sample. The large Ni particles size observed on Ni-imp can be attributed by substantial nucleation and rapid aggregation during the reduction. Weak interaction between TUD-1 and the surface Ni during reduction is suggested to be the main reason. CH_4 conversions at 550°C on Ni-DHT, Ni-graf, and Ni-imp are 17.9%, 5.4%, and 4.0%, respectively; Ni-DHT exhibits the lowest initiation temperature followed by Ni-graf and Ni-imp. A higher conversion is observed for CO_2 compared to CH_4. The presence of RWGS reaction is suggested to be the main reason contributing to such higher CO_2 conversion. For all catalyst, H_2 selectivity increases exponentially with increasing temperature, in agreement with previous thermodynamics studies where high temperature favors the formation of H_2 through various reactions such as reforming, WGS reaction, carbon gasification, and methane cracking [116]. Excess CO is observed (H_2/CO ratio less than unity) over the entire range of temperature investigated. CO was found to be in greater excess at low temperature. This can be due to the occurrence of side reaction such as RWGS and methanation reaction at low temperature which consumes H_2 [116]. Both Ni-DHT and Ni-graf exhibit analogous H_2 selectivity while Ni-imp shows obviously lower selectivity toward H_2. Moreover, Ni-graf exhibits noticeably lower H_2/CO ratio compared to Ni-DHT, implying that less carbon is formed on Ni-DHT catalyst. Severe sintering occurs on Ni-imp during reaction and active nickel sites aggregate into large particles that are catalytically less active. Although the direct synthesized Ni catalyst performed the best during the catalytic activity test, the grafted catalyst surpassed all other Ni catalysts in the long-term activity and stability evaluation. Despite the presence of a large amount of carbon deposition, Ni-graf TUD-1 catalyst exhibited high activity and strong resistance against catalyst deactivation due to the presence of more easily accessible Ni active sites and the nature of the moderate reactivity of surface carbonaceous species. The catalytic stability of the prepared samples was investigated at 750°C. Both Ni-graf and Ni-DHT are relatively stable at 750°C for 72 h of time on stream (TOS) with Ni-graf being more stable than Ni-DHT. The better anchoring effect of grafted nickel compared to framework incorporated nickel can be said to have contributed to the higher stability in Ni-graf sample. Ni-imp exhibits the poorest activity and stability, nearly no activity (in terms of CH_4 conversion) is observed during stability analysis on Ni-imp sample after 5 h of TOS. The deactivation of Ni-imp catalyst may be attributed to the metal sintering which is confirmed by the TEM observation. It can be observed that nickel particle size is a crucial factor affecting activity of the catalyst. Both Ni-graf and Ni-DHT catalysts are able to maintain high H_2 selectivity and H_2/CO ratio during 72 h of TOS. Ni-imp exhibits moderate H_2 selectivity and H_2/CO ratio which deteriorate completely after 5 h. A slight decrease in H_2 selectivity and H_2/CO ratio (stabilized after 25 h of TOS) is also observed on Ni-DHT. This decline can be due to the further nucleation of Ni particle at the reaction condition which is stabilized after 25 h of TOS. Although carbon deposition can be another reason for the observed decline in H_2 selectivity, no further decrease in the selectivity is observed despite both CH_4 and CO_2

conversion maintained above 50% and 60%, respectively. As observed, a decrease in Ni particle size has resulted in a larger amount of coke formation over Ni-graf and Ni-DHT. Although a large amount of carbonaceous species formed over these two catalysts, their catalytic activity and stability are well maintained. This is most likely due to the low probability of the blocking of the active sites by carbon deposition or the absence of pore-mouth plugging over these catalysts; noticeably, there are more exposed surface active sites over Ni-graf catalyst. The smaller particle size is very beneficial for the carbon dioxide reforming of methane [114, 115]. Graphite peak (from XRD results) was observed on Ni-DHT and Ni-graf, and not in the Ni-imp catalyst after the stability reaction. This behavior has been attributed to the low activity of the Ni-imp catalyst 114, 115]. Ni-graf catalyst with higher activity and stability is found to exhibit a higher intensity of graphite peak, suggesting more carbon is deposited on it compared to Ni-DHT catalyst. According to Raman results, both the amorphous carbon and graphite have been formed during the reaction [114, 115]. It must be noted that no obvious correlation exists between the amount of carbon and the catalytic activity in stability test [115]. Although a negligible amount of carbon is formed on Ni-imp, both the activity and stability are severely poor. While more carbon was deposited on Ni-DHT and Ni-graf, both catalysts show good activity and stability, suggesting that higher activity is accompanied by more carbon deposition [114, 115]. Indeed, most of carbonaceous species, both α-carbon and β-carbon, can participate as the reactive intermediates in the production of syngas. For Ni-imp, the deactivation of this catalyst is not related with carbon formation but due to the significant sintering of nickel particles, which coincides with the lower metal dispersion and larger particle size. The smaller Ni particle is beneficial to the catalytic performance (both activity and stability). Similar trend that correlated particle size and catalytic performance has also been reported [117, 118]. The high activity observed on Ni-graf and Ni-DHT should be caused by the improved dispersion of metal particles under reaction conditions. The large amount of carbon formation can be attributed to the relatively easier decomposition of CH_4 over small nickel particles. The facility of CH_4 decomposition coupled with carbon gasification is beneficial for the formation of CO and H_2 via an alternative reaction pathway [116]. However, when the decomposition of CH_4 and the Boudouard reaction proceed faster than carbon gasification, accumulation of carbon on the catalyst occurs [116]. The decrease in Ni particle size has resulted in a larger amount of coke formation over Ni-graf and Ni-DHT. Although a large amount of carbonaceous species formed over these two catalysts, their catalytic activity and stability are well maintained, especially for the former catalyst. This is most likely due to the low probability of the blocking of the active sites by carbon deposition or the absence of pore-mouth plugging over these catalysts; noticeably, there are more exposed surface active sites over Ni-graf catalyst [114, 115]. However, the 12.5% Ni/SBA-15 catalyst, prepared using an incipient wetness impregnation method, showed highly stable activity at 800°C for 600 h [119]. After reaction for 710 h, the conversion of CH_4 and CO_2 decreased by about 50% and 25%, respectively. In this case, the coking was the main reason for the deactivation of the Ni/SBA-15 catalysts. At higher reaction temperatures, the mesoporous structure of SBA-15 was not destroyed and the pore walls of SBA-15 could prevent the aggregation of the nickel species. It was reported that mesoporous-alumina-supported Ni-based catalysts showed good catalytic performance and significantly decreased coke formation by modification of these catalysts with tungsten [120].

12.4 Conclusions

The conversion of MCP with hydrogen at atmospheric pressure was selected as an index reaction for Fe-TUD-1. For low iron content, Si/Fe ratios of 85 and 65, the Fe-O-Si bond of the isolated tetrahedral iron ions in the silica TUD-1 framework is dominant. For the high iron content, Si/Fe ratio of 45, the isolated Fe sites of Si-O-Fe bonds coexist with the Fe-O-Fe clusters. It was observed that the reductive treatment does not cause a noticeable change in the mode of iron ion stabilization in the TUD-1 framework; in this case no partial or total breaking of the Si-O-Fe framework linkages take place (UV–vis spectroscopy). The catalytic studies indicate that Fe-TUD-1 with various Si/Fe ratios exhibited outstanding ring-opening selectivity. Among the ring-opening products, *n*-H was formed exclusively. The formation of *n*-H has been explained by the presence of metallocyclobutane intermediate, followed by the endocyclic C–C bond rupture of the MCP ring which leads to the formation of a carbene–olefin, and giving a π-adsorbed vinyl group via a 1,2-hydride shift. The active sites responsible for the endocyclic C–C bond rupture between substituted secondary–tertiary carbon atoms seem to be the tetrahedrally coordinated/atomically isolated sites on the mesoporous support, while the small clusters seem to be responsible for the successive C–C bond rupture. The selectivity toward the cracking reaction is generally explained by the difficulty in desorbing products formed on the catalyst surface. The results show that one atom of the Fe-O-Si species may react in the MCP conversion, favoring the single C–C rupture and increasing the selectivity to ring opening of MCP. On these Fe isolated species and tetrahedrally coordinated, the Fe carbide cannot be formed. In contrast, two atoms of Fe-O-Fe species may react in the conversion of MCP, favoring the consecutive reactions and decreasing the selectivity of the ring opening of MCP. On these species the Fe carbide has been formed. Although the conversion of MCP is relatively low as compared with noble metal catalysts [46, 48, 92], these empirical results in reductive media for Fe-TUD-1 catalysts are appropriate for generating ring-opening products at the secondary C–tertiary C with atom efficiency.

The production of syngas from DRM is a sustainable and effective way to catalytically convert methane. Mesoporous materials with nanosized pore diameters have attracted considerable attention in the syngas production. It has been observed that a high dispersion of metallic Ni over mesoporous nanostructured supports limits coke formation and preventing metal particles from sintering. Moreover, the mesoporous supports can provide catalysts with high performance, due to providing more edges and corners. For transition metal oxides, the active phases in mesoporous nanostructures can localize *d*-electrons within the thin walls between pores, and therefore, the materials are endowed with interesting catalytic properties. Numerous fundamental studies have also shown that the unprecedented activity of metal catalysts is primarily indebted to their particle size, the degree of coordinative unsaturation of the metal atoms, and the collaboration of the metal particles and supports. Therefore, downsizing the particle size to the nanometer or subnanometer scale is highly desirable for designing novel catalysts. As observed, the physical, chemical, and structural characteristics of these mesoporous catalysts and their catalytic reactivity are dependent on their preparation methods. These results on mesoporous catalysts create new challenges for the technology and catalysis.

Acknowledgments: Financial support by the CNRS France is gratefully acknowledged by Ioana Fechete. This work was partly supported by the Kuwait Institute for Scientific Research, Petroleum Research and Studies Center, Petroleum Refining, and by a DEMETER grant from the National Research Agency (ANR) of France. I. Fechete gratefully acknowledges the wonderful discussion of F. Garin.

References

1. Dinolfo PH, Hupp Joseph T. Chem Mater 2001;13:3113.
2. Mathur S, Shen H, Altmayer J, Mathur S, Shen H, Altmayer J. Rev Adv Mater Sci 2007;15:16.
3. Warren SC, Wiesner U. Pure Appl Chem 2009;81:73.
4. Macquarrie DJ. Phil Trans R Soc Lond A 2000;358:419.
5. Kamperman M, Wiesner U. In: Lazzari M, Liu G, Sebastien L, editors. Block copolymers in nanoscience. Weinheim: Wiley-VCH, 2006:309–335.
6. Al Othman Zeid A. Materials 2012;5:2874.
7. Soler-Illia GJ, Innocenzi P. Chem – Eur J 2006;12:4478, (a) Soler-Illia GJ, Crepaldi E, Grosso D, Sanchez C. Curr Opin Coll Int Sci 2003;8:109.
8. Fechete I, Vedrine JC. Nano-oxide mesoporous catalysts in heterogeneous catalysis. In: Van de Voorde Marcel, Sels Bert, editors. Nanotechnology in catalysis: applications in the chemical industry, energy development, and environment protection. Wiley, Chap. 4, 2017;57–89.
9. Beck JS, Vartuli JC, Roth WJ, Leonowicz ME, Kresge CT, Schmitt KD, Chu CT, Olson DH, Sheppard EW, McCullen SB, Higgins JB, Schlenker JL. J Am Chem Soc 1992;114:10834.
10. Beck JS, Vartuli JC. Curr Opin Solid State Mater Sci 1996;1:76.
11. Ying Y, Mehnert CP, Wong MS. Angew Chem Int Ed 1999;38:56.
12. Merkache R, Fechete I, Bernard M, Turek P, Al-Dalama K, Garin F, Appl Catal A 2015;504:672.
13. Jin J, Hines WA, Kuo Ch-H, Perry DM, Poyraz AS, Xia Y, Zaidi T, Nieh MP, Suib SL. Dalton Trans2015;44:11943.
14. Ishchenko OM, Krishnamoorthy S, Valle N, Guillot J, Turek O, Fechete I, Lenoble D. J Phys Chem C 2016;120:7067.
15. Scott BJ, Wirnsberger G, Stucky GD. Chem Mater 2001;13:3140.

16. Alberti S, Soler-Illia GJ, Azzaroni O. Chem Commun 2015;51:6050.
17. Fechete I, Védrine JC. Molecules 2015;20:5638.
18. Linden M, Schacht S, Schuth F, Steel A, Unger KK. J Porous Mater 1998;5:177.
19. Bensacia N, Fechete I, Moulay S, Hulea O, Boos A, Garin F. C.R. Chim 2014;17:869–880.
20. Ariga K, Vinu A, Yamauchi Y, Ji Q, Hill JP. Bull Chem Soc Jpn 2012;85:132.
21. Pellicer E, Sort J. Nanomaterials 2016;6:15.
22. Bensacia N, Moulay S, Debbih-Boustila S, Boos A, Garin F. Environ Eng Manage J 2014;13:2675.
23. Schuth F. Angew Chem Int Ed 2003;42:3604.
24. Somorjai GA, Rioux RM. Catal Today 2005;100:201.
25. Fechete I, Jouikov V. Electrochim Acta 2008;537:7107.
26. Fechete I, Simon-Masseron A, Dumitriu E, Lutic D, Caullet P, Kessler H. Rev Roum Chim 2008;53:55.
27. Fechete I, Caullet P, Dumitriu E, Hulea V, Kessler H. Appl Catal A 2005;280:245.
28. Dumitriu E, Fechete I, Caullet P, Kessler H, Hulea V, Chelaru C, Hulea T, Bourdon X. Stud Surf Sci Catal 2002;142 A:951.
29. Dees MJ, Bol MH, Ponec V. Appl Catal 1990;64:279.
30. Vaarkamp M, Miller JT, Modica FS, Koningsberger DC. J Catal 1996;163:294.
31. Alvarez WE, Resasco DE. J Catal 1996;164:467.
32. Djeddi A, Fechete I, Ersen O, Garin F. C.R. Chim 201316:433; (a) Djeddi A, Fechete I, Garin F. Appl Catal A 2012;413–414:340; (b) Djeddi A, Fechete I, Garin F. Catal Commun 2012;17:173.
33. Teschner D, Paal Z. React Kinet Catal Lett 1999;68:25.
34. Galperin LB, Bricker JC, Holmgren JR. Appl Catal A 2003;239:297.
35. Kubicka D, Kumar N, Arvela PM, Tiitta M, Niemi V, Karhu H, Salmi T, Murzin DY. J Catal 2004;227:313.
36. O'Cinneide A, Gault FG. J Catal 1975;37:311.
37. Maire G, Plouidy G, Prudhomme JC, Gault FG. J Catal 1965;4:556.
38. Dartiques J-M, Chgambellan A, Gault FG. J Am Chem Soc 1976;98:856.
39. Gault FG, Amir-Ebrahimi V, Garin F, Parayne P, Weisang F. Bull Soc Chim Belg 1979;88:475.
40. Kramer R, Zuegg H. J Catal 1983;80:446.
41. Paal Z. Catal Today 1988;2:595.
42. Fulop E, Gnutzmann V, Paal Z, Vogel V. Appl Catal 1990;66:319.
43. Van Senden JC, Broekhoven EH, Wreesman CT, Ponec V. J Catal 1984;87:468.
44. Chow M, McVicker GB. J Catal 1988;112:303.
45. Liberman AL. Kinet Katal 1964;5:128.
46. Gault FG. Adv Catal 1981;30:1.
47. Zimmer H, Paal Z. J Mol Catal 1989;51:261.
48. Paal Z, Tetenyi P. Nature 1977;267:234.
49. Glassl H, Hayek K, Kramer R. J Catal 1981;68:397.
50. Anderson JB, Burch R, Cairns JA. Appl Catal 1986;28:255.
51. Anderson JB, Burch R, Cairns JA. J Catal 1987;107:351.
52. Clarke JK, Dempsey RJ, Baird Th. J Chem Soc Faraday Trans 1990;86:2789.
53. Tauster SJ. Acc Chem Rev 1987;20:389.
54. Bond GC, Burch R. In: Bond GC, Webb G, editors. Catalysis, vol 6. London: Royal Society of Chemistry, 1983:27.
55. Resasco DE, Haller GL. J Catal 1983;82:279.
56. Fenoglio RJ, Nunez GM, Resasco DE. J Catal 1990;121:77.
57. Tauster SJ, Fung SC, Garten RL. J Am Chem Soc 1978;100:170.
58. Tauster SJ, Fung SC. J Catal 1978;55:29.
59. Horsley JA. J Am Chem Soc 1979;101:2870.

60. Schwab GM. Adv Catal 1978;27:1.
61. Solymosi F. Catal Rev Sci Eng 1967;1:233.
62. Djeddi A, Fechete I, Garin F. Top Catal 2012;55:700.
63. Dees MJ, Ponec V. J Catal 1989;115:347.
64. Williams FL, Nason D. Surf Sci 1974;45:377.
65. Burton JJ, Hyman E. J Catal 1975;37:114.
66. Tomanek D, Mukhergee S, Kumar V, Bennemann KH. Surf Sci 1982;114:11.
67. Fechete I, Debbih-Bourstila S, Merkache R, Hulea O, Lazar L, Lutic D, Balasanian I, Garin F. Environ Eng Manage J 2012;11:1931.
68. Fechete I, Ersen O, Garin F, Lazar L, Rach A. Catal Sci Technol 2012;3:444.
69. Vralstad T, Øye G, Stocker M, Sjoblom J, Microporous Mesoporous Mater 2006;104:10.
70. Li Y, Feng Z, Lian Y, Sun K, Zhang L, Jia G, Yang Q, Li C. Microporous Mesoporous Mater 2005;84:41.
71. Fechete I, Donnio B, Ersen O, Dintzer T, Djeddi A, Garin F. Appl Surf Sci 2011;257:2791.
72. Tsoncheva T, Rosenholm J, Linden M, Kleitz F, Tiemann M, Ivanova L, Dimitrov M, Paneva D, Mitov I, Minchev C. Microporous Mesoporous Mater 2008;112:327.
73. Bakala PC, Briot E, Salles L, Brégeault J-M. Appl Catal A 2006;300:92.
74. Fechete I, Gautron E, Dumitriu E, Lutic D, Caullet P, Kessler H. Rev Roum Chim 2008;53:49.
75. Dias APS, Rozanov VV, Waerenborgh JC, Portela MF. Appl Catal A 2008;345:185.
76. Boulaoued A, Fechete I, Donnio B, Bernard M, Turek P, Garin F. Microporous Mesoporous Mater 2012;155:131.
77. Qin S, Zhang C, Xu J, Wu B, Xiang H, Li Y. J Mol Catal A: Chem 2009;304:128.
78. Dumitriu E, Hulea V, Fechete I, Catrinescu C, Auroux A, Lacaze J-F, Guimon C. Appl Catal A 1999;181:15.
79. Novak Tusar N, Ristic A, Cecowski S, Arcon I, Lazar K, Amenitsch H, Kaucica V. Microporous Mesoporous Mater 2007;104:289.
80. Hamdy MS, Mul G, Jansen JC, Ebaid A, Shan Z, Overweg AR, Maschmeyer Th. Catal Today 2005;100:255.
81. Gervasini A, Messi C, Carniti P, Ponti A, Ravasio N, Zaccheria F. J Catal 2009;262:224.
82. Szostak R. molecular sieves: principles of synthesis and identification. New York: Van Nostrand Reinhold, 1989.
83. Haddoum S, Fechete I, Donnio B, Garin F, Lutic D, Chitour CE. Catal Commun 2012;27:141.
84. Brunauer S, Deming LS, Deming E, Teller E. J Am Chem Soc 1940;62:1723.
85. Shan Z, Gianotti E, Jansen JC, Peters JA, Marchese L, Maschmeyer T. Chem Eur J 2001;7:1437.
86. Kumar MS, Schwidder M, Grunert W, Bruckner A. J Catal 2004;227:384.
87. Park JW, Chon H. J Catal 1992;133159.
88. Perez-Ramirez J, Kapteijn F, Bruckner A. J Catal 2003;218:234.
89. Perez-Ramirez J, Suthosh M, Bruckner A. J Catal 2004;223:13.
90. Weckhuysen BM, Wang D, Rosynek MP, Lunsford JH. Angew Chem Int Ed 1997;36:2374.
91. Chen H-Y, Sachtler WM. Catal Today 1998;42:73.
92. Dokjampa S, Rirksomboon T, Phuong Do TM, Resasco DE. J Mol Catal A 2007;274:231.
93. Miki Y, Yamadaya S, Oba M. J Catal 1977;49:278.
94. Alayoglu S, Aliaga C, Sprung C, Somorjai GA. Catal Lett 2011;141:914.
95. Garin F, Gault FG. J Am Chem Soc 1975;97:4466.
96. Garin F, Gault FG. In: Prins R, Schuit GC, editors. Chemistry and chemical engineering of catalytic processes. The Netherlands Germantown, Maryland, USA: Sijthoff & Noordhoff, Alphen aan den Rijn, 1980:351.
97. Touroude R, Gault FG. J Catal 1974;32:288.

98. Fechete I, Vedrine JC, Chim CR. 2016;19:1203. (a) Upare DP, Lee CW. Fuel Process Technol 2014;126:243.
99. Pena MA, Gomez JP, Fierro JL. Appl Catal A 1996;144:7. (a) Rostrup-Nielsen JR. Catal Today 1993;18:305.
100. Zhang J, Wang H, Dalai AK. J Catal 2007;249:300.
101. Bradford MC, Vannice MA. Catal Rev Sci Eng 1999;41:1.
102. Fan MS, Abdullah AZ, Bhatia S. ChemCatChem 2009;1:192.
103. Erdohelyi A. J Catal 1993;141:287.
104. Zhang Z. J Catal 1996;158:51.
105. Carrara C, Munera J, Lombardo EA, Cornaglia LM. Top Catal 2008;51:98.
106. Wang S, Lu G, Millar G. Energy Fuels 1996;0624:896.
107. Guczi L, Erdohelyi A. Catalysis for alternative energy generation. New York: Springer, 2012.
108. Dibbern HC, Olesen P, Rostrup-Nielsen JR, Tottrup PB, Udengaard NR. Hydrocarbon Process 1986;65:71.
109. Udengaard NR, Hansen JH, Hanson DC, Stal JA. Oil Gas J 1992;90:62.
110. Neumann P, Teuner SC, Von Linde F. Oil Gas Eur Mag 2001;27:44.
111. Liu H, Tao K, Xiong C, Zhou S. Catal Sci Technol 2015;5:405.
112. Damyanova S, Pawelec B, Arishtirova K, Fierro JL, Sener C, Dogu T. Appl Catal B 2009;92:250.
113. Liu D, Lau R, Borgna A, Yang Y. Appl Catal A 2009;358:110
114. Liu D, Lau R, Jia X, Borgna A, Yang Y. In: Suib Steven L, editor. New and future developments in catalysis. ISBN: 978-0-444-53882-6, Elsevier, Ch. 11, 2013:297.
115. Quek X-Y, Liu D, Cheo WN, Wang H, Chen Y, Yang Y. Appl Catal B 2010;95:374.
116. Haghighi M, Sun ZQ, Wu JH, Bromly J, Wee HL, Ng E, Wang Y, Zhang DK. Proc Combust Inst 2007;31:1983.
117. Gallego GS, Mondragon F, Barrault J, Tatibouet JM, Batiot-Dupeyrat C. Appl Catal A 2006;311:164.
118. Hou ZY, Gao J, Guo JZ, Liang D, Lou H, Zheng XM. J Catal 2007;250:331.
119. Zhang M, Ji S, Hu L, Yin F, Li C, Liu H. Chin J Catal 2006;27:777.
120. Arbag H, Yasyerli S, Yasyerli N, Dogu T, Dogu G. Top Catal 2013;56:1695.

Petra Göring, Monika Lelonek

13 Highly ordered porous materials

13.1 Introduction: Why Highly Ordered?

Porous materials are of scientific and technological importance due to the presence of controllable dimensions at nanometer scale. Research efforts in this field have been driven by the rapidly emerging applications such as biosensors, drug delivery, gas separation, energy storage and fuel cell technology [1–3]. This research offers exciting new opportunities for developing new strategies and techniques for the synthesis and applications of these materials. Perfect control of the structure parameters of porous materials is of fundamental importance in order to tailor and verify their properties.

In addition to the self-assembly processes of organic materials, which are mainly based on different wetting properties and thus surface energies, there is a second large material system in which the self-assembly is based on electrochemical instabilities. This is the case with the so-called valve metals and some semiconductors, such as silicon and III–V compound semiconductors [4, 5]. The most studied systems today are silicon and alumina.

Electrochemical anodization of pure aluminum or silicon enables the growth of highly ordered nano- and macroporous structures with a controlled morphology. These porous materials possess perfect ordered arrays of pores with monodisperse pore diameters and a high aspect ratio. The currently accessible pore diameters (D_p) of nanoporous alumina range from 15 to 400 nm, those of macroporous silicon from 370 nm up to a few microns as indicated in Figure 13.1. The length of pores can be varied from a few hundreds of nanometers to a few hundreds of microns. Both template systems contain highly ordered arrays of straight pores with sharp diameter distribution (less than 10%) and uniform depth.

Aluminum oxide has been used since the beginning of the last century due to its chemical and corrosion resistance properties. Mainly in 1995, Masuda et al. investigated highly ordered arrays of porous alumina initiated by self-organization. The ordered pore domains are in the micron range only. Furthermore, nanoporous alumina has the disadvantage that no long-range order is available, and grain boundaries and point defects occur. Nevertheless, to obtain perfectly straight pores the anodization is typically done in a two-step process [6, 7]. A one-step anodizing process is commonly used for manufacturing protective porous alumina layer on aluminum, whereas the two-step anodizing process results in much more ordered nanopore arrays. Selectively dissolving of the first oxide layer and subsequently re-anodizing in a second oxidation process will achieve this. The depth of the pores is closely related to the duration of anodizing, and defined anodizing times improve the pore regularity.

https://doi.org/10.1515/9783110547221-013

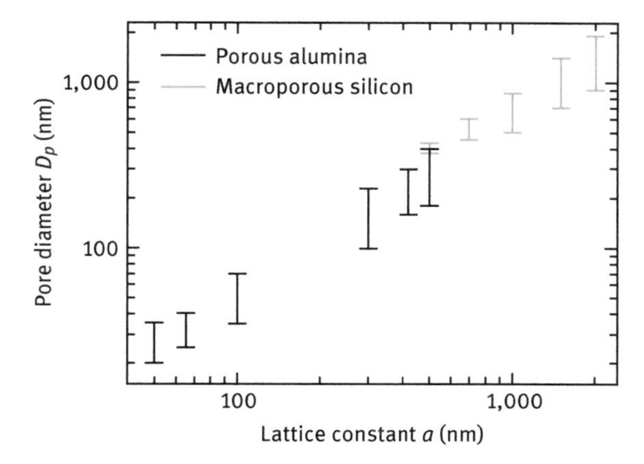

Figure 13.1: Overview of the pore diameters (D_p) and lattice constants (D_{int}) of the currently available highly ordered templates of nanoporous alumina (grey bars) and macroporous silicon (black bars).

The fabrication of macroporous silicon is based on a classical top-down approach. Starting from bulk material, regular pore structures are formed in an electrochemical etching process. The underlying fabrication is a controllable and parallell-processable method that makes the production of macroporous silicon scalable to large areas and therefore economically attractive.

The fundamentals of the electrochemical etching process of macroporous silicon have originated in the 1990s by Lehmann and Föll [8]. Perfectly ordered pore arrays in the micrometer range are producible using a photolithographic pre-patterning. These macropores are perfectly directional and separated pores in contrast to meso- or nanopores with a more or less sponge-like structure.

13.2 Macroporous Silicon

Macroporous silicon produced by electrochemical etching of lithographically pre-structured silicon wafers [8, 9] is a promising material for a variety of novel devices. Potential applications of macroporous silicon have been proposed in the fields of microprocessing, gas measurement, photonic crystals, biotechnology and many others. Examples are short-time optical filters [10], two-dimensional (2D) and three-dimensional (3D) photonic crystals [11–13] and optical and capacitive sensors [14, 15].

The etching of silicon and therefore the "growth" of pores with a certain diameter covers several orders of magnitude. According to the International Union of Pure and Applied Chemistry nomenclature for porous materials [16], structures with a pore width below 2 nm are called microporous. The mesoporous structures range from 2 to 50 nm. Bigger pores are referred to as macropores (Figure 13.2).

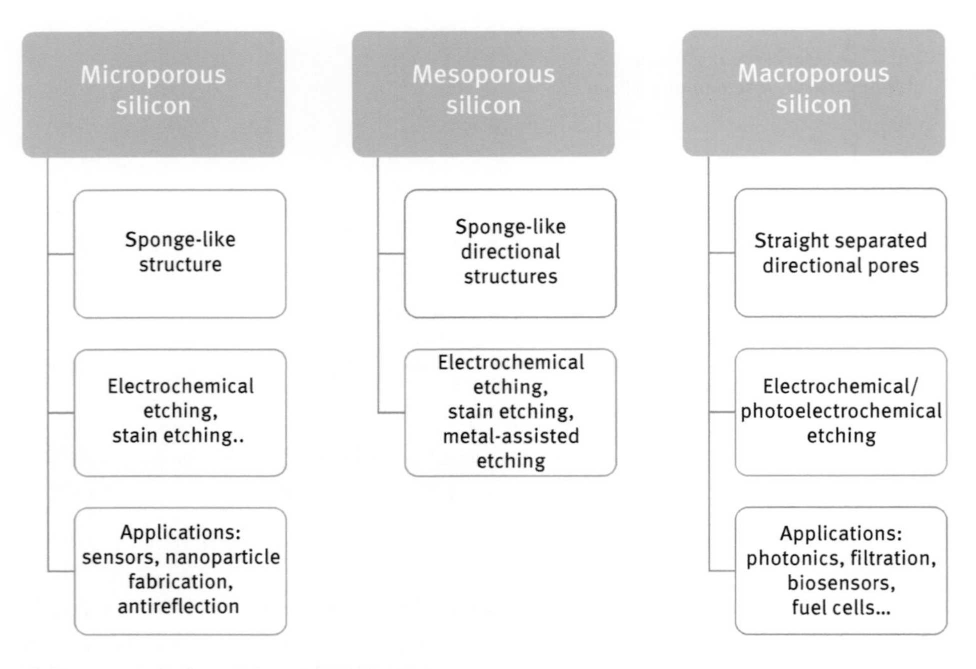

Figure 13.2: Overview of the porous silicon types.

This kind of pores can be obtained under a variety of conditions and with differing morphologies. The key parameters are the electrolyte type (aqueous, organic, oxidant), the hydrofluoric acid (HF) concentration, the surfactant, the Si doping type and level (n, n+, p, p+) and in some cases the illumination (backside or front side illumination). Detailed reviews about their formation are available, for example, [8, 17].

In this chapter, we focus on electrochemically etched macropores in n-type silicon.

13.2.1 Introduction to Silicon Electrochemistry

Electrochemical etching of silicon is a very versatile method for the fabrication of macroporous silicon.

The electrochemical dissolution of silicon is due to the stability of its oxide in aqueous solutions only possible with acidic solutions of HF. Silicon requires being anodic biased to dissolve efficiently. Micropore formation in p-type silicon has been known for about 50 years [18]. It was shown only in 1997 that macropores could be also generated under the same conditions in p-type silicon [19].

The pore evolution and morphology obtained by electrochemical etching depends on several factors. On the one hand, the current density and electrode potential determine the amount of etched silicon but have only a little influence in advance of the pore front. That is defined as the etch velocity. On the other hand, factors such as the electrolyte concentration, temperature and optional additives affect the etch velocity as well as the pore shape and limit the practical operation range of the electrical parameters. Furthermore, the bulk characteristics of the silicon wafer like doping type and density, crystal orientation and carrier lifetime strongly determine the morphology of the pores and feasible pore dimensions [20]. It has been found in the porous layer formation regime that currents through the electrochemical cell have a similar shape to that shown in Figure 13.3. (the current–voltage profile of an n-type silicon–electrolyte junction). Two competing mechanisms act on the silicon surface: formation of SiO_2 and subsequent dissolution by HF as well as direct dissolution of silicon. The *I*/*U* curve reaches a local maximum for a defined voltage U_{PS} with a corresponding current density J_{PS}. The critical current density J_{PS} marks the transition from divalent to tetravalent dissolution.

In particular, porous silicon formation occurs if the current density is lower than J_{PS}. For current densities below J_{PS} divalent dissolution of silicon takes place:

$$Si + 4HF_2^- + h^+ \rightarrow SiF_6^{2-} + 2HF + e^- \quad (13.1)$$

Thereby two charges per dissolved silicon atom can be measured in the electric circuit: one defect electron (h^+) moving from the silicon to the electrolyte and one electron (e^-) moving in the opposite direction through the interface.

Current densities exceeding the critical value J_{PS} will involve four minority charge carriers. A tetravalent dissolution occurs:

$$Si + 2H_2O + 4h^+ \rightarrow SiO_2 + 4H^+ \quad (13.2)$$

The HF electrolyte removes instantaneously the formed silicon dioxide – electropolishing takes place. Both the divalent and the tetravalent reactions lead to a dissolution

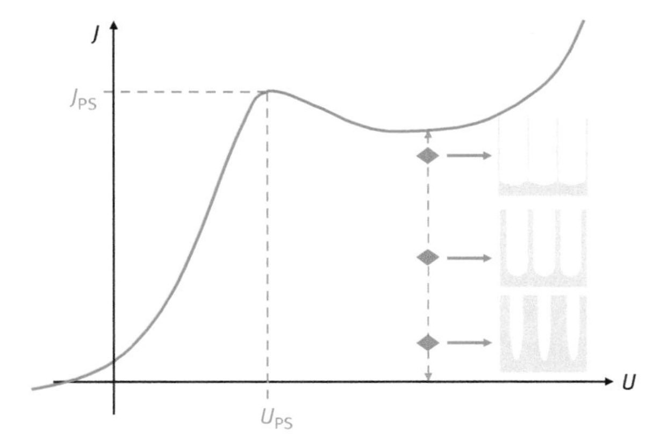

Figure 13.3: *I*–*U* characteristic curve of an illuminated reverse-biased Si-HF contact. The critical current density J_{PS} marks the transition from divalent to tetravalent dissolution. The limitation of the overall current density to values below J_{PS} by illumination from the backside influence directly the diameter of the pores.

of silicon, but only the first reaction (13.2) results in pores into silicon due to the strong dependence on the crystal orientation of the divalent dissolution. The chemical processes involved in the dissolution of silicon are quite complex and depend on the local current intensity and fluoride ion concentration.

More insight on the chemical–physical aspects of silicon dissolution can be found in the works of Lehmann, Zhang [20] and Kolasinski [21].

The investigation of the physicochemical processes at the silicon/HF interface plays an important role to understand the pore growth and therefore the local dissolution of silicon. These processes depend essentially on the doping of the silicon wafer and thus on the properties of the space charge region (SCR) at the interface. The formation of microporous silicon is based on quantum confinement effects [22, 23]. The doping and crystal orientation do not influence the pore growth. Such a layer of microporous silicon often covers the walls of meso- or macropores. The generation and properties of an SCR in an electrochemical etching setup is essential during the mesoporous and macroporous pore growth. Mesopores are formed mainly by tunneling processes within the SCR [24] while macropores are generated resulting in thermionic emission (for p-type doping) [25] or the collection of minority charge carriers (for n-type doping). The growth direction of meso- and macropores is dependent on crystal orientation.

The underlying mechanisms of macropore formation in n-type silicon will be discussed in the following section in more detail.

13.2.2 Electrochemical Macropore Formation in n-Type Silicon

For the release of anodic-poled silicon at the semiconductor electrolyte boundary layer, the existence of holes is required [9]. In the case of photoelectrochemical etching the silicon is therefore additionally illuminated from the rear side (i.e., the side facing away from the electrolyte) in order to produce electron–hole pairs by absorption of the photons. The generated electrons are directly 'sucked off' by the anodic polarity while the holes diffuse through the semiconductor and thus reach the boundary layer. In addition, the current–voltage characteristic is shifted by the generation of holes in the illuminated n-type silicon, so that in the case of maximum illumination it assumes the form of the characteristic curve for p-doped silicon.

The illumination of n-doped silicon also results in a controllable degree of freedom for the etching process since the current density at the semiconductor–electrolyte junction can now be regulated independently from the applied voltage via the illumination intensity. Figure 13.4 shows the structure for photoelectrochemical etching according to Lehmann et al. [21].

For n-type silicon, the condition $J < J_{PS}$ is always obtainable by the choice of the appropriate illumination intensity independently from the voltage as long as the

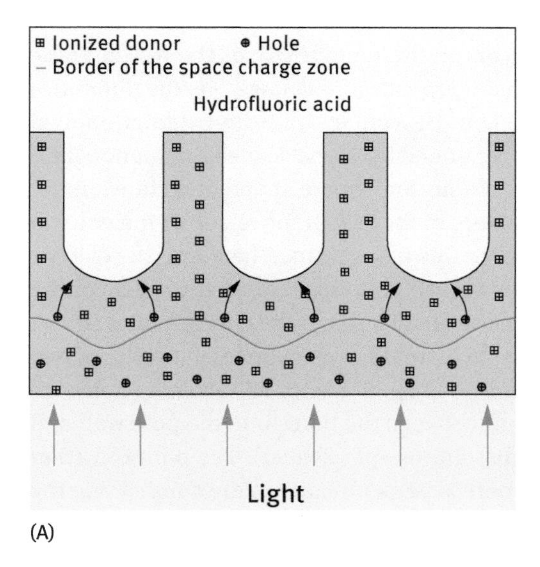

(A)

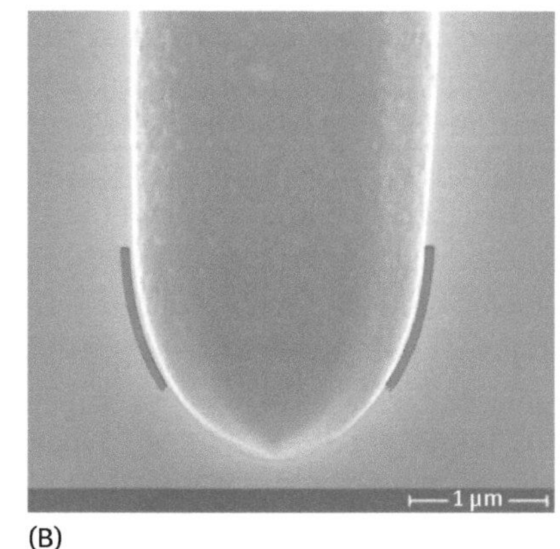

(B)

Figure 13.4: (A) Etching of macropores in n-type silicon under backside illumination. The holes generated by absorption of the light at the wafer backside diffuse through the wafer to the etching front. There they are consumed at the pore tips for the etching process. (B) Scanning electron microscope (SEM) picture of a silicon pore bottom. At the lowest region of the pore bottom electropolishing takes place due to the high hole density while the outer edge of the pore bottoms is divalent dissolved.

voltage remains in the range $U_{oc} < U < U_{through}$. (U_{oc} = idling voltage, $U_{through}$= breakthrough voltage). In particular, it is possible that U is selected in the range $U_{PS} < U < U_{through}$, whereas $J < J_{PS}$ remains. In this parameter range, stable growth of macropores is known for a longer time [9].

In the case of macropores, the development of the space charge zone at the Si/HF interface must be considered. The (100)-oriented n-type silicon wafer is in contact with HF at one side. The positive pole of the voltage source is placed on the backside of the n-type silicon wafer (anode). The negative pole forms the cathode made by platinum wire in the HF. The back side of the wafer is illuminated with light, thereby electrons at the wafer backside will raise from the valence band into the conduction band and generate simultaneously holes in the valence band. The electrons of the conduction band are extracted to the positive pole of the voltage source while the holes drift through the wafer to the opposite etching front. The development of SCR of the Si/HF interface follows the geometry of the pore tips.

The width of this region W_{SCR} for the case of a planar silicon interface can be estimated to

$$W_{SCR} = \sqrt{\frac{2\varepsilon\varepsilon_0 U}{eN_D}} \tag{A}$$

where ε is the dielectric constant of silicon, ε_0 the permittivity of the free space, e the elementary charge and N_D the density of dopants. The voltage U is the difference between the built-in potential of the silicon–HF contact and the external applied voltage. Furthermore, the charge carriers generated by the backside illumination limit the chemical reaction. The SCR almost conforms to the pore shape. It is shown in literature [20] that with the associated curvature of the SCR in the region of the pore tip a reduction of the SCR thickness at the pore tips also occurs. Therefore, the electric field is bigger in the region of the strongly curved pore tips than in the region of the pore walls. Since the electric field lines are perpendicular to the curved pore surface, the incoming holes are focused to the pore tip. In case of an overlap of the SCR of two adjacent pores, the entire pore wall belongs to the SCR. Hence all incoming holes are focused on the pore tips; no holes can drift between the pores into the pore wall and the pore wall is thus protected against dissolution (passivated). The pore is further etched at the tip only. The growth of the pore is perpendicular to the depth along the (100)-crystal direction.

Although the etching process is not a self-ordering process, it is a self-organizing one. For a given doping density, applied voltage and backside illumination, an average porosity and pore diameter will arise.

The growth of macropores in silicon follows a certain short-range order due to the formation of the SCR between neighbored pores but a long-range order that requires strict periodicity does not exist. Therefore, the lithography has to match the intrinsic material parameters. However, it is possible to initiate pore growth selectively at defined positions via artificial preparation of suitable ordered nucleation on the silicon surface. For this purpose, an anisotropic etching solution is used to generate etching pits in the form of inverted pyramids in the (100) silicon surface (Figure 13.5).

A defined top surface photomask determines the structure and arrangement of the required pits. For this purpose, first the silicon wafer is coated with a photoresist (Figure 13.5A, B). Photolithography is used to expose the mask (Figure 13.5C) and to transfer the structure into the silicon surface (Figure 13.5E). In addition to the periodically ordered trigonal or cubic pore arrangements by specifically omitting certain

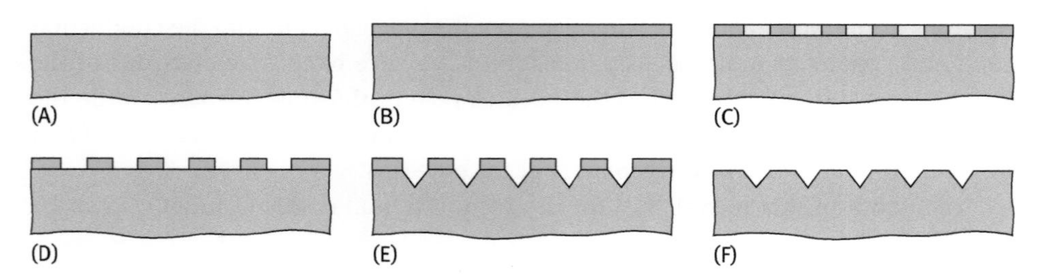

Figure 13.5: Macroporous silicon by photoelectrochemical etching: lithography and KOH pre-structuring. (A) plane Si; (B) photoresist + silicon dioxide; (C) photolithography; (D) HF dip; (E) KOH etching; (F) final pre-structured wafer surface.

nuclei, point or line defects can also be produced for the generation of resonant structures or photonic waveguides.

In 1990, Lehmann and Föll have demonstrated ordered macropore growth in n-doped (100)-oriented silicon [8] for the first time, and 3 years later Lehmann published his phenomenological, macroscopic growth model based on space charge effects [9]. The predictions derived from that model describe the growth of cylindrical macropores at steady growth conditions in an excellent manner and enable precise control of the production.

Starting material is a pre-patterned single crystal silicon wafer as shown in Figure 13.5. It is brought into contact with the structured front side with HF and illuminated at the back. In addition, an anodic voltage is applied between silicon (contacted at the backside) and the electrolyte (contact by means of a platinum wire). Depending on the intensity, the backlighting generates different electron–hole pairs. Due to the externally applied anodic voltage and supported by the back contact, the electrons are immediately sucked off. The generated defect electrons, driven by the concentration gradient, flow through the entire silicon wafer to the interface between silicon and HF.

An extensive SCR (see Section 13.2.1) has formed on this boundary surface. This static electric field is mainly deflected to the pore tips. Therefore, the SCR passivates the pore walls against the penetration of electronic holes and thus protects them from electrochemical dissolution. The electrochemical dissolution of the silicon takes place only at the pore tips according to eqs (13.1) and (13.2). Figure 13.6 shows typical example of such two-dimensional macropores that can be produced with lattice constants of 0.5–12 μm. They are characterized by a very flat etching front, that is, deep pores, pores with practically identical diameters, low surface roughness and great perfection and precision.

The silicon standards available at 6″ wafer scale at SmartMembranes are listed in Table 13.1. The structure parameters have some degrees of freedom, but the maximum or minimum main values are fixed as listed.

The standard spacings (a = pitch) are currently 1.5, 4.2 and 12 μm. Other pitches are reasonable as well but in needs of a new pre-patterned surface and thus a new

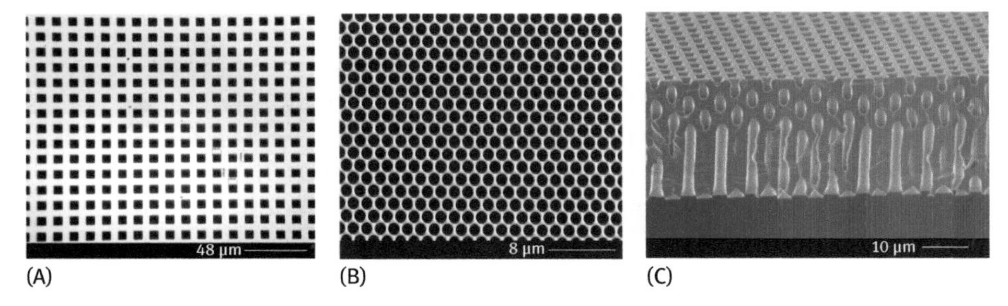

Figure 13.6: Standard electron microscopic pictures of silicon membranes: (A) anisotropic pore shape, (B) isotropic pore shape and (C) lifted membrane.

Table 13.1: Standard pore structure parameters for macroporous silicon – standards available at SmartMembranes.

Interpore distance	1.5 μm	4.2 μm	12 μm
Pore diameter	1 μm	2.5 μm	5–6 μm
Possible pore widening	Up to 1.2 μm	Up to 3.5 μm	Up to 10 μm
Pore arrangement	Trigonal	Trigonal	Cubic
Porosity	40–60%	20–60%	40–50%
Membrane thickness, standards	50, 200 μm (liftoff)	50, 200 μm (liftoff)	50, 200 μm (liftoff) 350, 500 μm (flat)
Membrane thickness, custom	15–200 μm	15–500 μm	15–500 μm
Membrane size	up to Ø 130 mm	up to Ø 130 mm	up to Ø 130 mm
Standard tolerance of ±10%	✓	✓	✓

lithographic mask. Additional various postprocessing steps are possible, such as substrate liftoff to generate membranes thinner than the bulk material (Figure 13.6A), anisotropic or isotropic pore shaping (Figure 13.6B, C) and laser dicing in membrane sizes regarding the customer needs. All processes are applicable on 6″ wafer size, but also scalable in future.

13.3 Nanoporous Alumina

The history of electrochemical oxidation of aluminum has already started in the beginning of the last century. Anodic treatment of aluminum was developed to obtain protective and decorative films on its surface [26]. Bengough and Stuart's patent in 1923 is being the first patent for protecting Al and its alloys from corrosion by means of an anodic treatment [27]. In 1936, Caboni invented the famous coloring method consisting of two sequential processes: anodization in sulfuric acid, followed by the application of an alternating current in a metal salt solution [28]. Between 1970 and 1990, studies led by Thompson and Wood explained the growth mechanisms of aluminum oxide films. A similar publication by O'Sullivan and Wood is well known for the anodization of aluminum to receive porous alumina structures [29] in a disordered way as shown in Figure 13.7A. Based on a two-step anodization process, a self-ordered (or high-ordered) porous alumina membrane with 100 nm interpore distance was synthesized by Masuda and Fukuda in 1995 [30]. This discovery was a breakthrough in the preparation of polydomain nanoporous alumina templates with a very narrow size distribution and high aspect ratios (Figure 13.7B). Two years later, they combined the aluminum anodization method with novel nanoimprint technologies, which allowed for the first time the preparation of a monodomain pre-structured porous alumina as shown in Figure 13.7C [31].

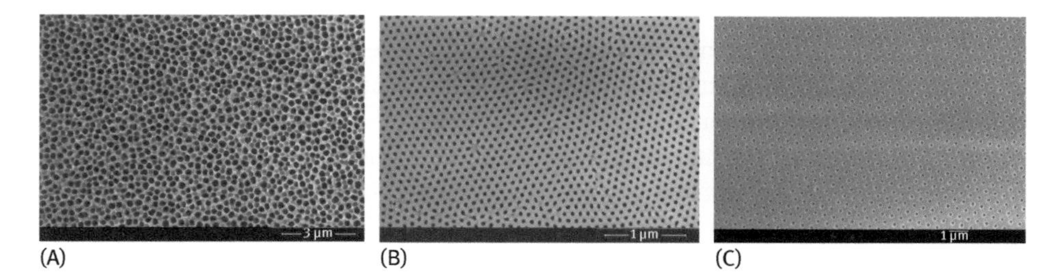

Figure 13.7: Overview of nanoporous alumina structures: (A) disordered, (B) self-ordered and (C) pre-structured.

Self-ordered nanoporous alumina, which is the main topic of this chapter, is produced by anodic oxidation of pure aluminum under defined process conditions. Those high-ordered alumina structures gained importance in the past years because of their wide range of commercial applications. They are often used for biomimetics or as templates for novel nanocomposites [32]. Also a fabrication of several types of devices such as filter modules, catalyst components and biological appliances as well as electronic, magnetic and optical devices are of more technological interest [33, 34]. The formation of alumina nanostructures (Figure 13.8) is based on a self-organization in which the pores grow perpendicular to the surface [35]. Pore size, the interpore distance and the pore length can be defined by a wide range of combinations of different process parameters such as temperature, etching time, applied voltage, current

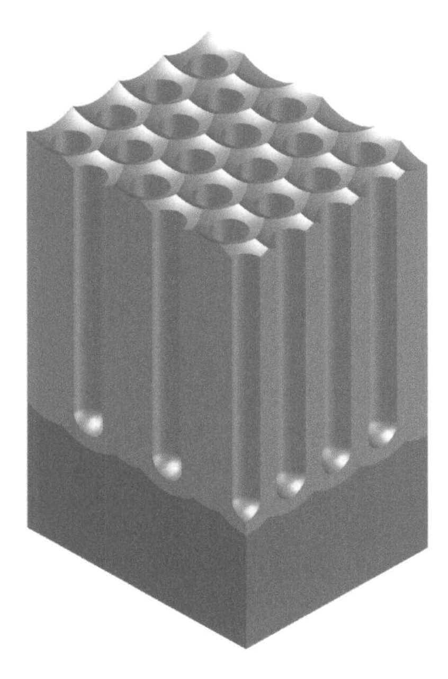

Figure 13.8: Self-ordered nanoporous alumina on aluminum support.

density as well as the concentration and mixture of the electrolyte. To create a two-side open membrane, the residual aluminum on the bottom has to be selectively dissolved and a chemical pore bottom etching has to be performed. All those parameters and additional post-processes make the fabrication very sensitive to changes, especially in terms of high throughput and size scaling.

13.3.1 Introduction to Alumina Electrochemistry

Depending on several factors, in particular the electrolyte, two types of anodic films can be produced: barrier-type films can be formed in completely insoluble electrolytes, for example, neutral boric acid, ammonium borate, tartrate and ammonium tetraborate in ethylene glycol. Porous-type films can be created in slightly soluble electrolytes such as sulfuric, phosphoric, chromic and oxalic acid [36].

The process takes place in an electrochemical cell, which consists of a two-electrode system, for example, a platinum (Pt) mesh acting as the counter electrode whereas aluminum is the anode. When voltage is applied, aluminum is being oxidized and hence the alumina layer is produced at the anode (eq. (13.3)). On the cathode side hydrogen evolves (eq. (13.4)):

$$2Al + 3H_2O \rightarrow Al_2O_3 + 6H^+ + 6e^- \tag{13.3}$$

$$6H^+ + 6e^- \rightarrow 3H_2 \tag{13.4}$$

The pore formation [37, 38] can be described in the following four steps (Figure 13.9):

- Step 1: once the voltage is applied, a barrier oxide layer is formed on the surface.
- Step 2: since the barrier oxide layer always shows irregularities due to the surface roughness of the aluminum foil or sheet used, the electric field focuses locally at those fluctuations.
- Step 3: that focus induces a field-enhanced or/and temperature-enhanced dissolution in the formed oxide and thus leads to the growth of pores.
- Step 4: pore growth with an equilibrium of field-enhanced oxide dissolution at the oxide/electrolyte interface and oxide growth.

At the equilibrium (step 4) negatively charged ions (O^{2-}/ OH^-) from the electrolyte migrate through the oxide layer to the pore bottoms, whereas Al^{3+} ions from the metal drift through the oxide layer into the solution at the oxide/electrolyte interface [42]. While Al^{3+} ions that reach the oxide/electrolyte interface contribute to oxide formation in case of barrier oxide growth (Figure 13.10), other Al^{3+} ions can be lost to the electrolyte solution [39].

The current density passing across the oxide is described as follows [40]:

$$j = j_a + j_c + j_e \tag{13.5}$$

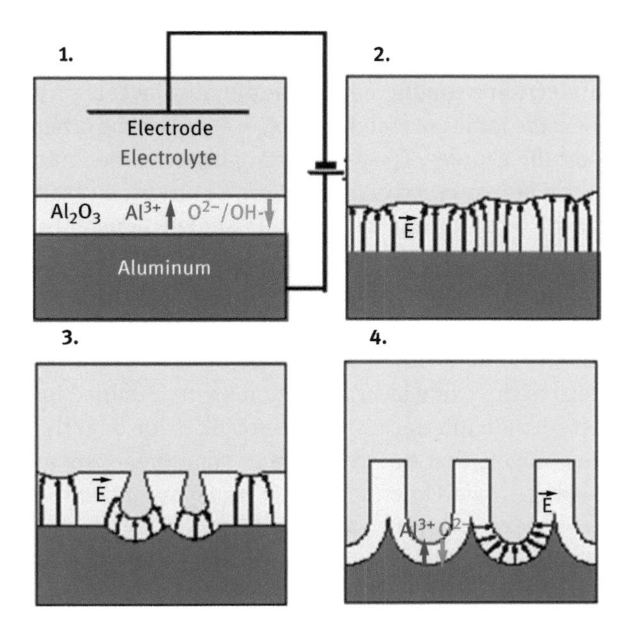

Figure 13.9: Schematic diagram of the pore formation at the beginning of the anodization: step 1: formation of barrier oxide on the entire area; step 2: local field distributions caused by surface fluctuations; step 3: creation of pores by field-enhanced or/and temperature-enhanced dissolution; step 4: stable pore growth [47, 49].

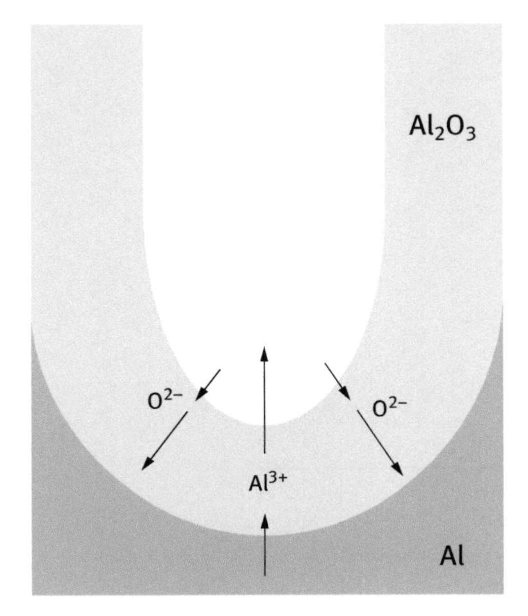

Figure 13.10: Ion diffusion at the interfaces during oxide formation.

where j_a, j_c and j_e are the anion-contributing, cation-contributing and electron-contributing current densities. The electronic conductivity in the alumina layer is very low due to the passivation properties; the ionic current density ($j_i = j_a + j_c$) on the other hand contributes mostly to transport the charges. The relationship between the ionic current, j_i, and the electric field, E, can be expressed in terms of the Guntherschultze–Betz equation [49], whereas β and j_0 are temperature and metal-dependent constants:

$$j_i = j_0 \exp(\beta E) \tag{13.6}$$

For the alumina the electric field E, j_0 and β are in the range of 106–107 V/cm, 1×10^{-16} to 3×10^{-2} mA/cm^2 and 1×10^{-7} to 5.1×10^{-6} cm/V, [39]. Based on the Guntherschultze–Betz equation, the rate-limiting steps of the oxide layer formation are determined by the ionic transport either at the metal/oxide interface, within the bulk oxide or at the oxide/electrolyte interface [44]. This means that the oxide grows simultaneously at both interfaces, for example, at the metal/oxide interface by Al^{3+} transport and at the oxide/electrolyte interface by oxygen ion transport [33, 40].

The transient of the potentiostatic current density shows the formation of barrier-type or porous-type alumina (see Figure 13.11) [41–44]. At the beginning of the anodization, both transients behave similarly. In case of the barrier film formation, the current density j_b decreases exponentially since the barrier oxide current is dominated by the ionic current j_i. For nanoporous alumina growth the current density profile according to the formation steps mentioned above is observed [45]. In the beginning, the current density j_p decreases as well (step 1 in Figures 13.9 and 13.11). In step 2 it passes through minimum value. Thereupon it again increases to reach a

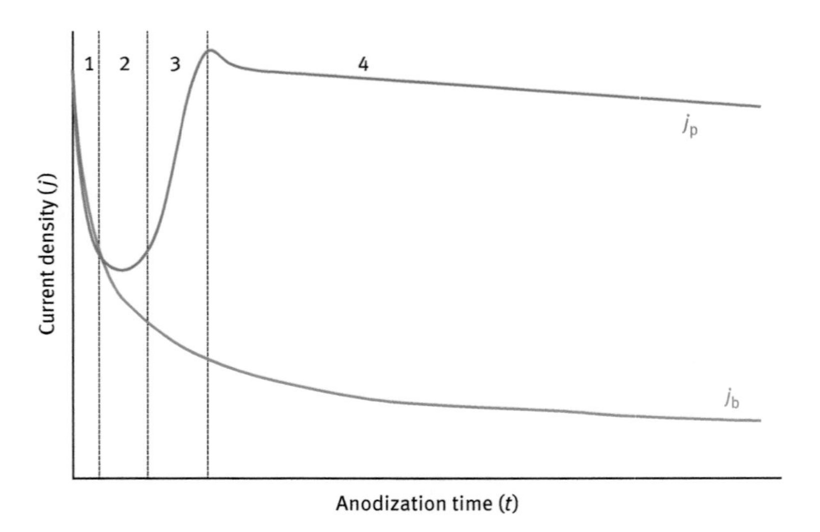

Figure 13.11: Schematic diagram of current density curves during the initial growth at constant voltage. j_b indicates the current density for the barrier film and j_p for the porous film formation.

maximum value (step 3). Since some pores begin to stop growing due to competition among the pores, the current slightly decreases. Finally, a constant current density remains (step 4). During a long etching time the current density may continuously decrease. This is due to diffusion limits in the long pore channels [47–49].

13.3.2 Self-Ordered Regimes via Two-Step Anodization

The pore diameter and the interpore distance can be varied by using different voltages or electrolytes such as oxalic, sulfuric and phosphoric acid [46, 47]. Self-ordered porous alumina structures can only be obtained under specific conditions. For example, structures with pore spacing of 65, 105, 125 and 480 nm are fabricated at 25 V in sulfuric acid, at 40 or 50 V in oxalic acid and at 195 V in phosphoric acid, respectively [48–50]. The high order of the pores is only created where the voltage is adapted to the right electrolyte and concentration, as shown in Figure 13.12.

The applied potential U is one of the most important factors to adjust self-assembly of porous alumina. The interpore distance, D_{in}, is linearly proportional to the applied potential with a proportionality constant k of approximately $2.5 \leq k$ (nm/V) ≤ 2.8 [33]:

$$D_{int} = kU \tag{13.7}$$

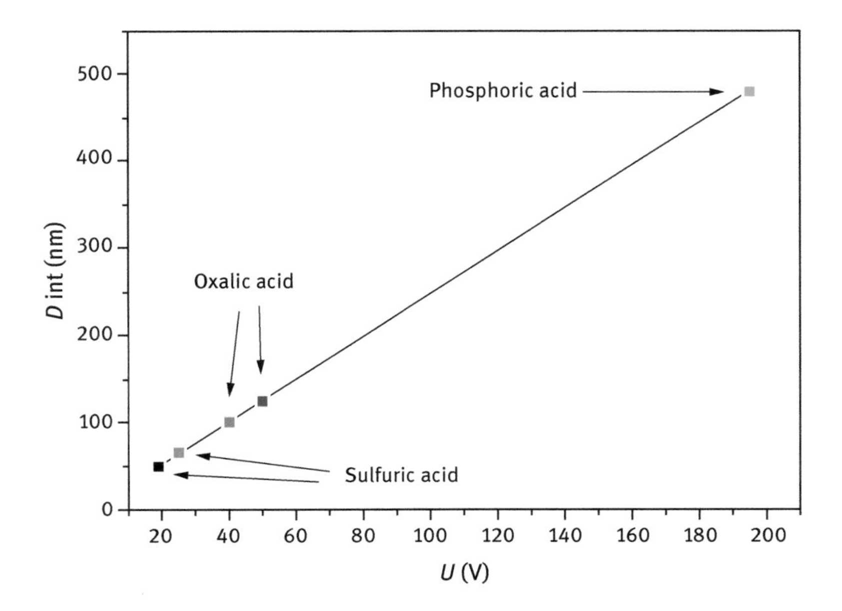

Figure 13.12: Effect of anodizing potential (U) on the interpore distance (D_{int}) for high-ordered porous alumina formed in different acid electrolytes. The known self-ordering regime takes place at the line where the equilibrium of dissolution and oxide creation exists.

In addition, the thickness of the barrier layer can be approximately estimated as half of the interpore distance ($D_{int} = 2\ t_b$, where t_b is the barrier-layer thickness). Since the interpore distance, D_{int}, is given by the applied voltage, only the pore diameter can be adjusted to offer a wide range of porosities. In a stable self-ordered etching process, the resulting pore diameter is similar to the applied voltage, for example, at 25 V in sulfuric acid the pore diameter is 25 nm. It is well known that those standard pore diameters follow the 10% porosity rule [51]. To achieve higher porosities, the pore diameter can be chemically widened in an additional step while keeping D_{int} constant [52].

The self-organized arrangement of pores in high-ordered hexagonal arrays can be explained by a repulsive interaction between the pores during growth due to volume expansion [53]. As a result, during the first oxidation as explained previously hexagonally close-packed arrays are obtained at the interface between the porous alumina layer and the aluminum substrate, whereas the top side remains unordered (Figure 13.13A and B). On a very rough surface also loss or bifurcation of pores during the first anodization may occur [54, 55]. Then, the porous alumina film is selectively dissolved in a wet chemical treatment [56]. Patterns that are replicas of the hexagonal pore array are preserved on the fresh aluminum surface (Figure 13.13B). This allows the preparation of pores with a high regularity by a subsequent second anodization under the same conditions as the first anodization (Figure 13.13C). If needed, the

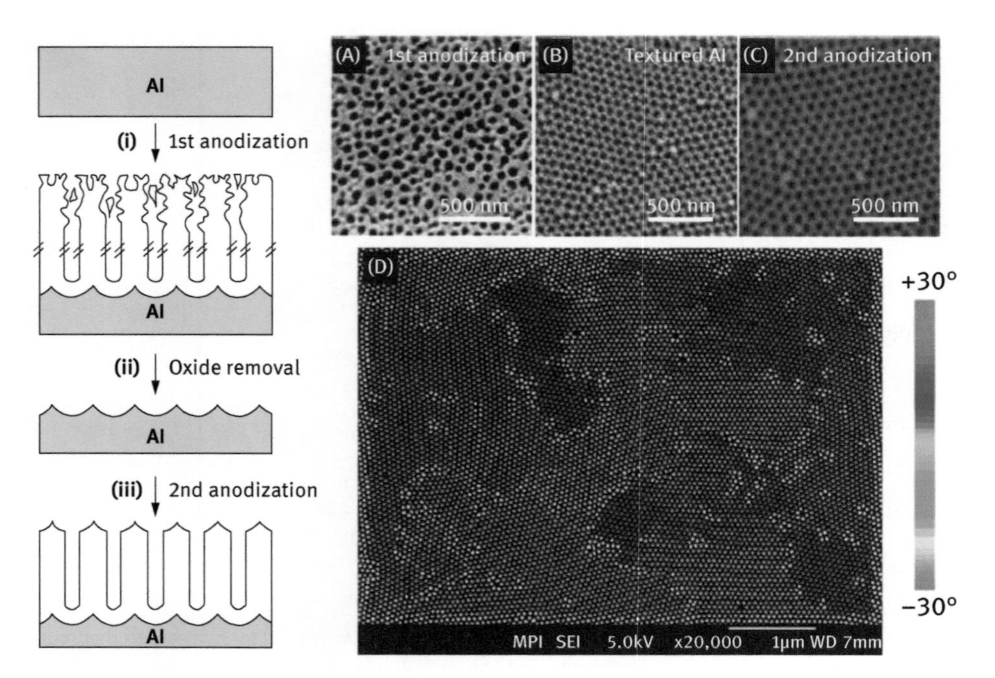

Figure 13.13: Two-step anodization procedure for a self-ordered nanoporous structure [55].

Table 13.2: Nanoporous alumina standard pore structure parameters – available at SmartMembranes.

Pitch/interpore distance	65 nm	125 nm	480 nm
Pore diameter	25 nm	40 nm	180 nm
Possible pore widening	Up to 45 nm	Up to 90 nm	Up to 400 nm
Pore arrangement	Trigonal	Trigonal	Trigonal
Porosity	10–45%	10–50%	10–50%
Membrane thickness, standards	50, 100 μm	50, 100 μm	50, 100 μm
Membrane thickness, custom	30–120 μm	30–120 μm	30–120 μm
Membrane size	Up to A4	Up to A4	Up to Ø 150 mm
Standard tolerance of ±10%	✓	✓	✓

resulting pores can be isotropically widened by chemical etching to increase the pore size and porosity at a constant interpore distance.

Since the aluminum used is a multicrystalline metal, the trigonal order of the self-ordered pores can only be reached in the single grains of the crystal structure. Those so-called monodomains are emphasized in Figure 13.13D with different colors.

The alumina membranes and templates available in a self-ordered regime at SmartMembranes are listed in Table 13.2. Besides those parameters, custom developments are possible for reaching other pitches and porosities depending on the order needed.

13.4 Potential Applications of Both Porous Material Systems

Based on SmartMembranes' experience the membranes have a large application portfolio in many areas of materials science as well as in nano- and microproduction (Table 13.3).

Nowadays the main focus of our membrane production lies in the needs of defined market segments – so-called niche market segments (Figure 13.14).

Applications of macroporous silicon have been shown in the fields of microsystem technology, gas sensors, photonic crystals, biotechnology and many others. Examples include short-pass optical filters, 2D and 3D photonic crystals, optical and capacitive immunosensors for monitoring immune complex formation. Macroporous silicon provides a platform for lots of interesting and innovative applications mostly in biotechnology. Silicon membranes with straight passages of different dimensions (pore size and length) have been proposed, for instance, for selective bioorganism detection [57]. Here an initial use of macroporous silicon is the Ratchet membrane [58], where the pores are characterized with asymmetrically modulated diameter in depth. This system can be suitable for efficient and selective continuous separation of sensitive biological materials such as viruses and cell fragments. Miniaturized biochips with involved high-ordered macroporous silicon arrays have also proved to

Table 13.3: Application portfolio of high-ordered porous alumina and silicon.

Application
Separation and protection against contamination (dust, bacteria, viruses, etc.)
Sterile and viral filtration
Solid-state sensors, multifunctional sensors for gases
Lab-on-chip systems, diagnostics
Bioanalysis
Microreactors
Catalysis
Template synthesis for novel composites

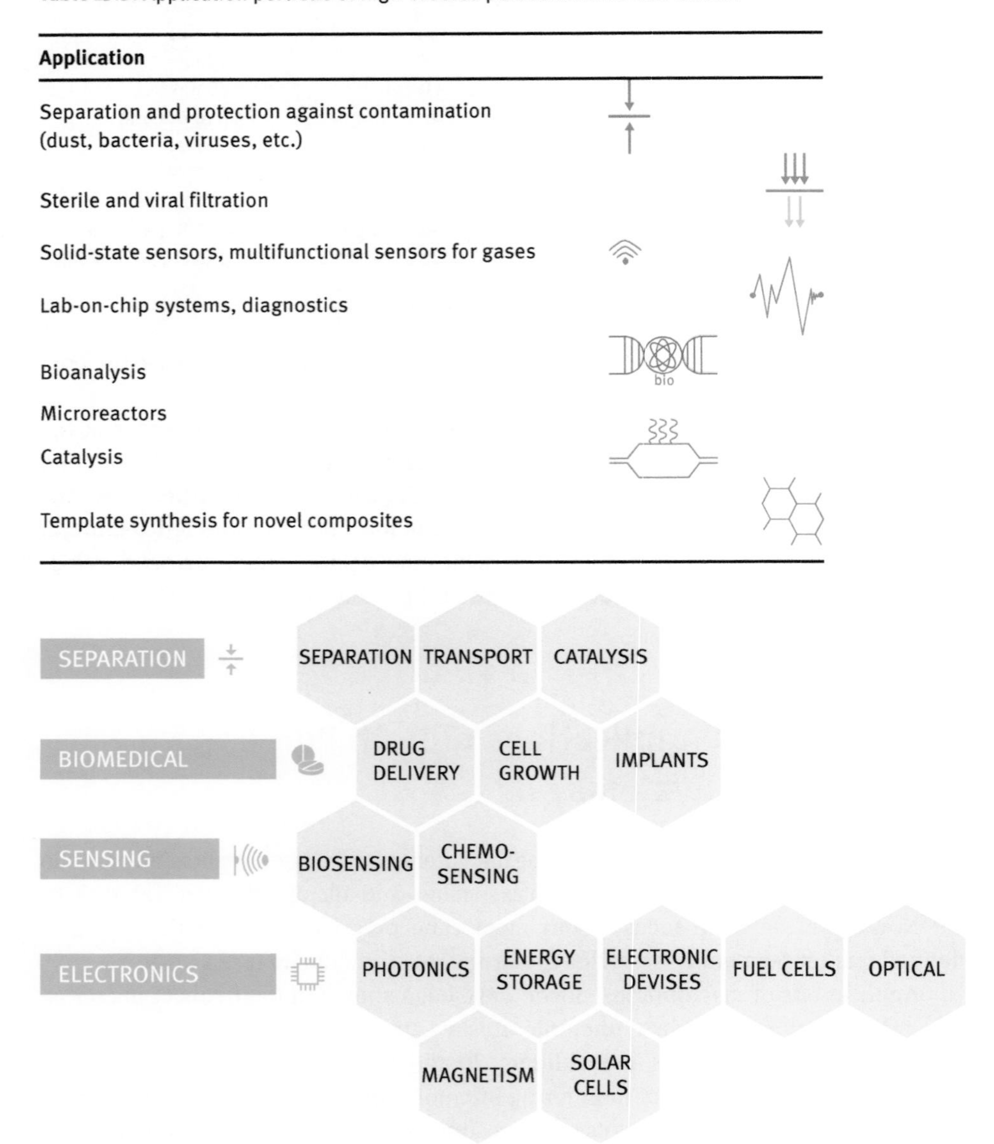

Figure 13.14: Application fields of the nano- and macroporous membranes of SmartMembranes.

be advantageous for the use of the detection and recognition of molecular binding events [59, 60]. All of the above examples utilize several unique features of macroporous silicon, for example, the ability to generate perfect periodic pore arrays of defined arrangements and high aspect ratio, large area-to-volume ratio and full process compatibility with silicon microtechnology. Most applications benefit from the remarkable periodicity and straightness of macropore arrays which are essentially 3D systems.

In contrast to other state-of-the-art porous materials and processes, nanoporous alumina structures offer a very high aspect ratio (up to 4,000) while still keeping all dimensions constant. Due to the precise membrane parameters, the surface area and also the exact flow rates through the membranes can be exactly calculated and adjusted on nanometer scale. These are the unique selling points, which lead to various new markets and applications.

13.4.1 Elected Applications of Macroporous Silicon Membranes and Templates

One of the innovative application examples is the multifunctional biochip sensor "TipChip," which was developed in cooperation with the company Axela Inc. (Canada). This flow-through chip is based on a lab diagnostic method for the determination of DNA and protein species. In this case, complex elaboration methods are replaced by substance-selective separations in the pores (transition from 2D to 3D structures with defined surface modification of the inner pore walls). The single-use consumable has been designed for routine and focused multiplex analysis for nucleic acids or proteins. The TipChip is a disposable device consisting of a 6.5 mm square chip mounted on a plastic tube. The chip is made of porous silicon with >200,000 microchannels incorporated in that area. A single capture probe site occupies approximately 70 microchannels. This approach facilitates the interaction between target molecules and immobilized probes, resulting in three to four times faster hybridization of oligonucleotides or protein binding (Figure 13.15). A highly selective multifunctional biosensor system could be realized by using an electrosensory measurement technology.

Another interesting application is the collaboration with Neah Power Systems Inc. (USA) for the development of a direct methanol fuel cell (Figure 13.16). In this case, the macroporous silicon acts as the electrode material. The 3D microchannel silicon structure has many beneficial attributes in contrast to the traditional proton exchange membrane (PEM) cell configuration. The enormous enlargement of the total surface area, the adjustable porosities and the straight and uniform air pore channels do not only allow high flow rates without the formation of blockages but also enable a clear miniaturization with the same or improved performance. The drawing showing a cross-sectional view of a single cell in Figure 13.16 illustrates the differences of the two designs. Power generation in the 3D structure of the Neah electrode design along with the liquid electrolyte coverage of that surface results in significantly larger surface area available for power generation compared to the corresponding 2D surface. Since electrical current is produced from the chemical reactions that occur when the fuel, electrolyte and catalyst interact, the PEM design has substantially less useable reaction sites.

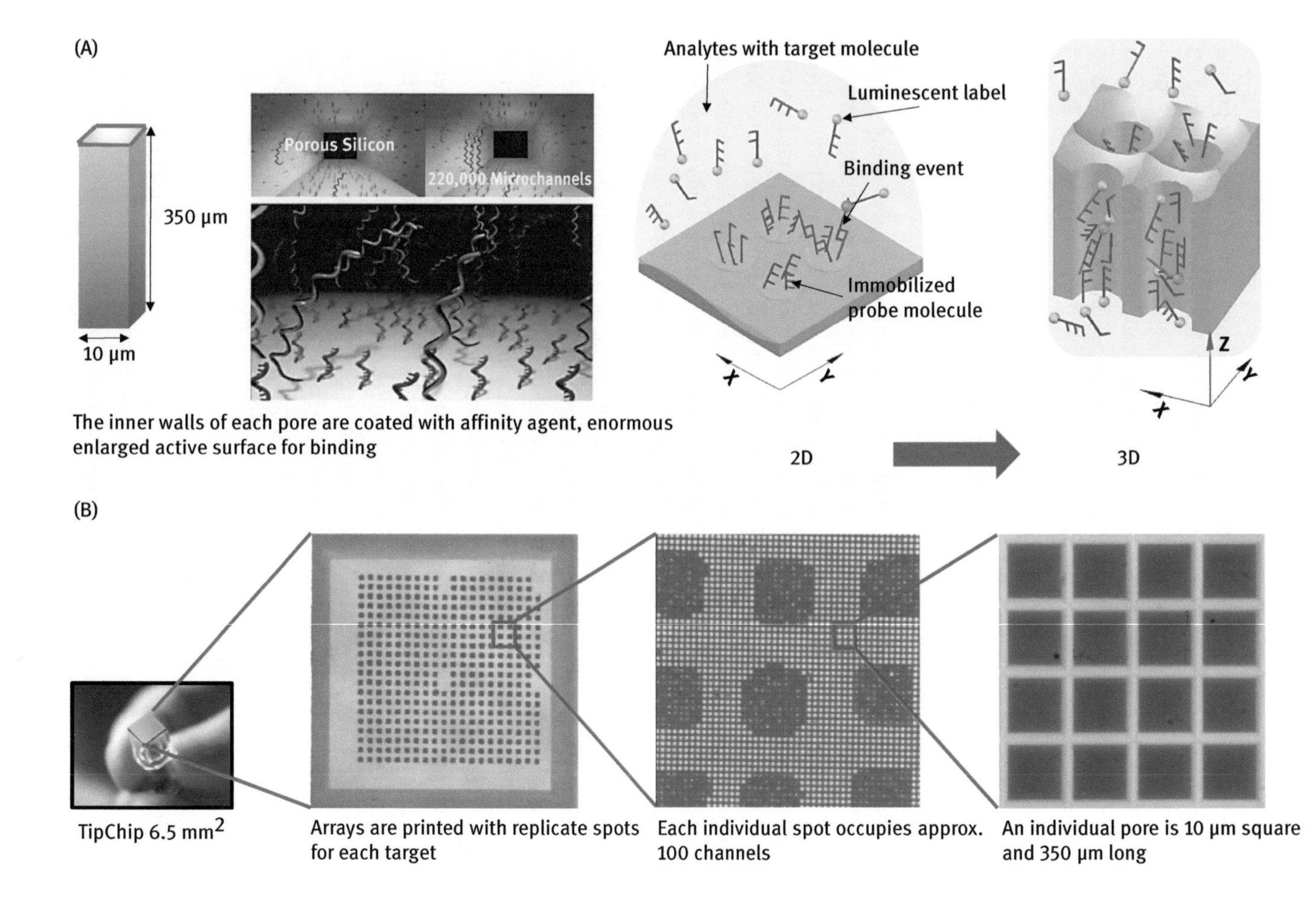

Figure 13.15: Three-dimensional flow-through TipChip Technology (Axela Inc.): (A) mode of operation – transfer from 2D to 3D substrate and (B) TipChip – functionalized macroporous silicon membrane.

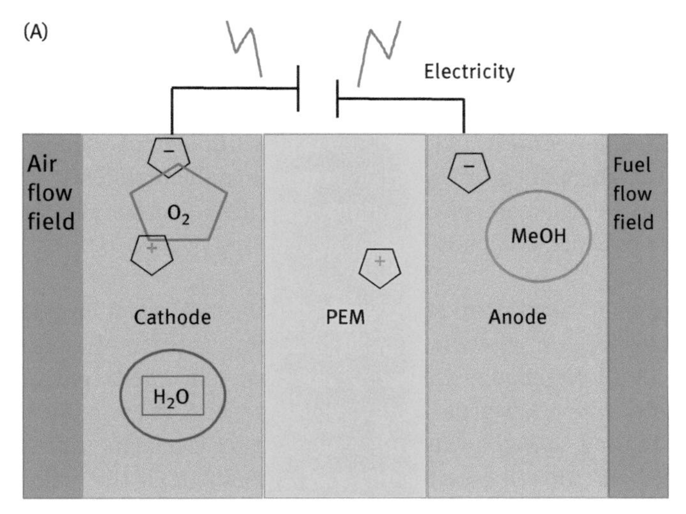

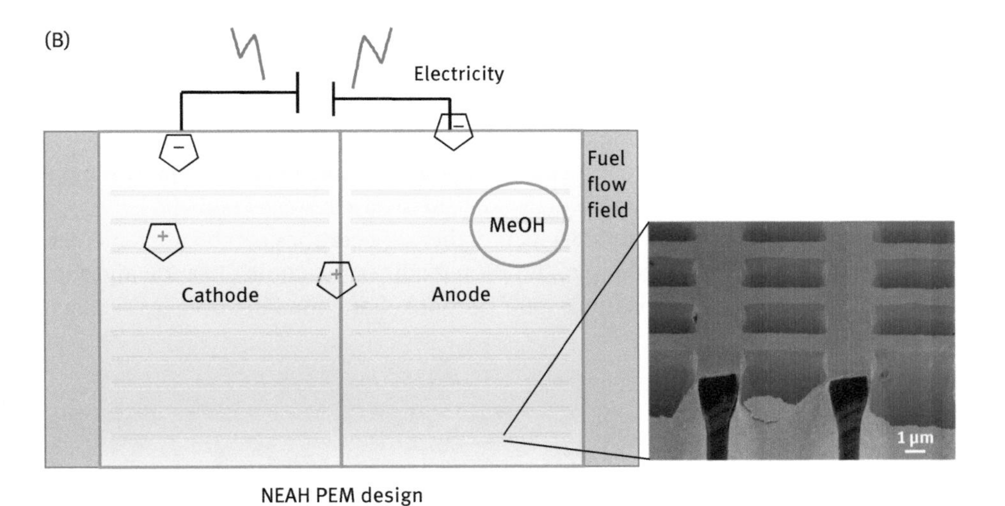

Figure 13.16: Side-view comparison of the PEM versus Neah cell designs (not to scale).

The liquid system drives other differences from a traditional PEM-based design. The gas diffusion layer needed in PEM systems is not a requirement any longer as the ordered pore structure of the silicon electrode allows a uniform fuel/electrolyte distribution. Ultimately, the micro electrochemical systems (MEMS)-based design techniques available for silicon structures will be capable of more and more functionality to be built directly into the electrode structure as the technology continues to mature.

13.4.2 Elected Applications of Nanoporous Alumina Membranes and Templates

The membranes at SmartMembranes are available either as flow-through membranes or one side closed templates. The main application of templates is the production of nanowires or nanorods from metals, metal oxides or polymers [61–63]. Examples for nanotubes produced from a polymer melt and metal wires achieved by electrodeposition are shown in Figure 13.17.

Based on those developments a new market has evolved while looking for inverse moth-eye molds as templates for printing tools in order to achieve antireflective and adhesive surfaces (Figure 13.18). Hereby the nanopores are created by anodic oxidation of an aluminum cylinder with a low aspect ratio, typically 1 or 2. The pore form is not straight as usual but rather cone shape having a slope or conical shape which is generated by a controlled variation of pore etching and widening [65]. The major advantage in contrast to other commercial solutions, such as interference or e-beam lithography or nickel shim applications, is the rather convenient production method at shorter times and favorable pricing. Moreover, this method offers a novel seamless printing cylinder which is not limited to the length or diameter of the tool [66]. This cylinder is then being used in printing machines to transfer the structure in either a resin which is quickly ultraviolet cured after the adoption of the inverse structure or directly on a melted foil following a cooling procedure.

Nanorods from metals are often used to create novel magnetic materials for magnetic memory in three dimensions which can lead to new storage media [67].

Nanoporous alumina membranes with higher aspect ratios can be implemented in flow-through devices, for example, filtration modules for sterile filtration and separation or for biosensing applications [68]. They are often found in novel gas sensors,

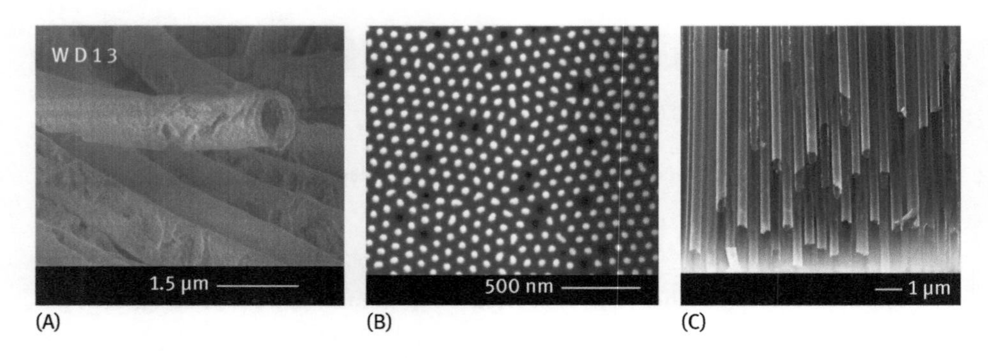

Figure 13.17: Nanowires and nanotubes from polymer melts (A) [64] and metal deposition (B, C) using the nanoporous membranes as templates; pictures contributed by Prof. Martin Steinhart, University of Osnabrück, Germany (A); Prof. Jörg Schilling, ZIK SiLi-nano®, MLU Halle, Germany (B); Institute NEEL, University Grenoble Alpes & CNRS, France (C).

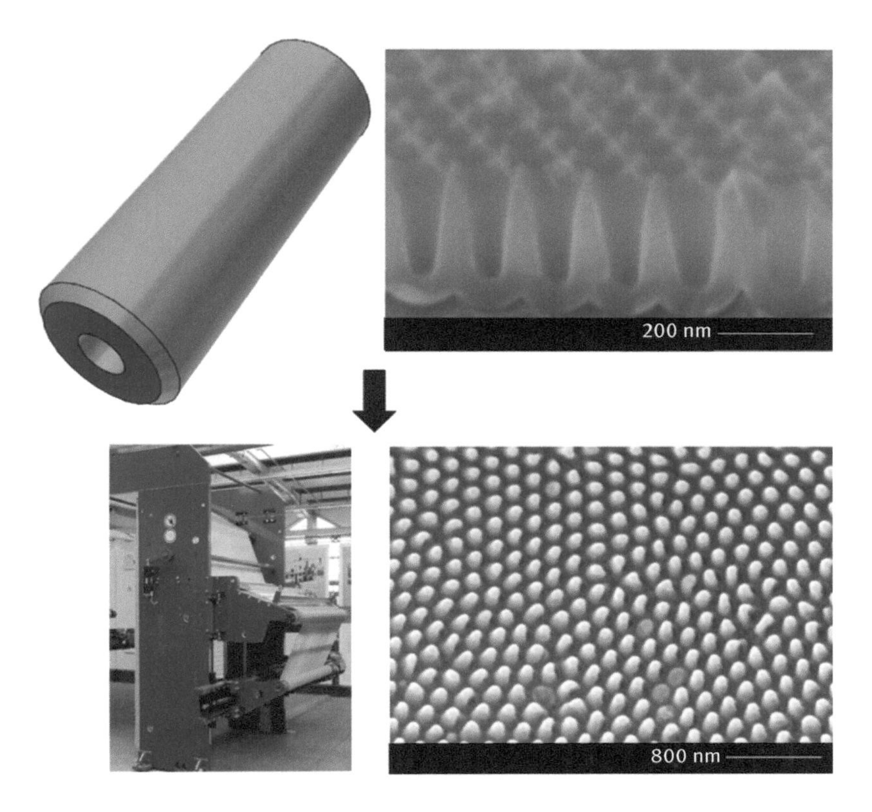

Figure 13.18: Porous alumina structure on aluminum cylinder for printing purposes on lacquers or foils.

for example, measuring moisture in gases [69, 70], for detecting ammonia [71] or hydrogen [72]. One of the main topics is using a thin porous oxide film as humidity sensors. Juhász and Mizsei created a humidity sensor with an integrated heating inside [73]. SmartMembranes also focuses on a new design as shown in Figure 13.19. The thin nanoporous alumina layer is prepared on a Si/SiO_2 standard semiconductor device with added electrodes for measurement of the capacitive changes. Once water is absorbed in the porous structure, a significant change in the capacity of the sensor is caused by the high dielectric constant of water in contrast to the one of air. The pore size has a direct influence on the sensitivity and the time-related responsive behavior. Heating is crucial to dry oxide layer and hence eliminate the drift caused by previous measurements.

The sensitivity of the thin nanoporous layer according to Juhász and Mizsei is one magnitude higher than other commercially available sensors due to the high surface area created by the porosity. Nanoporous alumina hence offers the option for new developments of miniaturized, highly sensitive devices with strongly improved performance.

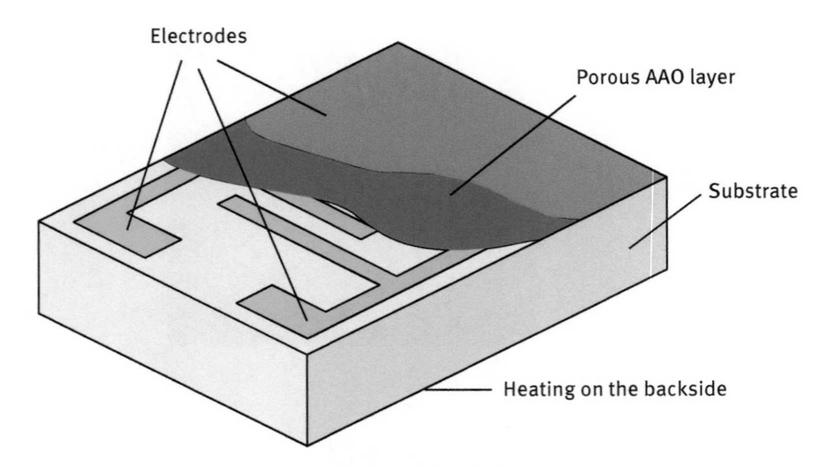

Figure 13.19: Schematic prototype of a humidity sensor with an integrated heating using nanoporous alumina as the active layer.

13.5 Conclusion

A basic introduction to the fabrication as well as applied aspects of macroporous silicon and nanoporous alumina were presented. Their outstanding material properties make both systems ideal candidates for great number of applications. The highly reproducible geometries can be fabricated using an inexpensive and well-controllable etching process.

The morphology can be tuned over a broad range (2 nm to a few microns) by modification of the anodization conditions as well as defined material properties.

Characteristics that occur due to the nanostructuring of bulk material, for example luminescence, dependence of the refractive index on the porosity, biodegradability and bioactivity, render porous silicon and alumina a material that can be exploited in optics, sensor technology, biomedicine and many more.

The focus of product development in the actual market in recent years is sensing and biomedical applications (gas sensors, biosensors, tissue engineering, controlled drug delivery and diagnostics).

References

1. Abd-Elnaiem AM, Gaber A. Parametric study on the anodization of pure aluminum thin film used in fabricating nano-pores template. Int J Electrochem Sci 2013;8:9741.
2. Pai Y-H, Tseng C-W, Lin G-R. Size-dependent surface properties of low-reflectivity nanoporous alumina thin-film on glass substrate. J Electrochem Soc 2012;159:E99.

3. Araoyinbo AO, Rahmat A, Derman MN, Ahmad KR. Room temperature anodization of aluminum and the effect of the electrochemical cell in the formation of porous alumina films from acid and alkaline electrolytes. Adv Mater Lett 2012;3:273.
4. Lohrengel M. Thin anodic oxide layers on aluminum and other valve metals: high field regime. Mater Sci Eng 1993;11:243.
5. Chazalviel, J-N, Wehrspohn R, Ozanam F. Electrochemical preparation of porous semiconductors: from phenomenology to understanding. Mater Sci Eng B 2000;69–70:1
6. Masuda H, Fukuda K. Cylindrical and spherical membranes of anodic aluminum oxide with highly ordered conical nanohole arrays. Science 1995;268:1466.
7. Kima TY, Jeong SH. Highly ordered anodic alumina nanotemplate with about 14 nm diameter. Korean J Chem Eng 2008;25:609.
8. Lehmann V, Föll H. Formation mechanism and properties of electrochemically etched trenches in n-type silicon. J Electrochem Soc 1990;137:653–9.
9. Lehmann V. The physics of macropore formation in low doped n-type silicon. J Electrochem Soc 1993;140:2836–43.
10. Lehmann V, Stengl R, Reisinger H, Detemple R, Theiss W. Optical shortpass filters based on macroporous silicon. Appl Phys Lett 2001;78:589–91.
11. Birner A, Wehrspohn R, Gösele U, Busch K. Silicon-based photonic crystals. Adv Mater 2001;13:377–88.
12. Schilling J, Wehrspohn RB, Birner A, Müller F, Hillebrand R, Gösele U, Leonard SW, Mondia JP, Genereux F, van Driel HM, Kramper P, Sandoghdar V, Busch K. A model system for two-dimensional and three-dimensional photonic crystals: macroporous silicon. J Opt A: Pure Appl Opt 2001;3:121–32.
13. Matthias S, Müller F, Jamois C, Wehrspohn RB, Gösele U. Large-area three-dimensional structuring by electrochemical etching and lithography. Adv Mater 2004;16:2166–70.
14. Betty C, Lal R, Sharma DK, Yakhmi JV, Mittal JP. Macroporous silicon based capacitive affinity sensor-fabrication and electrochemical studies. Sens Actuator B 2004;97:334–43.
15. Angelucci R, Poggi A, Dori L, Tagliani A, Cardinali GC, Corticelli F, Marisaldi M. Permeated porous silicon suspended membrane as sub-ppm benzene sensor for air quality monitoring. J Porous Mater 2000;7:197–200.
16. Everett DH. IUPAC, manual of symbol and terminology for physicochemical quantities and units, appendix, definitions, terminology and symbols in colloid and surface chemistry, part I. Pure Appl Chem 1972;31(4):577.
17. Chazalviel J-N, Ozanam F. Macropores in p-type silicon. In: Wehrspohn RB, editor. Ordered porous nanostructures and applications, Chapter 2. New York: Springer, 2005:15–35.
18. Uhlir A. Bell Syst Technol J 1956;35:333.
19. Wehrspohn R, Chazalviel J-N, Ozanam F, Solomon I. Thin Solid Films 1997;5:297.
20. Zhang XG. Morphology and formation mechanisms of porous silicon. J Electrochem Soc 2004;151:C69.
21. Kolasinski KW. Etching of silicon in fluoride solutions. Surf Sci 2009;603:1904–11
22. Lehmann V. Gösele U. Porous silicon formation: a quantum wire effect. Appl Phys Lett 1991;58(8):856–8.
23. Canham LT. Silicon quantum wire array fabrication by electrochemical and chemical dissolution of wafers. Appl Phys Lett 1990;57(10):1046–8.
24. Lehmann V, Stengl R, Luigart A. On the morphology and the electrochemical formation mechanism of mesoporous silicon. Mater Sci Eng B 2000;69–70:11–22.
25. Lehmann V, Rönnebeck S. The physics of macropore formation in low-doped p-type silicon. J Electrochem Soc 1999;146(8):2968–75.

26. Anodic oxidation of aluminum and its alloys, In Information Bulletin, vol. 14. London: The Aluminum development association, 1948.
27. British Patent 223, 994, 1923.
28. Italian Patent 741, 753, 1936.
29. O'Sullivan JP, Wood GC. Morphology and mechanism of formation of porous anodic films on aluminium. Proc R Soc Lond Ser A – Math Phys Sci 1970;317:511–43.
30. Masuda H, Fukuda K. Ordered metal nanohole arrays made by a two-step replication of honeycomb structures of anodic alumina. Science, 1995;268:1466–8.
31. Masuda H, Yamada H, Satoh M, Asoh H, Nakao M, Tamamura T. Highly ordered nanochannel-array architecture in anodic alumina. Appl Phys Lett 1997;71:2770–2
32. Hanaoka T-A, Heilmann A, Kröll M. Kormann H-P, Sawitowski T, Schmid G, Jutzi P, Klipp A, Kreibig U, Neuendorf R. Appl Organometal Chem 1998;12:367.
33. Asoh H, Ono S. Electrocrystallization in nanotechnology. Weinheim: Wiley-Vch, 2007:138–66.
34. Lee W, Park S-J. Chem Rev 2014;114: 7487–556.
35. Jessensky O, Müller F, Gösele U. Appl Phys Lett 1998;72:1173.
36. Diggle JW, Downie TC, Coulding CW. Anodic oxide films on aluminum. Chem Rev 1969;69:365–405.
37. T. Våland, Heusler KE. J Electroanal Chem 1983;149:71.
38. Parkhutik VP, Shershutsky VI. J Phys D: Appl Phys 1992;25:1258.
39. Lohrengel M. Thin anodic oxide layers on aluminum and other valve metals – high-field regime. Mater Sci Eng Rep 1993;11:243–94.
40. Thompson GE, Xu Y, Skeldon P, Shimizu K, Han SH, Wood GC. Anodic oxidation of aluminum. Philos Mag B 1987;55:651–67.
41. Li F. Nanostructure of anodic porous alumina films of interest in magnetic recording. Tuscaloosa: The University of Alabama, 1998.
42. Jessensky O. Untersuchungen zum Porenwachstum in 6H-Siliziumkarbid und anodischem Aluminiumoxid. Halle: Martin-Luther-Universität Halle-Wittenberg, 1997.
43. Nielsch K. Hochgeordnete ferromagnetische Nano-Stabensembles: Elektrochemische Herstellung und magnetische Charakterisierung. Halle: Martin-Luther-Universität Halle-Wittenberg, 2002.
44. Wehrspohn RB, Li AP, Nielsch K, Müller F, Erfurth W, Gösele U. Highly ordered alumina films: pore growth and applications. In: Hebert KR, Lillard RS, Mac Dougall BR, editors. Oxide films in the electrochemical society proceeding series. vol. PV 2000-4, pp. 271. Pennington: Marcel Dekker, 2000.
45. Choi J. Fabrication of monodomain porous alumina using nanoimprint lithography and its applications. Halle: Martin-Luther-Universität Halle-Wittenberg, 2003.
46. Asoh H, Nishio K, Nakao M, Yokoo A, Tamamura T, Masuda H. J Vac Sci Technol B 2001;19:569.
47. Li AP, Müller F, Birner A, Nielsch K, Gösele U. J Appl Phys 1998;84:6023.
48. Masuda H, Hasegwa F, Ono S. Self-ordering of cell arrangement of anodic porous alumina formed in sulfuric acid solution. J Electrochem Soc 1987;144:L 127–30.
49. Masuda H, Yada K, Osaka A. Self-ordering of cell configuration of anodic porous alumina with large-size pores in phosphoric acid solution. Jpn J Appl Phys Part 2: Lett 1988;37:L 1340–2.
50. Li AP, Müller F, Birner A, Nielsch K, Gösele U. Hexagonal pore arrays with a 50–420 nm interpore distance formed by self-organization in anodic alumina. J Appl Phys 1988;84:6023–6.
51. Nielsch K, Choi J, Schwirn K, Wehrspohn RB, Gösele U. Self-ordering regimes of porous alumina: the 10% porosity rule. Nano Lett 2002;2(7):677–80.
52. Zhang J, Kielbasa JE, Carroll DL. Controllable fabrication of porous alumina templates for nanostructures synthesis. Mater Chem Phys 2010;122:295–300.
53. Jessensky O, Müller F, Gösele U. J Electrochem Soc 1998;145:3735.
54. Kopp O, Lelonek M, Knoll M. The influence of pore diameter on bifurcation and termination of individual pores in nanoporous alumina. J Phys Chem C 2011;115(16):7993–6.

55. Lelonek M, Kopp O, Knoll M. Pore bifurcation, growth and pore termination in nanoporous alumina with concave and convex surfaces. Electrochim Acta 2009;54(10):2805–9.
56. Schwartz GC, Platter V. Anodic process for forming planar interconnection metallization for multilevel LS I. J Electrochem Soc 1975;122:1508–16.
57. Letant SE, Hart BR, van Buuren AW, Terminello LJ. Functionalized silicon membranes for selective bio-organism capture. Nat Mater 2003;2:391–5.
58. Matthias S, Müller F. Asymmetric pores in a silicon membrane acting as massively parallel Brownian ratchets. Nature 2003;424:53–7.
59. Lehmann V. Barcoded molecules. Nat Mater 2002;1:12–13.
60. Lehmann V. Trends in fabrication and applications of macroporous silicon. Phys Stat Sol A 2003;197:13–15.
61. Steinhart M, Wehrspohn RB, Gösele U, Wendorff JH. Nanotubes by template wetting: a modular assembly system. Angew Chem Int Ed 2004;43:1334–44.
62. Sauer G, Göring P, et al. Surface-enhanced Raman spectroscopy employing monodisperse nickel nanowire arrays. Appl Phys Lett 2006;88:023106.
63. Steinhart, Göring P, et al. Coherent kinetic control over crystal orientation in macroscopic ensembles of polymer nanorods and nanotubes. Phys Rev Lett 2006;97:027801.
64. Poly(vinylidene difluoride) (Aldrich, M_w = 180 000 g/mol, M_n = 71 000 g/mol) infiltrated into self-ordered AAO (D_p = 400 nm, T_p = 50 μm), with permission from Steinhart M, Senz S, Wehrspohn RB, Gösele U, Wendorff JH. Curvature-directed crystallization of poly(vinylidene difluoride) in nanotube walls. Macromolecules 2003;36: 3646–51.
65. Choi K, Park SH, Song YM, Lee YT, Hwangbo CK, Yang H, Lee HS. Nano-tailoring the surface structure for the monolithic high-performance antireflection polymer film. Adv Mater 2010;22:3713–18.
66. Masuda H, et al. Antireflection polymer surface using anodic porous alumina molds with tapered holes. Chem Lett 2007;36(4):530.
67. FP7 European project aiming at developing advanced magnetic materials suitable for designing a data storage solution in three dimensions (3D), http://mem3d.eu/
68. Santos A, Kumeria T, Losic D. Review: nanoporous anodic alumina: a versatile platform for optical biosensors. Materials 2014;7:4297–320.
69. F. Casanova1, Chiang C, C.-P. Li, Roshchin I, Ruminski A, Sailor M, Schuller I. Gas adsorption and capillary condensation in nanoporous alumina films. Nanotechnology 2008;19:315709.
70. Basu S, Chatterjee S, et al. Study of electrical characteristics of porous alumina sensors for detection of low moisture in gases. Sens Actuators B 2001;79: 182–6.
71. Dickey EC, Varghese OK, Ong KG, Gong D, Paulose M, Grimes CA. Room temperature ammonia and humidity sensing using highly ordered nanoporous alumina films. Sensors 2002;10: 91–110.
72. Rumiche F, Wang HH, Hu WS, Indacochea JE, Wang ML. Anodized aluminum oxide (AAO) nanowell sensors for hydrogen detection. Sens Actuators B 2008;134:869–77.
73. Juhász L, Mizsei J. A simple humidity sensor with thin film porous alumina and integrated heating. Proc Eurosensors 2010;XXIV:701–4.

Atasi Dan, Bikramjit Basu, Subhajit Roychowdhury, Kanishka Biswas and Baldev Raj†

14 Nanotechnology and energy conversion: A solution using spectrally selective solar absorbers and thermoelectrics

14.1 Introduction

The world is experiencing a global energy crisis which is deeply affecting most of the countries in the world and threatening the existence of civilization [1, 2]. Nanotechnologies [3] as key and cross-sectional technologies introduce numerous technological breakthroughs in energy sectors to provide substantial contribution in renewable energy supply. The inventions in nanotechnology have paved the way to move beyond conventional energy generation approaches in a more productive, sustainable, environment-friendly and cost-effective manner. It should be emphasized that nanotechnology has a remarkable possibility to reduce the dependence on depletable source of conventional energy. Also, nanotechnology can extensively enhance the efficiency, storage and conservation of the green energy-based systems [4]. Nanoengineered materials can also play an important role in the development of high energy density systems, renewable and low-cost Li-ion [5], Li-S [6], Li-air [7] and other type batteries (e.g., supercapacitors [8] and hydrogen storage materials [9, 10]). The safety of Li-ion batteries can be improved by adopting several approaches including use of safety vents [11], positive temperature coefficient elements, shutdown separators [12], less flammable electrolytes [13] and redox shuttles [14]. The processing and packaging, safety, environmental fallout and so on are next frontiers toward large-scale commercialization of Li-ion batteries [15]. Also, nanotechnology-based innovations enable the development of state-of-the-art cost-effective components as well as systems with outstanding energy efficiency and noteworthy performance. Recently, nano-enabled coatings are also interesting research topics for energy applications [16–18]. Such coatings incorporate nanostructured and nanofunctional materials [19], such as nanoengineered steels [20, 21], nano-enhanced lightweight for transport applications [22], nanostructured surfaces and nanocoatings, nanolubricants [23] and nanosensors [24]. The surface area related property improvement in nanomaterials is also utilized in other functional applications, such as nanopores and their morphology control in double-layer supercapacitors [25], superhydrophobic surfaces [26] for corrosion mitigation and improving performance of steam condensers, specificity of nanoparticles along with understanding taxonomic and physiological types of microbial species for enhancing efficiency of conversion of biowaste and biomass to gaseous energy recovery [27].

https://doi.org/10.1515/9783110547221-014

Catalysts with shell and core material structures with nano-thin coating of noble metal can contribute in realizing cost-effectiveness of vehicle emission control systems, fuel cells of electrolyzers and so on [28].

Solar thermal technology, also known as concentrated solar power (CSP) [29], is an advanced technology that has a great potential to usher in universal energy access. Such systems have emerged as a significant player in renewable energy production by focusing solar irradiation using a number of mirrors into a spectrally selective absorber to heat up a liquid, solid or gas to several hundreds of degrees Celsius, is then used in a downstream process for electricity generation [30]. Recently, thermoelectric materials also have been investigated extensively as it can directly and reversibly convert the temperature difference to electricity. A constant source of thermal energy can generate temperature gradient in thermoelectric legs. Solar thermoelectric generator (STEG) [31] consists of thermoelectric materials in between a solar absorber and a heat sink to produce a temperature gradient which in turn generates electricity. STEG systems are achieving significant interest in both concentrated [32] and nonconcentrated [33, 34] solar power systems (see Figure 14.1).

Solar collectors that are used in nonconcentrated systems can be stationary, and tracking mechanism is not required in such systems while a strong optical setup is needed in concentrated systems to focus solar radiation. It is also important to mention that in nonconcentrating collectors the temperature reaches from ambient to 240 °C, whereas the operating temperature for concentrating collector is almost up to 1,500 °C. The solar selective absorber coatings on collectors in CSP systems generate heat energy by utilizing the entire solar spectrum whereas photovoltaic cells convert only a small portion of solar energy into electric power. A lot of research is going on to combine the CSP and thermoelectric technology. Figure 14.2 shows a schematic diagram of STEG systems. STEG systems have many advantages compared to other conventional electrical power generator systems, since it does not require any moving parts and produce electricity from solar energy. Additionally, STEG technology can be used for both small- and large-scale applications. However, these hybrid systems have relatively low conversion efficiencies, so their applications have been usually

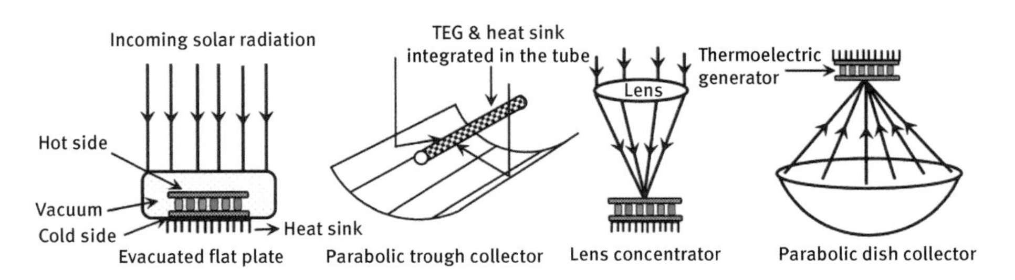

Figure 14.1: Thermoelectric generator setup in different solar thermal systems [33].

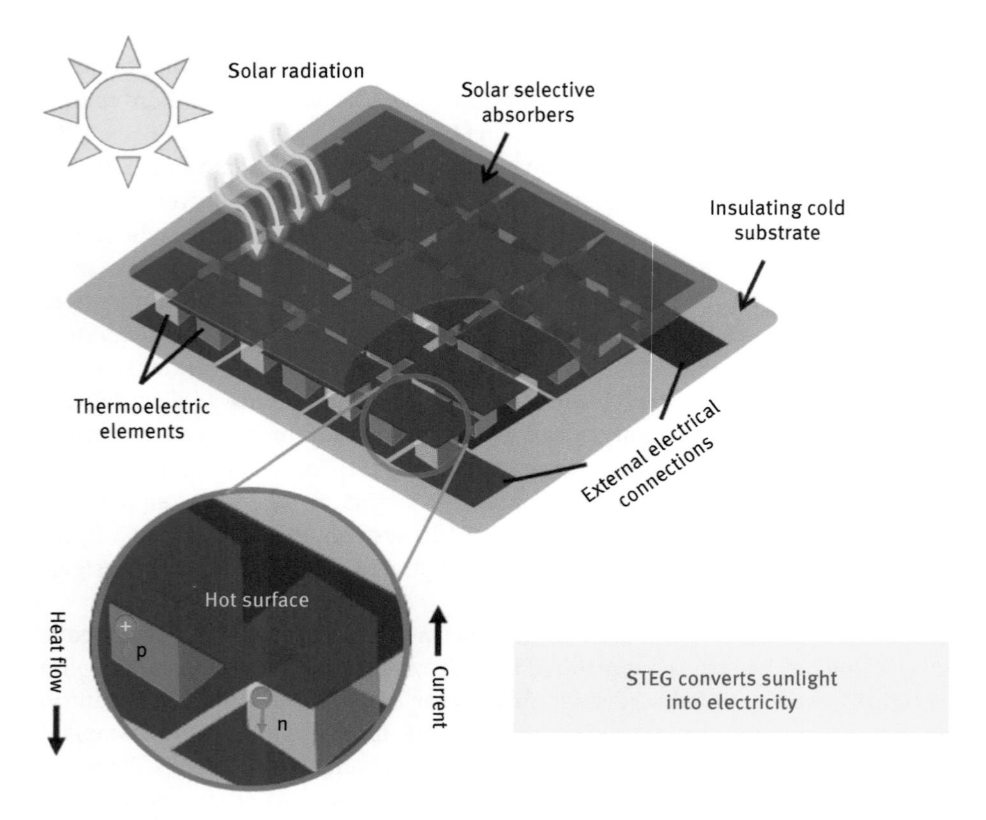

Figure 14.2: Conversion of solar energy to electricity using STEGs.

limited to the specific situations where reliability is a major consideration such as in military applications and aerospace [35].

The methodologies and achievements of nanotechnology make us optimistic to set our goals high in developing nanostructured thermoelectric materials that empower energy to be recuperated from waste heat in the form of electricity. The products based on thermoelectric materials provide next revolution by perceiving a promise to fabricate cost-effective and pollution-free energy conversion systems. In parallel, nanocrystalline and nanolayered spectrally selective coatings with high solar absorptance and low thermal emittance will address the transformation of the solar energy into more useful form of electrical energy in the promising STEG systems. In this chapter, the authors have ensured coverage of different classes of solar selective coatings and thermoelectric materials to demonstrate the importance of nanomaterials and nanotechnology in renewable energy applications and economic growth. This chapter also demonstrates the legitimate analysis on issues pertaining to acknowledge the role of nanotechnology in improving long-term durability of such systems. In addition, the future directions have also been discussed, which can be appreciated and utilized to maximize harnessing of nanotechnologies for green and renewable energy.

14.2 Solar Thermoelectric Generator

Solar thermal technologies along with thermoelectric devices have been used for electrical power generation since the nineteenth century. STEGs need a thermoelectric generator, a collector and a heat sink. Incident solar flux on the thermoelectric generator depends on collector options, such as evacuated flat plate, parabolic troughs, Fresnel lenses and parabolic dishes (Figure 14.1). For an evacuated system, the maximum efficiency of an STEG can be expressed as follows [36]:

$$\eta_{\text{STEG}} = \left[\tau_g \alpha \eta_{\text{op}} - \frac{\varepsilon\sigma(T_H^4 - T_C^4)}{CR \times G(\lambda)}\right] \times [\,\eta_{\text{TEG}}\,] \tag{14.1}$$

where τ_g, α, η_{op}, ε, σ $G(\lambda)$ and η_{TEG} are the transmittance of the glass enclosure, absorptance of the selective surface to the solar flux, optical concentration efficiency, effective emittance of the absorber and the envelope, Stephen Boltzmann constant, the incident solar flux and efficiency of STEG, respectively.

Telkes [31] was the pioneer for using thermoelectric material in solar thermal technology in 1954. Combination of p-type ZnSb (Sn, Ag, Bi), an n-type 91% Bi + 9% Sb and a double-paned flat plate collector in an STEG with a 70°C temperature difference provided efficiency of approximately 0.63% [31]. The efficiency of STEG enhances to 3.35% after using concentrated system with lens for a temperature difference of 523 K. Kraemer et al. [37] used flat panel collector system inside an evacuated glass chamber which achieved a peak efficiency of 4.6% at 1 kW/m^2. Amatya et al. [38] developed concentrated STEG system with 3% efficiency. However, Goldsmid et al. [39], Omer et al. [40] and Suter et al. [41] also developed concentrated STEG systems, but achieved low system efficiencies. Recently, Arturo et al. [42] performed an experimental study on a solar concentrating system based on thermoelectric generators. They have used six serially connected Bi_2Te_3-based thermoelectric generators. A solar tracking system with concentrator made of a mosaic set of mirrors was used. The maximum electric efficiency of the system is 5% with a 50°C temperature difference between hot and cold sides of thermoelectric generator [42].

The efficiency of STEGs can be further enhanced by the use of high-performance nanostructured thermoelectric materials and highly efficient spectrally selective solar absorbers. In this chapter, we will discuss about the development of three different types of spectrally selective absorbers. We will also demonstrate the synthesis of highly efficient thermoelectric materials.

14.3 Spectrally Selective Absorber Materials

Spectrally selective absorber coating is one of the key components of all the solar thermal systems [43]. These coatings should be capable of capturing maximum amount of solar radiation with a very less infrared emission. Therefore, one should aim at designing a thin film to have the highest optical absorptance ($\alpha \geq 0.95$) in solar spectrum (0.25–2.5 μm) and a minimum emittance ($\varepsilon \leq 0.05$) in the infrared range (2.5–25 μm) to avoid radiative heat losses [44]. However, the main challenge of the absorber lies with the performance of the coating in adverse environment and at high temperature.

14.3.1 Intrinsic Absorbers

As the name suggests, intrinsic absorbers such as V_2O_5 [45], LaB_6 [46], Fe_3O_4 [47] and Al_2O_3 [48], and few ultra-high temperature ceramics such as ZrB_2 [49], ZrC, TaC, HfC [50], SiC, ZrB_2 and HfB_2 [51] possess selectivity as inherent property. These materials have been researched as potentially suitable candidates for solar selective applications. Recently, Chen et al. [52] have investigated the intrinsic selective property of three types of carbon nanotubes (N-CNT, P-CNT and T-CNT), which were electrophoretically developed on aluminum using kinetically stable CNT aqueous suspensions. A absorptance of 0.79 and emittance of 0.14 were achieved by T-CNT absorber. N- and P-CNT absorbers have a better spectral selectivity, with $\alpha = 0.90$ and $\varepsilon = 0.14$ for N-CNT absorber, $\alpha = 0.90$ and $\varepsilon = 0.13$ for P-CNT absorber. In a different study, the selective properties ($\alpha = 0.76$–0.85 and $\varepsilon = 0.05$–0.1) of chemically synthesized Fe_3O_4 nanoparticles were explored by González et al. [47], while deposited by dip coating method on Cu substrates. Doping enhances the intrinsic selectivity by two ways: first, the donor atoms give rise to electron plasma, and second, these donors act as a scattering center, which enhances the optical path of solar radiation and leads to maximum absorptance. For example, the doping of W in VO_2 matrix ($V_{1-x}W_xO_2$) makes it suitable as solar thermal absorber [53].

14.3.2 Multilayer Absorbers

In recent years, the coatings with the magnetron sputtered multilayer structures have attracted attention in high-temperature applications. A multilayer coating consists of several layers of transition metal oxides, nitrides, oxynitrides, carbides and/or silicides. For example, Ning et al. [54] have developed Mo/ZrSiN/ZrSiON/SiO_2 coating by magnetron sputtering which has a high solar absorptance of 0.94 and a low thermal emittance of 0.06. The optimized thicknesses of the layers from substrate to surface were 180, 51, 65 and 110 nm, respectively. The optical properties of the coating were stable at 500°C in vacuum for 500 h. Recently, a W/WAlN/WAlON/Al_2O_3-based spectrally selective absorber has been developed by our group on stainless steel substrate by direct current (DC) and radio frequency (RF) magnetron sputtering. Such coating with a sky blue appearance has a high absorptance of 0.958 and a low emittance of 0.08 [55]. The coating has also a wide angular absorptance up to an incidence angle of 58° [56]. It is worthwhile to mention that the fabrication of a spectrally selective multilayer coating by magnetron sputtering is extremely challenging while maintaining thermal stability of the layered structure for prolonged duration. In this coating, W layer serves the role of diffusion barrier to reduce the emittance while WAlN layer acts as main absorber layer and WAlON as the semiabsorber layer. The top dielectric Al_2O_3 layer with the lowest refractive index (n = 1.65) is used as antireflection layer. The layer thicknesses were approximately 87, 50, 23 nm, respectively, for WAlN, WAlON and Al_2O_3 layers with a total thickness of approximately 160 nm.

Figure 14.3 represents an overview of the spectrally selective properties of the entire stack. The coating also exhibited an outstanding thermal stability at 500°C in air for 150 h without any noticeable change in solar absorptance and thermal emittance (α/ε = 0.918/0.11). Long-term thermal stability tests confirmed that the expected service lifetime of the coatings is more than 25 years [57, 58]. Figure 14.4 represents the spectral performance along with morphological changes as well as chromatic appearance in high temperature. Several other multilayer coatings, such as TiAlSiN/TiAlSiON/SiO_2 [59], NbAlN/NbAlON/Si_3N_4 [60], NbMoN/NbMoON/SiO_2 [61], AlSiN/AlSiON/$AlSiO_y$ [62] and TiAlCrN/TiAlN/AlSiN [63] have been investigated. The concept of gradation in metallic property and refractive index from substrate to top surface in these multilayer coatings help to capture maximum sunlight by destructive interference mechanism [64].

14.3.3 Dielectric/Metal/Dielectric Absorbers

In order to absorb solar radiation, a different approach is adopted to sandwich a metallic layer (Cr, Mo, Al, Pt, etc.) between two stable nanocrystalline and amorphous dielectric layers (Cr_2O_3, MgO, Al_2O_3, etc.). Such a coating structure is known as dielectric/metal/dielectric absorber and can be employed to improve the absorption by the use of surface plasmon polaritons.

In response to the growing demand for the industrial use of Cr_2O_3 or black chrome [65, 66] in the field of solar selective coating, Barshilia et al. [67] first investigated Cr_xO_y/Cr/Cr_2O_3-based coating using asymmetric bipolar-pulsed DC generators. They observed that Cr_xO_y/Cr/Cr_2O_3 (28 nm/13 nm/64 nm) has a very high selectivity with an absorptance of 0.899–0.912 and a low emittance of 0.05–0.06. Nuru et al. [68] developed MgO/Zr/MgO-based solar-selective coating on Zr-coated stainless steel (SS) substrate. The absorptance and the emittance of the film were 0.92 and 0.09, respectively.

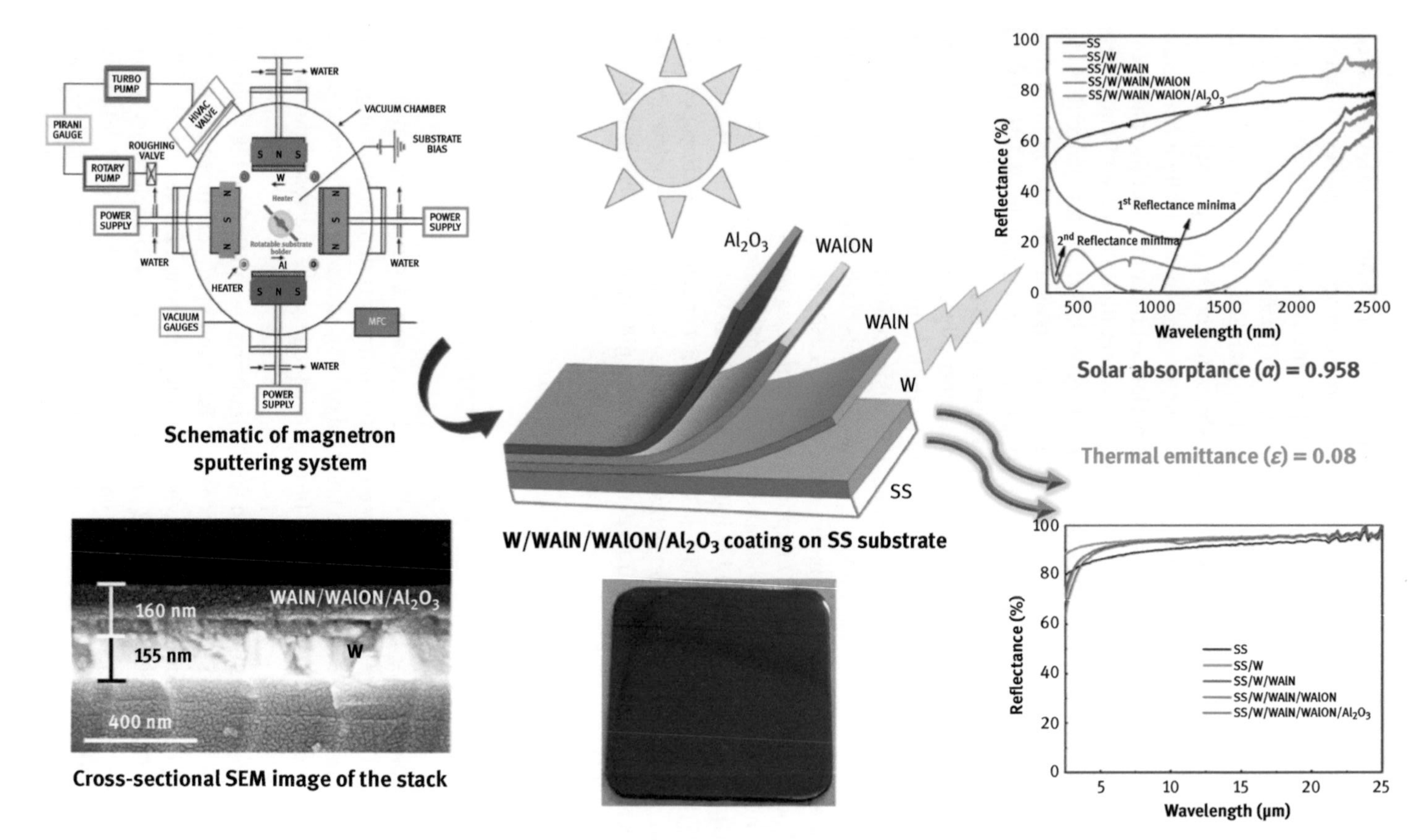

Figure 14.3: Spectrally selective properties of $W/WAlN/WAlON/Al_2O_3$-based absorber coating deposited by magnetron sputtering on stainless steel substrate [56].

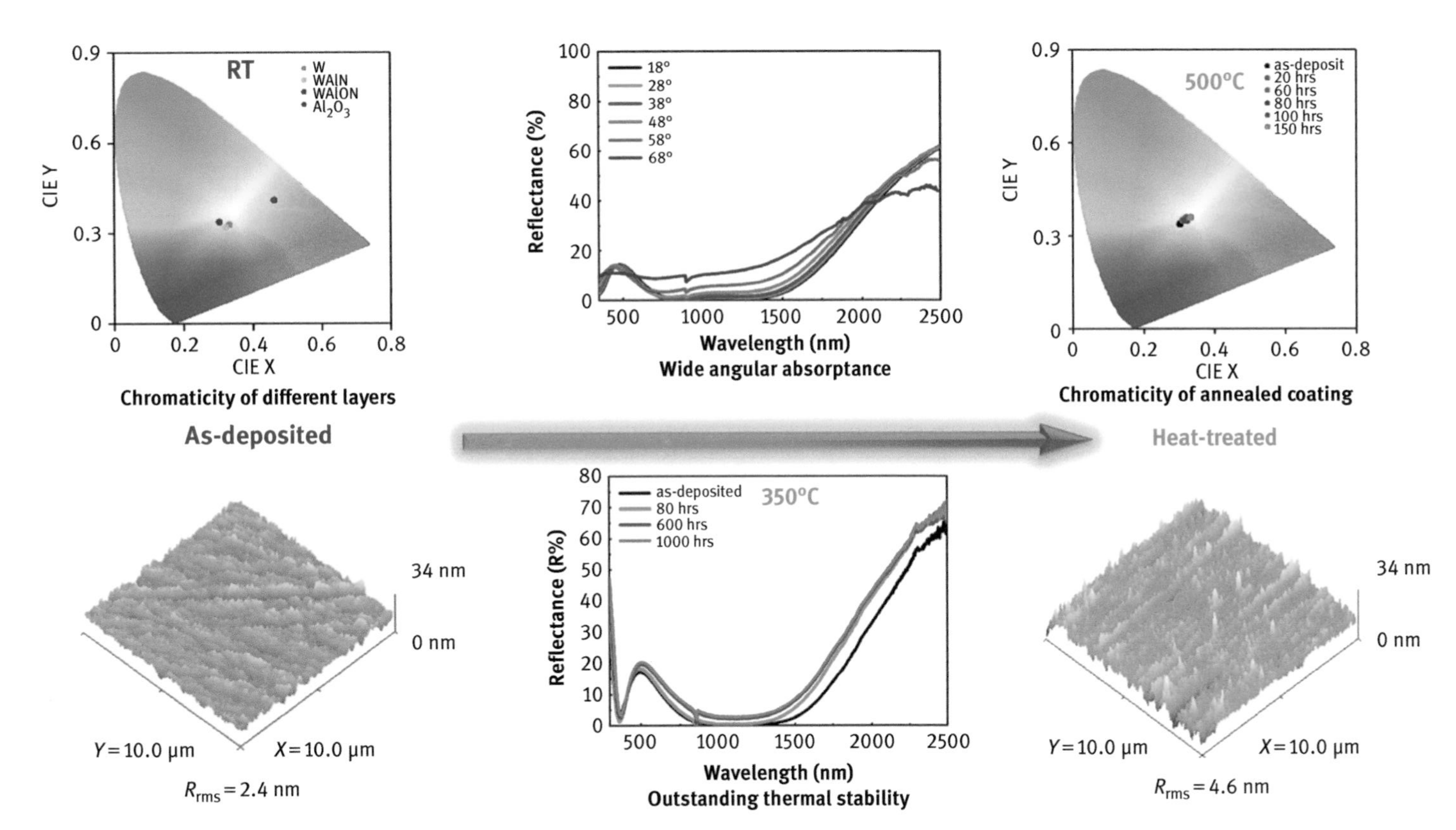

Figure 14.4: Thermal stability of W/WAlN/WAlON/Al_2O_3-based absorber coating.

In a separate study, Tsai et al. [69] demonstrated the optical properties of Al_xO_y (70 nm)/Ni (20 nm)/Al_xO_y (70 nm) coating, which exhibited highest selectivity with an absorptance of 0.932 and emittance of 0.038. It is worth mentioning that the `target power and oxygen flow rate during sputtering also influence the structure, surface morphology and selective properties of the thin film. The transmission electron microscopic images in Figure 14.5(A)–(C) indicate the loose and rough coating morphology, deposited at lower oxygen flow rate (2 sccm), while the coating fabricated at higher oxygen flow rate (8 sccm) appears to be dense with a very smooth surface. Additionally, an increase in target power from 150 to 250 W provides a film with looser and rougher surface. The reason behind such morphological changes can be attributed to the change in Al_xO_y deposition rate as it increases with an increase in sputtering power and decreases with an increase in oxygen flow rates. It is also important to note that loose coating with voids exhibits lower transmittance, that is, higher absorptance than that of dense coating as the voids and loose pattern assist in enhancing the light absorption (see Figure 14.5(D)). Different types of spectrally selective coatings along with optical properties and thermal stability have been summarized in Table 14.1.

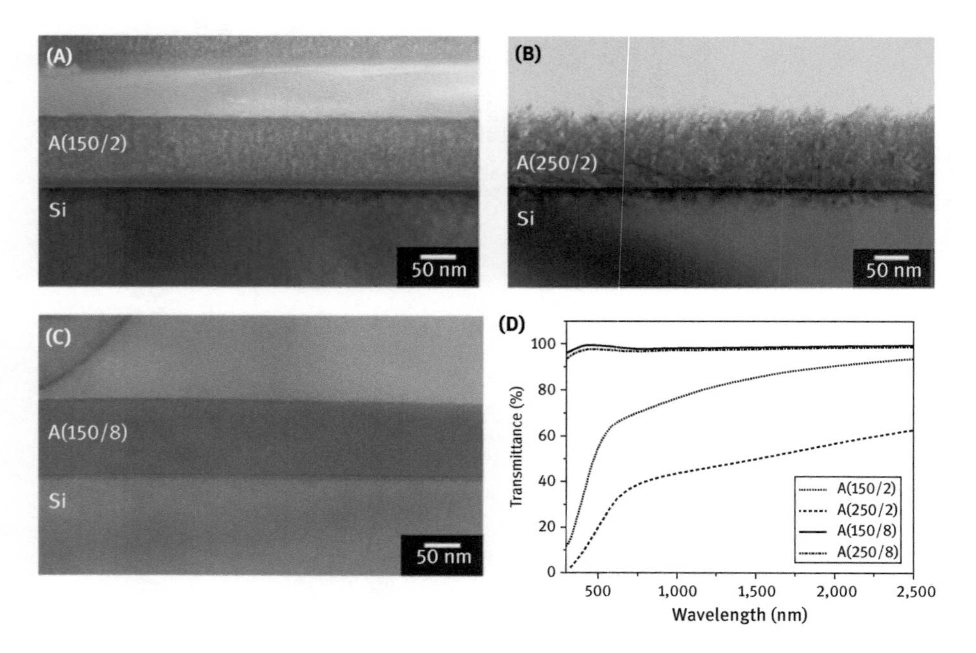

Figure 14.5: Transmission electron microscopic cross-sectional images of Al_xO_y coatings deposited using different DC powers and oxygen flow rates; A (*x*/*y*) represents Al_xO_y coating (DC power/oxygen flow rate) (A) A (150/2), (B) A (250/2) and (C) A (150/8); (D) transmittance spectra of as-deposited Al_xO_y coatings, fabricated in different deposition conditions [69].

Table 14.1: Optical properties and thermal stability of different types of spectrally selective absorber coating.

Solar absorbers	Type	Deposition method	Absorptance (α)	Emittance (ε)	Thermal stability in air		References
					Temperature (°C)	Duration (h)	
N-CNT	Intrinsic	Electrophoretic deposition	0.90	0.14	500	–	[53]
P-CNT	Intrinsic	Electrophoretic deposition	0.90	0.13	500	–	[53]
Fe_3O_4	Intrinsic	Dip coating	0.76–0.85	0.05–0.1	–	–	[48]
$Si_{0.8}Ge_{0.2}$	Semiconductor	Drop casting	0.90–0.95	<0.3	750	1	[79]
Mo/ZrSiN/ZrSiON/SiO_2	Multilayer	Sputtering	0.94	0.06	500 (vacuum)	500	[55]
W/WAlN/WAlON/Al_2O_3	Multilayer	Sputtering	0.958	0.08	500	150	[58]
Mo-SiO_2	Cermet	Sputtering	0.94	0.05	450	–	[80]
Ni-nanochain-Al_2O_3	Cermet	Spin coating	>0.90	<0.10	–	–	[81]
1D CuO nanofibers and nanoneedles	textured	Chemical oxidation	0.95	0.07	–	–	[82]
V-groove gratings coated with aperiodic metal-dielectric stacks	textured	Rigorous coupled-wave analysis	>0.94	<0.06	–	–	[83]
Cr_xO_y/Cr/Cr_2O_3	Dielectric/metal/dielectric	Sputtering	0.899–0.912	0.05–0.06	250	250	[84]
MgO/Zr/MgO	Dielectric/metal/dielectric	Sputtering	0.92	0.09	250	24	[68]
Al_xO_y /Ni/Al_xO_y	Dielectric/metal/dielectric	Sputtering	0.932	0.038	400	12	[69]

14.4 Surface Engineering of Superhydrophobic Coatings for Solar Energy-Harvesting Systems

The superhydrophobic coatings are developed on a wide variety of materials, including titanium [70–74]. However, these hybrid systems have relatively low conversion efficiencies, so their applications have been usually limited to the specific situations where reliability is a major consideration such as in military applications and aerospace chrome moly steels, [75, 76] glass slides [77] and marine steels [78]. Alumina on ZnO for near-complete transmission on the optimized coatings is a promising superhydrophobicity surface. The accumulation of dust and sand on solar power reflectors and photovoltaic cells is one of the main efficiency drags for solar power plants, capable of reducing reflectivity up to 50%. The cover glass of the solar cell coated with the film becomes superhydrophobic, in which water rolls off easily from the surface, carrying dirt and grime with it and allowing more solar energy to be absorbed. Though currently there are several coating techniques available, most of them are based on vapor deposition that require high vacuum and are cost intensive. Therefore, low-cost, environmental-friendly coatings based on superhydrophobicity that is amenable to scale up in addition to providing multifunctional properties for solar reflectors is necessary to optimize energy efficiency.

Corrosion Science and Technology Division, IGCAR, has been involved in the development of lotus effect-based superhydrophobic coatings for the past 7 years. One of the authors (Baldev Raj and his coauthors) have developed superhydrophobic surfaces based on ZnO [70, 71] and aluminum oxide [72] coatings on glass substrates via solution-based approach for solar panel cover glass applications (Figure 14.6). The fabricated surface was amorphous with an interconnected porous network of nanoflakes. The static contact angle of the prepared coatings was 161° and exhibited superior self-cleaning behavior at a tilting angle less than 10°. The interconnected network of coatings exhibited average transmittance level of 95%.

The photovoltaic performance of commercially available solar cells covered with uncoated glass substrates and superhydrophobic glass substrates was measured under various conditions (such as fabricated, artificially contaminated and self-cleaned conditions) and the results were compared. The uncoated glass substrates and aluminum oxide-coated superhydrophobic glass substrates recovered the efficiency of sawdust-contaminated solar panels by 67% and 91%, respectively, thereby enabling the fabricated superhydrophobic glass substrates to be effectively useful for self-cleaning cover glass applications.

It is believed that the development of transparent, mechanically robust superhydrophobic coatings will eventually solve the challenging issues regarding dust accumulation in near future. Similar strategies can be adopted to self-clean the solar absorber surfaces for STEG systems. The combination of micro- and nanostructures, together with a hydrophobic chemistry, generates the phenomenon of superhydrophobicity with water droplets on such surfaces exhibiting contact angles above 150°.

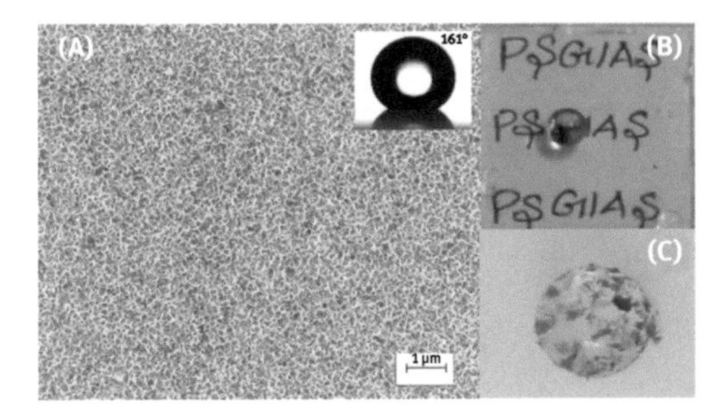

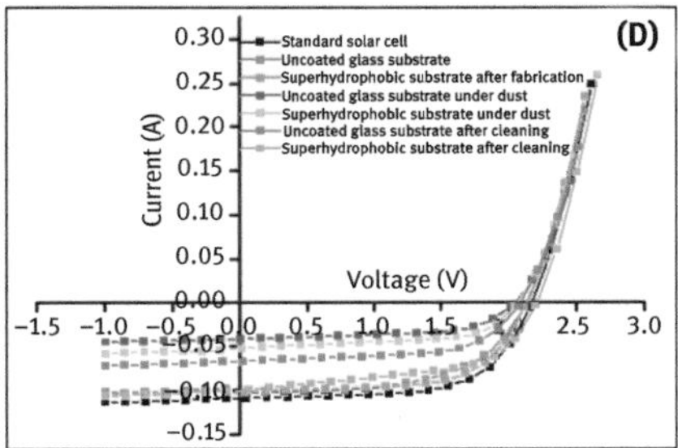

Figure 14.6: (A) Surface morphology of superhydrophobic alumina coating (inset shows water-surface contact angle 161°); (B) fully transparent superhydrophobic alumina surfaces; (C) self-cleaning nature of superhydrophobic ZnO surface; (D) solar simulation studies on transparent superhydrophobic alumina surfaces; the saw dust contaminated uncoated glass substrate recovered only 67% of efficiency of solar panel after cleaning with water. Aluminum oxide-coated superhydrophobic surfaces recovered 91% efficiency of solar panel after cleaning with water.

It will be exciting to enhance performance and applications of different types of absorber coatings evolution in terms of selective property requirements (α and ε), reliability of the materials and testing requirements. In particular, the constant demand on renewable energy can be satisfied with enormous research effort by making the STEG systems competitive with the conventional power generation. It can be expected that the combined efforts into the further advancement on fabrication procedures, structure and material optimization along with performance testing for practical application of spectrally selective absorbers will have noteworthy impact in developing superior STEG systems.

14.5 Thermoelectric Materials

Recently, significant interest has grown for energy generation, conversion, storage, and management due to emerging global need in renewable energy sources. Enrichments to the existing energy supply must emerge from different renewable sources such as solar, wind, hydropower, biomass and others. Motivated by the requirement for efficient, clean and sustainable energy sources, thermoelectrics has become an essential part of the research portfolio seeking to identify new and efficient energy materials for power generation and cooling. The major share of the energy was consumed after generation by the use of coal, petroleum, natural gas, nuclear and coal. To run the transportation and industrial sector, combustions of coal, petroleum and gas are used. Importantly, ~60–65% of the utilized energy is being lost as waste heat. Thermoelectric materials are the all solid-state converters without any moving part which can directly and reversibly convert heat energy into electrical energy [85, 86].

Over the last two decades, there has been an escalated research interest in the field of thermoelectric materials and devices. The application of thermoelectric materials in the industrial and defense applications are creating enhanced activity in this field arising from demands of high-performance low-cost materials. More recently, research on solar-thermoelectric is gaining attention to utilize the infrared part (heat) of the solar spectrum, which can be converted to electricity by thermoelectric devices [34]. Novel applications of thermoelectrics are biothermal batteries to power heart pacemakers, wearable devices, coolers for optoelectronics and devices, durable coolers for automobiles and batteries and power generation for deep-space mission via radioisotope thermoelectric generators. Significant research has already been devoted to replace the alternator in cars with a thermoelectric generator. Thermoelectric generators have been installed in automobiles to capture waste heat from the exhaust and to transform it into useful electrical energy for automotive electrical systems and for increased fuel efficiency [85–87]. The deep-space applications of NASA's *Voyager* and *Cassini* missions are using radioactive thermoelectric generators [88]. Thermoelectric refrigeration is an environmentally green method of small-scale, infrared detectors, electronics and optoelectronics, localized cooling in computers and many other applications. Moreover, the applications include comfort seat coolers in luxury vehicle. Recently, Peltier coolers are used for the refrigeration of biological specimens [85]. Thus, in future energy management and technology, development of low-temperature thermoelectric refrigeration devices, as well as for the development of high-temperature materials for waste heat recovery are important and will play a significant role.

14.5.1 Thermoelectric Parameters

Despite great advantage for waste to electricity conversion using the thermoelectric device, the fundamental challenge of designing high-performance materials has been to increase thermoelectric figure of merit (*ZT*), which is defined as

$$ZT = \frac{\sigma S^2 T}{K} \tag{14.2}$$

Clearly, to have high *ZT*, the materials need to have either low thermal conductivity (κ), or high power factor (σS^2), or both at the same time. As these above-mentioned transport characteristics are interrelated, a number of parameters need to be optimized to maximize *ZT* [88–90]. Thermal conductivity, κ, is the measure of transfer of heat through a material by charge carrier and phonons. The thermal conductivity κ_{total} comprises two major components: (i) electronic contribution, carriers transporting heat (κ_{el}) and (ii) lattice contribution, lattice vibrations or phonons carrying the heat (κ_{lat}) and it can be expressed as $\kappa = \kappa_{el} + \kappa_{lat}$. The necessity of low thermal conductivity is another confliction in thermoelectric material design [88, 91–94].

14.5.2 Recent State-of-the-Art Thermoelectric Materials

A good thermoelectric material should have high electrical conductivity (property of metal), large Seebeck (property of semiconductor) and low thermal conductivity (property of glass). The main challenge lies in the field to decouple the electronic and phonon transport parts of the *ZT* [93]. In the following part, we will briefly discuss about the state-of-the-art recent thermoelectric materials and concept developed from our group and others.

14.5.2.1 Intrinsic Rattlers

Crystalline solids exhibiting ultralow thermal conductivity are centric to the development of thermoelectrics, refractories and thermal barrier coatings for efficient energy management. Thermal transport is indeed reckoned to be a limiting factor for realizing efficient thermoelectric materials. Extrinsic strategies such as alloying and nanostructuring have been proven effective to suppress lattice thermal conductivity (κ_{lat}) but may also deteriorate electrical mobility. Solids with intrinsically low (κ_{lat}) are, therefore, practically attractive being capable of offering nearly independent control over electrical transport. Biswas group has discovered very low lattice thermal conductivity (ca. 0.5 W/mK at 300 K) in Zintl-type $AInTe_2$ compounds ($A = Tl^+/In^+$), which decays to amorphous limit at elevated temperatures [95, 96]. These compounds feature interlocked rigid anionic and weakly bound cationic substructures. $TlInTe_2$ exhibits ultra-low lattice thermal conductivity, owing to rattling dynamics of weakly bound Tl cations. Large displacements of Tl cations along the *c*-axis, driven by electrostatic repulsion between localized electron clouds on Tl and Te atoms, are akin to those of rattling guests in caged systems like skutterudites. Heat capacity of $TlInTe_2$ exhibits a broad peak at low temperatures due to contribution from Tl-induced low-frequency Einstein modes as also evidenced from THz time-domain spectroscopy. First-principles calculations reveal a strong coupling between large-amplitude coherent optic vibrations of Tl-rattlers along the *c*-axis, with acoustic phonons which likely cause the low lattice thermal conductivity in $TlInTe_2$. Jana et al. [95, 96] proposed that large atomic displacement parameter associated with a specific set of atoms (i.e., hierarchical bonding) besides rightly oriented electron-lone pairs should serve as indicators of low thermal conductivity, and guide in exploring new materials for thermoelectric applications.

14.5.2.2 Lone Pair-Induced Bond Anharmonicity

Biswas group has discovered a new class of inexpensive Te-free cubic $I\text{-}V\text{-}VI_2$ semiconductors (I = Cu, Ag; V = Sb, Bi; and VI = S, Se) with high thermoelectric performance due to ultra-low thermal conductivity and enhancement of the electrical conductivity by p-type or n-type aliovalent doping [90, 97–102]. The electrostatic repulsion between the stereochemically active lone pair on group V element and the valence bonding charge of the chalcogen atom results in strong anharmonicity in the bonding arrangement. We have shown that substitution of p-type and n-type dopants in $AgSbSe_2$ and $AgBiSe_2$, respectively, enhances the electrical conductivity significantly. With ultra-low thermal conductivity and superior electronic transport, promising thermoelectric properties were achieved in $AgSbSe_2$ [90, 97, 98]. The electronic transport properties of $AgBiSe_{1.98}X_{0.02}$ (X = Cl, Br and I) is mainly controlled by temperature-dependent cation order–disorder phase transition [90]. Important thermoelectric properties with fascinating cation order–disorder transition have been discovered recently in kinetic cubic phase of $AgBiS_2$ and AgBiSeS synthesized by solution chemistry [100–102].

14.5.2.3 Tin Telluride

Pristine tin telluride (SnTe) is a p-type degenerate semiconductor with high p-type carrier density (~$10^{21}/cm^3$) due to its intrinsic Sn vacancy which results in small Seebeck coefficient and large electrical conductivity [103]. In SnTe, the energy difference between light and heavy hole valence bands is ~0.3–0.4 eV at 300 K, leading to negligible involvement of the heavy hole valence band in electrical transport of SnTe [103].

Significant attention has been given for enhancing the thermoelectric properties of SnTe via nanostructuring [104] and modifying the electronic structure to create resonance levels near the valence band, which indeed enhances the Seebeck coefficient [105]. High thermoelectric performance has been achieved through enhancement of the Seebeck coefficient via the synergistic effect of indium and silver co-doping in SnTe (ZT ~1), where In acts as a resonant dopant and Ag doping enables the valence band convergence [105, 106].

At room temperature, lattice thermal conductivity (k_{lat}) of pristine SnTe is to be ~2.88 W/m/K, whereas the minimum lattice thermal conductivity (k_{min}) of SnTe is ~0.5 W/m/K [104]. The introduction of second-phase nanoprecipitates or alloy formation is known to reduce the k_{lat} of SnTe moderately [107]. Recent research shows that the lattice thermal conductivity of SnTe can be reduced near to its theoretical minimum limit, k_{min}, via formation of spontaneous nanodomains of Sb-rich layered intergrowth $Sn_mSb_{2n}Te_{3n+m}$ compounds, which are actually natural heterostructures (Figure 14.7) [104]. This mechanism should provide new guidelines for achieving low lattice thermal conductivities in SnTe by nanoscale engineering.

14.5.2.4 Germanium Telluride

Germanium telluride (GeTe) and its derivatives have been regarded as potential thermoelectric materials since the 1960s and offer significant technological importance.

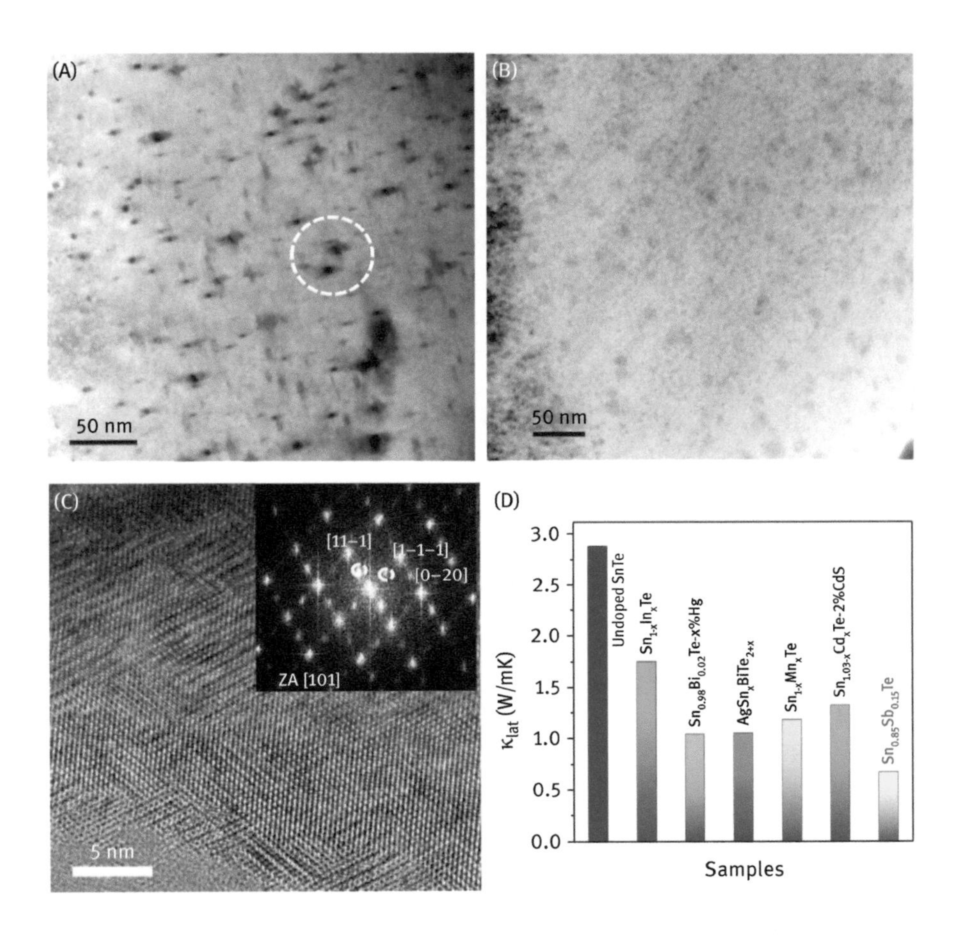

Figure 14.7: Low-magnification transmission electron microscopic (TEM) image of sample (A) $Sn_{0.96}Sb_{0.04}Te$ and (B) $Sn_{0.85}Sb_{0.15}Te$ showing nanoscale precipitates, (C) high-resolution TEM images of $Sn_{0.85}Sb_{0.15}Te$ showing the nanodomains of layered intergrowth nanostructures (inset of (C)) electron diffraction pattern indexed to the cubic rocksalt parent structure with white circles indicating superstructure ordering spots at ½ (h, k, l) with h, k, l values of (1, 1, 1). (D) Comparison of the room temperature κ_{lat} of $Sn_{0.85}Sb_{0.15}Te$ sample with the previously reported high-performance $Sn_{1-x}In_xTe$, Hg/Mn/Cd alloyed SnTe and SnTe-$AgBiTe_2$ which is very close to theoretical minimum lattice thermal conductivity of SnTe [104]. Copyright © 2016, Royal Society of Chemistry. Copyright © 2016, Royal Society of Chemistry.

GeTe undergoes a ferroelectric structural phase transition from paraelectric cubic phase (β phase, Fm−3m) to ferroelectric rhombohedral phase (α phase, R3m) at ~700 K. Pristine GeTe is less attractive for thermoelectric application because of its high p-type carrier density ($\sim10^{21}/cm^3$), which results in high electrical conductivity and a low Seebeck coefficient [108]. At 300 K, κ_{lat} for pristine GeTe is ~3 W/mK whereas theoretical limit of the minimum lattice thermal conductivity (κ_{min}) of GeTe is ~0.3 W/mK [109].

Recently, thermal conductivity of GeTe has been reduced by using different strategies in pseudobinary solid solutions of GeTe-$AgSbTe_2$ (TAGS-x) [108] and Sb/Bi-doped GeTe [110, 111]. Recently, Biswas group demonstrates low κ_{lat} (~0.7 W/mK) and an ultrahigh thermoelectric figure of merit ($ZT = 2.1$ at 630 K) in the Sb-doped pseudoternary $(GeTe)_{1-2x}(GeSe)_x(GeS)_x$ system (Figure 14.8) [109]. Entropy-driven solid solution point defects rather than enthalpy-driven conventional endotaxial nanostructuring is mainly responsible for low lattice thermal conductivity in Sb-doped $(GeTe)_{1-2x}(GeSe)(GeS)_x$ sample.

14.5.2.5 $AgSbTe_2$

$AgSbTe_2$ is a promising thermoelectric material for power generation application in mid-temperature range (400–700 K) owing to its anomalously low thermal conductivity (0.6–0.7 W/m/K) [112]. At room temperature, $AgSbTe_2$ crystallizes in a rock-salt structure (space group, Fm-3m). $AgSbTe_2$ exhibits strong bond anharmonicity due to the presence of $5s^2$ lone pair on Sb, which leads to low thermal conductivity [113]. Pseudobinary phase diagrams of Sb_2Te_3–Ag_2Te and Sb_2Te_3–Ag_2Te–Te reveals that $AgSbTe_2$ is a metastable compound and slowly decomposes to Ag_2Te and Sb_2Te_3 at 630 K [114]. However blocking the formation of Ag_2Te impurity is required to optimize thermoelectric properties of $AgSbTe_2$, due to its n-type conduction and it undergoes a structural transition at ~425 K. Recently, Biswas group showed that incorporation of Zn^{2+} at Sb^{3+} site in $AgSbTe_2$ suppresses the inevitable formation of Ag_2Te impurity by enhancing the solubility of Ag_2Te in $AgSbTe_2$ matrix, which indeed increases the

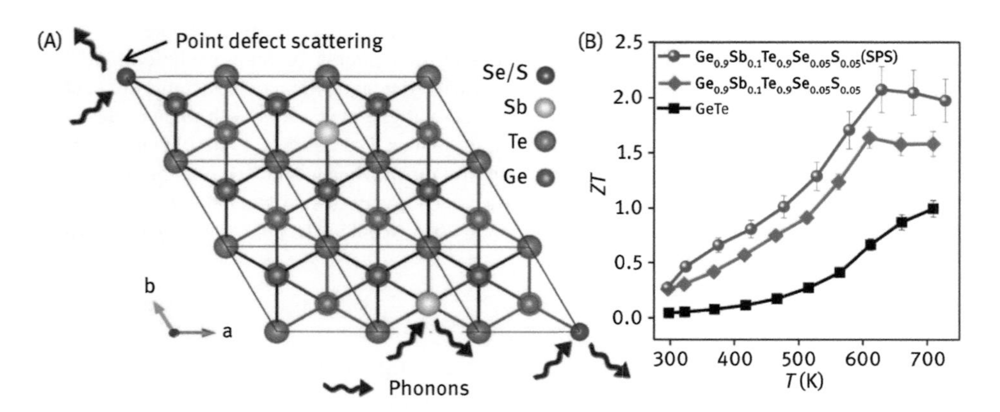

Figure 14.8: (A) Phonon scattering by entropically driven solid solution point defects, which is mainly responsible for low lattice thermal conductivity in $(GeTe)_{1-2x}(GeSe)_x(GeS)_x$, (B) temperature-dependent thermoelectric figure of merit (ZT) of GeTe and $(GeTe)_{1-2x}(GeSe)_x(GeS)_x$ (pristine & SPS sample) [109] Copyright © 2017, American Chemical Society.

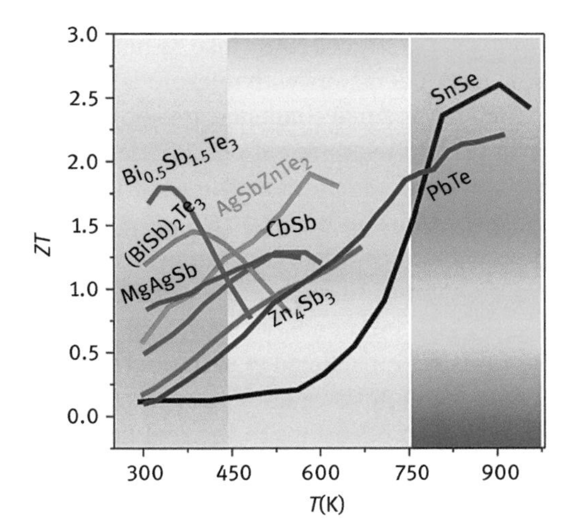

Figure 14.9: The p-type Zn-doped AgSbTe2 shows the highest *ZT* in the lower-medium temperature range (450–650 K) [115]. Copyright © 2017, American Chemical Society.

thermal and mechanical stability of $AgSbTe_2$. A maximum *ZT* of 1.9 at 585 K has been achieved for p-type $AgSb_{1-x}Zn_xTe_2$ samples, which is one of the highest value obtained in the lower-medium temperature range (450–600 K) (Figure 14.9) [115].

14.5.3 Topological Crystalline Insulator

Topological crystalline insulators (TCIs) are new classes of topological materials in which crystal mirror symmetry protects the surface state, whereas for topological insulators time-reversal symmetry protects the surface state [116]. A TCI-like $Pb_{0.6}Sn_{0.4}Te$ exhibits poor thermoelectric performance due to its vanishingly small bulk band gap [117]. Apart from various physical perturbation techniques, chemical doping can also perturb the local structure thereby the mirror symmetry which may tailor the electronic structure of TCI. Recently, Biswas group showed that Na (K) as a chemical dopant in $Pb_{0.6}Sn_{0.4}Te$ breaks the crystal symmetry locally and widens a bulk electronic band gap, resulting in high thermoelectric figure of merit value (*ZT*) ~1 at 856 K for Na (K)-doped $Pb_{0.6}Sn_{0.4}Te$ sample which is considerably higher than that of undoped $Pb_{0.6}Sn_{0.4}Te$ sample [117, 118]. The primary criteria for good thermoelectrics and topological insulators (TIs) are comparatively similar as both need heavy elements and narrow bandgap semiconductor. We can use same material either for electronics/spintronics as a topological material or for heat to energy conversion as thermoelectrics.

14.6 A Way Forward

Exploring the excellent properties of STEG systems further is likely to be a very complex task, which requires the use of both theoretical and experimental knowledge in interdisciplinary field. Detailed experimental case studies could also be helpful, as would deeper understanding on different operational factors with relevant effects on efficiency. Though improving the system efficiency is a real challenge, one hopes that this does not prevent the researchers from at least trying different possibilities.

Concerning other functional applications of nanomaterials, the absorption of solar radiation in entire spectrum can be enhanced by optimizing the size/morphology/dimension of the nanoparticles and nanocoatings. Also, the number of layers, roughness and thickness of individual layer, compositions, optical constants and so on are the essential components to enhance the solar energy absorption. Hence, in spectrally selective absorber, nanostructures are introduced to enhance the solar absorption by minimizing thermal heat loss. Though there are a number of spectrally selective absorber coatings, their practical applications in the real field still remain a challenge and are at risk due to various critical environmental factors. The pretreatments of the substrates and careful surface engineering of the coating with a roughness in nanometer scale can satisfy the criterion of self-cleaning. Apart from operational properties, the mechanical properties, for example, hardness or scratch resistance, could be modified extremely in the nanoscale region compared with the identical material with coarser microstructural characteristics. A year-round exposure of these coatings in various climatic conditions is required to demonstrate the environmental stability of these coatings. It is also expected that optical simulation and optimization of performance of multilayered coatings before experimental study can reduce the cost and parameterization time for fabricating spectrally selective coating. As a result, the overall production cost of STEG systems can also be minimized.

Bulk thermoelectric material-based devices with power output of 20 MW have efficiencies as low as 0.1% for temperature difference of 1.5 K in low-temperature domain. In order to compete with current dynamic technologies, there is an urgent need to reach the *ZT* value above 3. However, *ZT* hardly surpasses 2 in case of bulk engineered materials because of interdependences. Such circumstances obstruct the commercial applications of thermoelectric devices. A number of investigations are undergoing to synthesize promising chalcogenides such as PbTe, GeTe,β-Zn_4Sb_3, *skutterudites, clathrates* and Zintl phases to achieve higher *ZT* for utilizing in practical applications. There are additional difficulties to ensure electrical contact to the nanowires in a high-density arrangement with minimizing heat leakages. As alternative materials, the topological insulators and polymer thermoelectric are also potential competitors for commercial implantation of thermoelectric materials. Applications have been explored where thermoelectric materials can be used up to 1,000 °C. For high-temperature thermoelectrics, nanocomposites are advantageous as thermal conductivity in these materials is suppressed by the phonon scattering without hindering the electrical properties. Such consideration points will help to explore the vital role of STEG systems in the field of environmental study and sustainability.

14.7 Closure

The advancement of cleaner, cost-effective and more efficient energy production demands an interdisciplinary effort including various scientific domains and technological fields. Indeed, significant improvement has been achieved in the production of green energy followed by distribution at an affordable cost. During the last few decades, nanomaterials and nanotechnology have offered a number of promising and constructive innovations as well as research possibilities. Not only that, nanotechnology has been at the premise of numerous technological advances with lot of modern devices or applications for sustainable energy development while reducing their manufacturing and electricity production cost. The existing research and development infrastructure on nanotechnology in universities and industries as well as the modern, intriguing, encouraging approach of nanotechnology-based research have attracted bright young students, researchers, industries and policy makers. In order to enhance the competitiveness and innovation strength as well as to achieve the aspiration of the world community in the years up to 2025, consequential collaborations between researchers, manufacturing industry and government will be extremely advantageous.

The combination of solar thermal and/or thermoelectric systems, that is, STEGs are attractive technologies, which can not only able to generate power but also can serve a lead role for environmental protection. It can be commented that despite low efficiency and high cost, research on STEG is gaining a lot of interest in small-scale applications. Thus, we conclude that STEGs with the spectrally selective absorber coatings and thermoelectric materials and other components can be considered as a promising alternative to traditional renewable energy systems. Taking into account the robust advances in the fabrication of new and efficient nanostructured solar absorber coating and thermoelectric materials, it can be predicted that the efficiency and attractiveness of STEGs can excel than that of other conventional solar hybrid systems in near future.

Acknowledgments: This work was supported by SERB (EMR/2016/000651) and Technology Research Centre (TRC, Department of Science & Technology – subproject number: TRC-JNC/KB/4396) in JNCASR. S.R. thanks CSIR for research fellowship. This paper is based upon work supported in part under the US-India Partnership to Advance Clean Energy-Research (PACE-R) for the Solar Energy Research Institute for India and the United States (SERIIUS), funded jointly by the U.S. Department of Energy (Office of Science, Office of Basic Energy Sciences, and Energy Efficiency and Renewable Energy, Solar Energy Technology Program, under Subcontract DE-AC36-08GO28308 to the National Renewable Energy Laboratory, Golden, Colorado) and the Government of India, through the Department of Science and Technology under Subcontract IUSSTF/JCERDC-SERIIUS/2012 dated 22nd Nov. 2012. A.D. thanks DST for providing INSPIRE scholarship.

Appendix I

List of symbols

Symbol	Definition
η_{STEG}	Efficiency of solar thermoelectric generator
τ_g	Transmittance
α	Solar absorptance
ε	Thermal emittance
η_{op}	Optical concentration efficiency
σ	Stefan–Boltzmann constant
$G(\lambda)$	Incident solar radiation
λ	Wavelength
ZT	Thermoelectric figure of merit
κ	Thermal conductivity
σS^2	Power factor
T	Temperature
κ_{lat}	Lattice thermal conductivity
κ_{el}	Electronic thermal conductivity

References

1. Ekatah I, Samy M, Bampton R, Halabi A. The relationship between corporate social responsibility and profitability: the case of Royal Dutch Shell Plc. Corp Reput Rev 2011;14:249–61.
2. Raj B, Van de Voorde M, Mahajan Y. Nanotechnology for energy sustainability, 3 vol. Weinheim, Germany: John Wiley & Sons, 2017.
3. Moniz EJ. Nanotechnology for the energy challenge. Weinheim, Germany: John Wiley & Sons, 2010.
4. Liu C, Li F, Ma LP, Cheng HM. Advanced materials for energy storage. Adv Mater 2010;22:E28–E62.
5. Barpanda P, Djellab K, Recham N, Armand M, Tarascon J-M. Direct and modified ionothermal synthesis of $LiMnPO_4$ with tunable morphology for rechargeable Li-ion batteries. J Mater Chem 2011;21:10143–52.
6. Barshilia HC, Selvakumar N, Rajam KS, Rao DVS, Muraleedharan K, Biswas A. TiAlN/TiAlON/Si_3N_4 tandem absorber for high temperature solar selective applications. Appl Phys Lett 2006;89:191909-.
7. Lee JS, Tai Kim S, Cao R, et al. Metal–air batteries with high energy density: Li–air versus Zn–air. Adv Energy Mater 2011;1:34–50.
8. Vangari M, Pryor T, Jiang L. Supercapacitors: review of materials and fabrication methods. J Energy Eng 2012;139:72–9.
9. Doyle ML, Harris IR, Pratt AS, Willey DB. Hydrogen storage materials. In: Google Patents; 2000.
10. Schlapbach L, Züttel A. Hydrogen-storage materials for mobile applications. Nature 2001;414:353–8.
11. Kitoh K, Nemoto H. 100 Wh Large size Li-ion batteries and safety tests. J Power Sources 1999;81:887–90.
12. Zhang SS. A review on the separators of liquid electrolyte Li-ion batteries. J Power Sources 2007;164:351–64.
13. Abouimrane A, Belharouak I, Amine K. Sulfone-based electrolytes for high-voltage Li-ion batteries. Electrochem Commun 2009;11:1073–6.

14. Dahn JR, Jiang J, Moshurchak LM, Fleischauer MD, Buhrmester C, Krause LJ. High-rate overcharge protection of LiFePO4-based Li-ion cells using the redox shuttle additive 2, 5-ditertbutyl-1, 4-dimethoxybenzene. J Electrochem Soc 2005;152:A1283–A9.
15. Daniel C. Materials and processing for lithium-ion batteries. Jom 2008;60:43–8.
16. Wang Z, Wang Z, Madhavi S, Lou XWD. α-Fe 2 O 3-mediated growth and carbon nanocoating of ultrafine SnO 2 nanorods as anode materials for Li-ion batteries. J Mater Chem 2012;22:2526–31.
17. Si Y, Guo Z. Superhydrophobic nanocoatings: from materials to fabrications and to applications. Nanoscale 2015;7:5922–46.
18. Linic S, Christopher P, Ingram DB. Plasmonic-metal nanostructures for efficient conversion of solar to chemical energy. Nat Mater 2011;10:911.
19. Shakouri A, Zebarjadi M. Nanoengineered materials for thermoelectric energy conversion. In: Thermal nanosystems and nanomaterials. Berlin, Heidelburg: Springer,2009:225–99.
20. Sun C, Zheng S, Wei CC, et al. Superior radiation-resistant nanoengineered austenitic 304L stainless steel for applications in extreme radiation environments. Sci Rep 2015;5.
21. Zheng YB, Huang TJ. Surface plasmons of metal nanostructure arrays: from nanoengineering to active plasmonics. JALA 2008;13:215–26.
22. Parlak O, Mishra YK, Grigoriev A, et al. Hierarchical aerographite nano-microtubular tetrapodal networks based electrodes as lightweight supercapacitor. Nano Energy 2017;34:570–7.
23. Mohammadnejad M, Ghazvini M, Javadi FS, Saidur R. Estimating the exergy efficiency of engine using nanolubricants. Energy Educ Sci Technol Part A: Energy Sci Res 2011;27:447–54.
24. Lichtenstein A, Havivi E, Shacham R, et al. Supersensitive fingerprinting of explosives by chemically modified nanosensors arrays. Nature Commun 2014;5:4195.
25. Torad NL, Salunkhe RR, Li Y, et al. Electric double-layer capacitors based on highly graphitized nanoporous carbons derived from ZIF-67. Chem-A Eur J 2014;20:7895–900.
26. Latthe SS, Terashima C, Nakata K, Fujishima A. Superhydrophobic surfaces developed by mimicking hierarchical surface morphology of lotus leaf. Molecules 2014;19:4256–83.
27. Chan FL, Tanksale A. Review of recent developments in Ni-based catalysts for biomass gasification. Renewable Sustainable Energy Rev 2014;38:428–38.
28. Chalker J, Worthington M. New nanoparticle-catalysts can reduce need for precious metals. Chem Today Mag 2016:76–9.
29. Müller-Steinhagen H, Trieb F. Concentrating solar power. A review of the technology Ingenia Inform QR Acad Eng 2004;18:43–50.
30. Siddique ARM, Mahmud S, Van Heyst B. A review of the state of the science on wearable thermoelectric power generators (TEGs) and their existing challenges. Renewable Sustainable Energy Rev 2017;73:730–44.
31. Telkes M. Solar thermoelectric generators. J Appl Phys 1954;25:765–77.
32. Moh'd A A-N, Tashtoush BM, Khasawneh MA, Al-Keyyam I. A hybrid concentrated solar thermal collector/thermo-electric generation system. Energy 2017;134:1001–1012.
33. Sundarraj P, Maity D, Roy SS, Taylor RA. Recent advances in thermoelectric materials and solar thermoelectric generators – a critical review. RSC Advances 2014;4:46860–74.
34. Willars-Rodríguez FJ, Chávez-Urbiola EA, Vorobiev P, Vorobiev YV. Investigation of solar hybrid system with concentrating Fresnel lens, photovoltaic and thermoelectric generators. Int J Energy Res 2017;41:377–88.
35. Lenoir B, Dauscher A, Poinas P, Scherrer H, Vikhor L. Electrical performance of skutterudites solar thermoelectric generators. Appl Therm Eng 2003;23:1407–15.
36. Chen G. Theoretical efficiency of solar thermoelectric energy generators. J Appl Phys 2011;109:104908.
37. Kraemer D, Poudel B, Feng H-P, et al. High-performance flat-panel solar thermoelectric generators with high thermal concentration. Nature Mater 2011;10:532.

38. Amatya R, Ram RJ. Solar thermoelectric generator for micropower applications. J Electron Mater 2010;39:1735–40.
39. Goldsmid HJ, Giutronich JE, Kaila MM. Solar thermoelectric generation using bismuth telluride alloys. Sol Energy 1980;24:435–40.
40. Omer SA, Infield DG. Design and thermal analysis of a two stage solar concentrator for combined heat and thermoelectric power generation. Energy Convers Manage 2000;41:737–56.
41. Suter C, Tomeš P, Weidenkaff A, Steinfeld A. A solar cavity-receiver packed with an array of thermoelectric converter modules. Sol Energy 2011;85:1511–8.
42. Chávez Urbiola EA, Vorobiev Y. Investigation of solar hybrid electric/thermal system with radiation concentrator and thermoelectric generator. Int J Photoenergy 2013;2013, Article ID 704087:7 pp.
43. Selvakumar N, Barshilia HC. Review of physical vapor deposited (PVD) spectrally selective coatings for mid-and high-temperature solar thermal applications. Sol Energy Mater Sol Cells 2012;98:1–23.
44. Dan A, Barshilia HC, Chattopadhyay K, Basu B. Solar energy absorption mediated by surface plasma polaritons in spectrally selective dielectric-metal-dielectric coatings: a critical review. Renewable Sustainable Energy Rev 2017;79:1050–77.
45. Ehrenreich H, Seraphin BO. Symposium on the fundamental optical properties of solids relevant to solar energy conversion. NASA STI/Recon Technical Report N 1975;77:12538.
46. Touloukian YS, Powell RW, Ho CY, Nicolaou MC. Thermophysical properties of matter-the TPRC data series. Volume 10. Thermal diffusivity: thermophysical and electronic properties information analysis center Lafayette In, USA, 1974.
47. González F, Barrera-Calva E, Huerta L, Mane RS. Coatings of Fe_3O_4 nanoparticles as selective solar absorber. Open Surf Sci J 2011;3:131–5.
48. Ienei E, Isac L, Duta A. Synthesis of alumina thin films by spray pyrolysis. Rev Roum Chim 2010;55:161–5.
49. Randich E, Allred DD. Chemically vapor-deposited ZrB_2 as a selective solar absorber. Thin Solid Films 1981;83:393–8.
50. Sani E, Mercatelli L, Sansoni P, Silvestroni L, Sciti D. Spectrally selective ultra-high temperature ceramic absorbers for high-temperature solar plants. JRenewable Sustainable Energy 2012;4:033104.
51. Sani E, Mercatelli L, Francini F, Sans JL, Sciti D. Ultra-refractory ceramics for high-temperature solar absorbers. Script Mater 2011;65:775–8.
52. Chen Z, Boström T. Electrophoretically deposited carbon nanotube spectrally selective solar absorbers. Sol Energy Mater Sol Cells 2016;144:678–83.
53. Paone A, Geiger M, Sanjines R, Schüler A. Thermal solar collector with VO_2 absorber coating and $V_{1-x}W_xO_2$ thermochromic glazing–Temperature matching and triggering. Sol Energy 2014;110:151–9.
54. Ning Y, Wang W, Wang L, et al. Optical simulation and preparation of novel Mo/ZrSiN/ZrSiON/SiO_2 solar selective absorbing coating. Sol Energy Mater Sol Cells 2017;167:178–83.
55. Dan A, Jyothi J, Chattopadhyay K, Barshilia HC, Basu B. Spectrally selective absorber coating of WAlN/WAlON/Al_2O_3 for solar thermal applications. Sol Energy Mater Sol Cells 2016;157:716–26.
56. Dan A, Chattopadhyay K, Barshilia HC, Basu B. Angular solar absorptance and thermal stability of W/WAlN/WAlON/Al_2O_3-based solar selective absorber coating. Appl Therm Eng 2016;109:997-1002.
57. Dan A, Chattopadhyay K, Barshilia HC, Basu B. Colored selective absorber coating with excellent durability. Thin Solid Films 2016;620:17–22.

58. Dan A, Chattopadhyay K, Barshilia HC, Basu B. Thermal stability of WAlN/WAlON/Al_2O_3-based solar selective absorber coating. MRS Adv 2016:1–7.
59. Rebouta L, Capela P, Andritschky M, et al. Characterization of TiAlSiN/TiAlSiON/SiO_2 optical stack designed by modelling calculations for solar selective applications. Sol Energy Mater Sol Cells 2012;105:202–7.
60. Barshilia HC, Selvakumar N, Rajam KS, Biswas A. Spectrally selective NbAlN/NbAlON/Si_3N_4 tandem absorber for high-temperature solar applications. Sol Energy Mater Solar Cells 2008;92:495–504.
61. Wu Y, Wang C, Sun Y, et al. Optical simulation and experimental optimization of Al/NbMoN/NbMoON/SiO_2 solar selective absorbing coatings. Sol Energy Mater Sol Cells 2015;134:373–80.
62. Rebouta L, Sousa A, Andritschky M, et al. Solar selective absorbing coatings based on AlSiN/AlSiON/$AlSiO_y$ layers. Appl Surf Sci 2015;356:203–12.
63. Valleti K, Krishna DM, Joshi SV. Functional multi-layer nitride coatings for high temperature solar selective applications. Sol Energy Mater Sol Cells 2014;121:14–21.
64. Dan A, Biswas A, Sarkar P, Kashyap S, Chattopadhyay K, Barshilia HC, et al. Enhancing spectrally selective response of W/WAlN/WAlON/Al_2O_3 –based nanostructured multilayer absorber coating through graded optical constants. Sol Energy Mater Sol Cells 2018;176:157–66.
65. Lampert CM. Thermal degradation of a black chrome solar selective absorber coating: short term, Lawrence Berkeley National Laboratory. International Solar Energy Society Meeting, Atlanta, GA, May 28–June 1, 1979
66. Survilienė S, Češūnienė A, Juškėnas R, et al. The use of trivalent chromium bath to obtain a solar selective black chromium coating. Appl Surf Sci 2014;305:492–7.
67. Barshilia HC, Selvakumar N, Rajam KS, Biswas A. Structure and optical properties of pulsed sputter deposited Cr_xO_y/Cr/Cr_2O_3 solar selective coatings. J Appl Phys 2008;103:023507.
68. Nuru ZY, Msimanga M, Muller TFG, Arendse CJ, Mtshali C, Maaza M. Microstructural, optical properties and thermal stability of MgO/Zr/MgO multilayered selective solar absorber coatings. Sol Energy 2015;111:357–63.
69. Tsai TK, Hsueh SJ, Fang JS. Optical properties of Al_xO_y/Ni/Al_xO_y Multilayered absorber coatings prepared by reactive DC magnetron sputtering. J Electron Mater 2014;43:229.
70. Mahalakshmi PV, Vanithakumari SC, Gopal J, Mudali UK, Raj B. Enhancing corrosion and biofouling resistance through superhydrophobic surface modification. Curr Sci 2011;101:1328–36.
71. Mudali U, Vanithakumari S, Raj B. Superhydrophobic surface modification to enhance corrosion resistance of titanium in chloride and nitric acid media. Surface Modification Technologies XXV Conference Proceedings, edited by T S Sudarshan and Per Nylen 2011:211–21.
72. Vanithakumari SC, George RP, Mudali UK. Enhancement of corrosion performance of titanium by micro-nano texturing. Corrosion 2013;69:804–12.
73. Vanithakumari SC, George RP, Kamachi Mudali U. Influence of silanes on the wettability of anodized titanium. Appl Surf Sci 2014;292:650–7.
74. Vanithakumari SC, George RP, Kamachi Mudali U. Environmental stability and long-term durability of superhydrophobic coatings on titanium. J Mater Eng Perf 2017;26:2640–8.
75. Vizhi ME, Vanithakumari SC, George RP, Vasantha S, Mudali UK. Superhydrophobic coating on modified 9Cr-1Mo ferritic steel using perfluoro octyl triethoxy silane. Surf Eng 2016;32:139–46.
76. Ezhil Vizhi M, Vanithakumari SC, George RP, Vasantha S, Kamachi Mudali U. Superhydrophobic coating on Mod.9Cr-1Mo ferritic steel for enhancing corrosion resistance and antibacterial activity. Trans Indian Inst Met 2016;69:1311–8.
77. Sutha S, Vanithakumari SC, George RP, Mudali UK, Raj B, Ravi KR. Studies on the influence of surface morphology of ZnO nail beds on easy roll off of water droplets. Appl Surf Sci 2015;347:839–48.

78. Sampathkumar Chrisolite V, Prashant Y, Rani Pongachira G, Uthandi Mudali Kamachi M. Lotus effect-based coatings on marine steels to inhibit biofouling. Surf Innov 2015;3:115–26.
79. Tritt TM. Thermoelectric phenomena, materials, and applications. Ann Rev Mater Res 2011;41:433–48.
80. Shakouri A. Recent developments in semiconductor thermoelectric physics and materials. Ann Rev Mater Res 2011;41.
81. Bell LE. Cooling, heating, generating power, and recovering waste heat with thermoelectric systems. Science 2008;321:1457–61.
82. Snyder GJ, Toberer ES. Complex thermoelectric materials. Nat Mater 2008;7:105–14.
83. Guin SN, Chatterjee A, Negi DS, Datta R, Biswas K. High thermoelectric performance in tellurium free p-type $AgSbSe_2$. Energy Environ Sci 2013;6:2603–8.
84. Guin SN, Srihari V, Biswas K. Promising thermoelectric performance in n-type $AgBiSe_2$: effect of aliovalent anion doping. J Mater Chem A 2015;3:648–55.
85. Sootsman JR, Chung DY, Kanatzidis MG. New and old concepts in thermoelectric materials. Angew Chem Int Ed 2009;48:8616–39.
86. Biswas K, He J, Blum ID, et al. High-performance bulk thermoelectrics with all-scale hierarchical architectures. Nature 2012;489:414.
87. Tan G, Zhao L-D, Kanatzidis MG. Rationally designing high-performance bulk thermoelectric materials. Chem Rev 2016;116:12123–49.
88. Zeier WG, Zevalkink A, Gibbs ZM, Hautier G, Kanatzidis MG, Snyder GJ. Thinking like a chemist: intuition in thermoelectric materials. Angew Chem Int Ed 2016;55:6826–41.
89. Jana MK, Pal K, Waghmare UV, Biswas K. The Origin of Ultralow thermal conductivity in InTe: lone-pair-induced anharmonic rattling. Angew Chem Int Ed 2016;55:7792–6.
90. Jana MK, Pal K, Warankar A, Mandal P, Waghmare UV, Biswas K. Intrinsic rattler-induced low thermal conductivity in Zintl type $TlInTe_2$. J Am Chem Soc 2017;139:4350–3.
91. Guin SN, Chatterjee A, Negi DS, Datta R, Biswas K. High thermoelectric performance in tellurium free p-type $AgSbSe_2$. Energy Environ Sci 2013;6:2603–8.
92. Guin SN, Biswas K. Sb deficiencies control hole transport and boost the thermoelectric performance of p-type $AgSbSe_2$. J Mater Chem C 2015;3:10415–21.
93. Guin SN, Negi DS, Datta R, Biswas K. Nanostructuring, carrier engineering and bond anharmonicity synergistically boost the thermoelectric performance of p-type $AgSbSe_2$–ZnSe. J Mater Chem A 2014;2:4324–31.
94. Guin SN, Biswas K. Cation disorder and bond anharmonicity optimize the thermoelectric properties in kinetically stabilized rocksalt $AgBiS_2$ nanocrystals. Chem Mater 2013;25:3225–31.
95. Guin SN, Banerjee S, Sanyal D, Pati SK, Biswas K. Origin of the order–disorder transition and the associated anomalous change of thermopower in $AgBiS_2$ nanocrystals: a combined experimental and theoretical study. Inorg Chem 2016;55:6323–31.
96. Guin SN, Banerjee S, Sanyal D, Pati SK, Biswas K. Nanoscale stabilization of nonequilibrium rock salt BiAgSeS: colloidal synthesis and temperature driven unusual phase transition. Chem Mater 2017;29:3769–77.
97. Banik A, Shenoy US, Anand S, Waghmare UV, Biswas K. Mg alloying in SnTe facilitates valence band convergence and optimizes thermoelectric properties. Chem Mater 2015;27:581–7.
98. Banik A, Vishal B, Perumal S, Datta R, Biswas K. The origin of low thermal conductivity in $Sn_{1-x}Sb_xTe$: phonon scattering via layered intergrowth nanostructures. Energy Environ Sci 2016;9:2011–9.
99. Banik A, Shenoy US, Saha S, Waghmare UV, Biswas K. High power factor and enhanced thermoelectric performance of SnTe-$AgInTe_2$: synergistic effect of resonance level and valence band convergence. J Am Chem Soc 2016;138:13068–75.

100. Banik A, Biswas K. AgI alloying in SnTe boosts the thermoelectric performance via simultaneous valence band convergence and carrier concentration optimization. J Solid State Chem 2016;242:43–9.
101. Tan G, Shi F, Doak JW, et al. Extraordinary role of Hg in enhancing the thermoelectric performance of p-type SnTe. Energy Environ Sci 2015;8:267–77.
102. Rosi FD, Dismukes JP, Hockings EF. Semiconductor materials for thermoelectric power generation up to 700°C. Electr Eng 1960;79:450–9.
103. Samanta M, Biswas K. Low thermal conductivity and high thermoelectric performance in $(GeTe)_{1-2x}(GeSe)_x(GeS)_x$: competition between solid solution and phase separation. J Am Chem Soc 2017;139:9382–91.
104. Perumal S, Roychowdhury S, Biswas K. Reduction of thermal conductivity through nanostructuring enhances the thermoelectric figure of merit in $Ge_{1-x}Bi_xTe$. Inorg Chem Front 2016;3:125–32.
105. Perumal S, Roychowdhury S, Negi DS, Datta R, Biswas K. High Thermoelectric Performance and Enhanced Mechanical Stability of p-type $Ge_{1-x}Sb_xTe$. Chem Mater 2015;27:7171–8.
106. Ma J, Delaire O, May AF, et al. Glass-like phonon scattering from a spontaneous nanostructure in $AgSbTe_2$. Nature Nanotechnol 2013;8:445–51.
107. Nielsen MD, Ozolins V, Heremans JP. Lone pair electrons minimize lattice thermal conductivity. Energy Environ Sci 2013;6:570–8.
108. Ayral-Marin RM, Brun G, Maurin M, Tedenac JC. Contribution to the study of $AgSbTe_2$. Eur J Solid State Inorg Chem 1990;27:747–57.
109. Roychowdhury S, Panigrahi R, Perumal S, Biswas K. Ultrahigh Thermoelectric Figure of Merit and Enhanced Mechanical Stability of p-type $AgSb_{1-x}Zn_xTe_2$. ACS Energy Lett 2017;2:349–56.
110. Müchler L, Zhang H, Chadov S, et al. Topological insulators from a chemist's perspective. Angew Chem Int Ed 2012;51:7221–5.
111. Roychowdhury S, Shenoy US, Waghmare UV, Biswas K. Tailoring of electronic structure and thermoelectric properties of a topological crystalline insulator by chemical doping. Angew Chem 2015;127:15456–60.
112. Roychowdhury S, Sandhya Shenoy U, Waghmare UV, Biswas K. Effect of potassium doping on electronic structure and thermoelectric properties of topological crystalline insulator. Appl Phys Lett 2016;108:193901.
113. Moon J, Lu D, VanSaders B, et al. High performance multi-scaled nanostructured spectrally selective coating for concentrating solar power. Nano Energy 2014;8:238–46.
114. ÜÖè LZ, Diao X. New criterion for optimization of solar selective absorber coatings. Chin Opt Lett 2013:122–6.
115. Wang X, Li H, Yu X, Shi X, Liu J. High-performance solution-processed plasmonic Ni nanochain-Al_2O_3 selective solar thermal absorbers. Appl Phys Lett 2012;101:203109.
116. Kumar SK, Suresh S, Murugesan S, Raj SP. CuO thin films made of nanofibers for solar selective absorber applications. Sol Energy 2013;94:299–304.
117. Sergeant NP, Agrawal M, Peumans P. High performance solar-selective absorbers using coated sub-wavelength gratings. Opt Express 2010;18:5525–40.
118. Barshilia HC, Selvakumar N, Rajam KS, Biswas A. Structure and optical properties of pulsed sputter deposited $Cr_xO_y/Cr/Cr_2O_3$ solar selective coatings. J Appl Phys 2008;103:023507.

Matthias Werner, Wolfgang Wondrak and Colin Johnston

15 Nanotechnology and transport: Applications in the automotive industry

15.1 Introduction

Nanotechnology is becoming increasingly economically important worldwide. Today, numerous products already include nanotechnological components or they are made using nanotechnologies. Signs for a broad industrial process of transformation through nanotechnologies have been apparent from a scientific perspective since the 1980s. Today, a consensus prevails that nanotechnologies will have an impact on virtually all areas of life, and thus on the economy, in the mid- and long term. Unlike many other high technologies, nanotechnology has a cross-sectorial character and therefore possesses a very broad potential of applications in many areas of economy, including automotive industry [1, 2]. In 2003, different automotive applications have been forecasted [3], several of them have been well established on the market today.

Catalysts and air filter systems cause clean air inside and outside the car. Optical layers for reflection reduction on dashboards or hydrophobic and dirt repellent "easy-to-clean" surfaces on car mirrors are further examples of applications of nanotechnology in automobiles. Nowadays, profits amounting to billions are being generated using such high-end products. Nanotechnologies are thereby incorporated as components into the product or into production technologies.

In the production technology of future automotive engineering, nanotechnological adhesives have an enormous economic potential since they allow energy savings in assembly processes. An interesting application relates to adhesives that are modified with magnetic nanoparticles. The coupling of thermal energy in the form of microwave radiation induces the chemical reaction necessary for the gluing process.

Through nanoadditives in plastics, clearly improved processing properties in injection-molding machines can be achieved. Here, energy savings of up to 20% are possible.

The key topics in automotive industry have evolved from reduction of fuel consumption and emissions, safety, driver information and comfort into electric mobility, autonomous driving, connectivity and shared mobility (see Figure 15.1). Nanotechnology obtains a huge potential for innovations in these fields and is seen as one of the pacemakers for future automobiles [4] to sustain competitiveness. Nanotechnology may also provide substantial improvements in electric and hybrid cars, for example, regarding energy storage systems, drivetrain components, robust power device assemblies and thermal management.

https://doi.org/10.1515/9783110547221-015

Figure 15.1: Megatrends in the automotive industry [5].

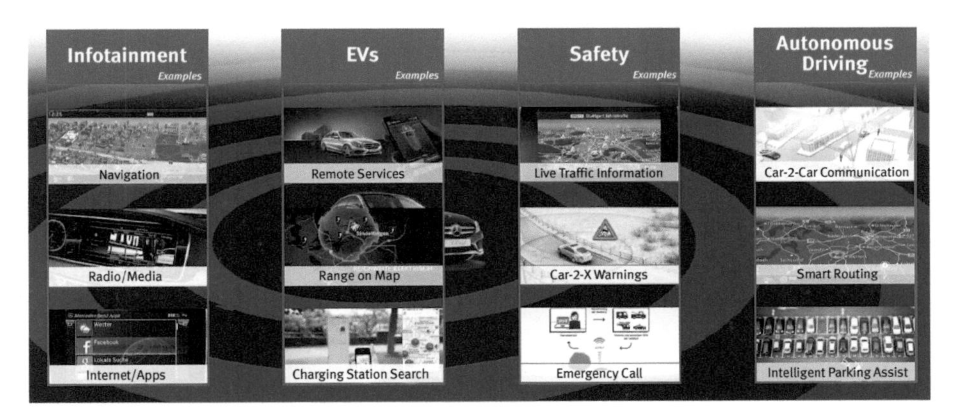

Figure 15.2: Functions enabled by connectivity [6].

Future cars can be regarded as digitized ecosystems, where information handling and exchange will enable new functions and services, as indicated in Figure 15.2. For these, connectivity with the environment, the Internet, traffic systems and GPS satellites is indispensible.

This chapter points out some applications of nanotechnology for the automotive industry and its suppliers that are either already existing or at the research stage.

15.2 Functionalities of Nanotechnology

Nanotechnologies can be applied in almost all industries and technologies because of their effects and functionalities. Given their cross-sectional capacity, nanotechnologies are especially important in automotive engineering, for recent reviews see [7, 8]. The functionalities of nanotechnological materials, products and processes discovered so far offer an application-oriented access to nanotechnologies for enterprises. These phenomena are closely related to the benefit and function of the product and thus to customer-oriented demand and provide the connection between nanotechnology and automotive engineering. Functionalities relevant for automotive engineering are presented below.

Table 15.1 gives a short overview of possible application fields of nanotechnological functionalities in automotive engineering. In future, we will be able to count on a number of further innovations that will concern all branches of the automotive industry and its suppliers. Some of the existing or emerging applications are illustrated in Figure 15.3.

Table 15.1: Current and future applications of nanotechnologies in automobiles.

Effects/ properties	Mechanical effects Hardness, friction, fracture toughness	Size effects High surface-to-volume relation	Electronic/ magnetic/thermal effects	Optical effects Color, transparency	Chemical effects Reactivity, selectivity, surface properties
Exterior	Scratch resistant or self-healing nano-lacquers, polymer windows			Electrochromic layers, angle-dependent colors	Cleaning and protection, photo-catalytic windows, window heating
Car body	Lightweight/ composite materials, carbon fiber reinforced elements		Novel adhesives		Corrosion protection
Interior	Cover hoods, panels, reinforced plastic parts, movable sun-shields	Nanofilters	Energy-independent sensor networks, low-cost sensors, heating elements, thermal management, textile electronics	Antiglare mirrors, touchscreens, antireflective instrument panel, thermal management, interior light designs, organic displays, flexible displays, panel lighting	Water- and soil-repellant coatings and textiles, antismell and antibacterial filters

(continued)

Table 15.1: (Continued)

Chassis and tires	Carbon black in tires, nanosteel		Magnetorheological damping, energy recuperating damping, electronic body control systems		
Combustion engine and drivetrain	Low-friction components, wear-resistant coatings, advanced ceramics. cylinders, pistons, valves, bearings	Catalysts, nanoporous exhaust filters	Piezoinjectors, energy harvesting		Catalysts, catalytic coatings, fuel additives
Electrics and electronics	Electronic packages, contacts, relays, connectors	New solder materials, nanosilver sintering or Cu sintering for die attach	Giant magnetoresistance (GMR) sensors, solar cells, memories, driver health monitoring, cooling media, quantum computing	Optoelectronic devices, lighting, optical computing and data transmission	
Electric cars	Gearbox, housings, lightweight electric motors, high-speed motors	Super caps, fuel cells, Li-ion batteries	Magnetic materials, power devices, capacitors, battery cooling, battery heating, failure prediction		

15.3 Application Areas

In the following, we describe some of the application fields in more detail. Thereby we follow the sequence of the categories of Table 15.1. In addition, we give a glance on the needs of automated driving, and how nanotechnology may contribute to safer and environmentally friendly cars.

Figure 15.3: Application fields of nanotechnology for automobiles, based on [8].

15.3.1 Car Body Shell

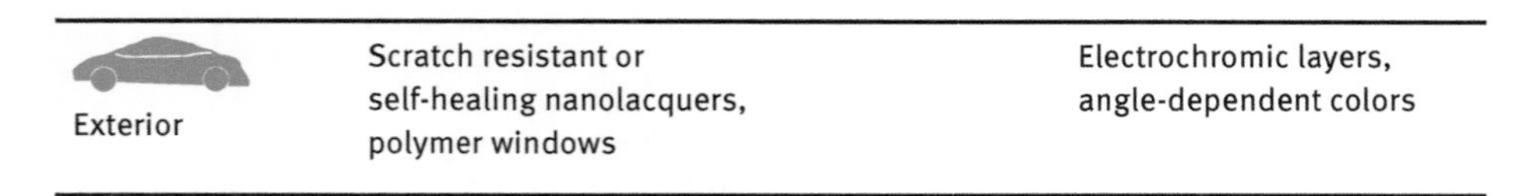

Exterior	Scratch resistant or self-healing nanolacquers, polymer windows	Electrochromic layers, angle-dependent colors

An unblemished car body shell should be guaranteed even after numerous car washes and several years of operation. Compared to conventional paint systems, nanovarnishes allow for higher scratch resistance and paint brilliance. The basis for this technological effect is embedded ceramic particles in the final varnish layer in the nanometer range. Nanoparticles such as Degussa's AEROSIL R9200 for car varnishes, which greatly account for the improvement in scratch resistance, are gaining importance very much. Traditionally, AEROSIL can also be found in other layers of the car body shell, where it is used for pigment stabilization, rheology control and corrosion resistance, for example. These are special types of silica that play an important role in innovative car paints. Their basis are nanostructured powders produced in a gaseous-phase synthesis in the flame and are therefore called pyrogenic constitute. Starting with silica tetrachloride, small, spherical single silica parts with a mean diameter in the range between 7 and 40 nm result from flame hydrolysis [9]. If the paint is liquid, these particles are initially randomly distributed in the solution. During the drying and hardening process they cross-link deeply with the molecular structure of the paint matrix (Figures 15.4 and 15.5).

The result is a very dense and ordered matrix on the paint surface. Thus the scratching resistance is tripled and the paint brilliance improves considerably. This

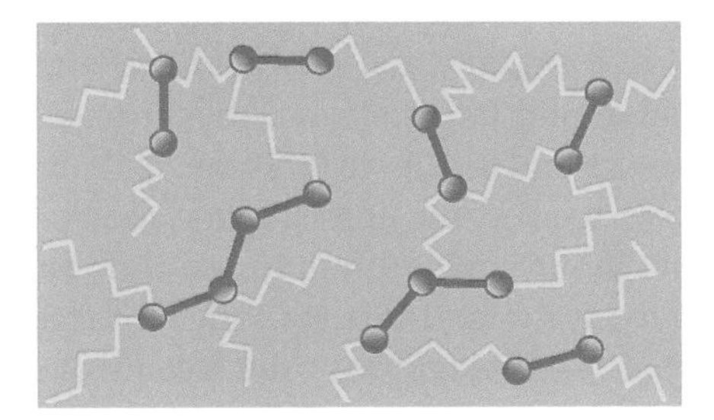

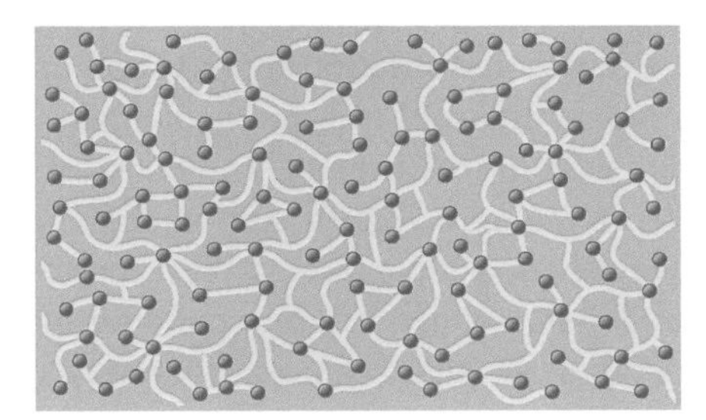

Figure 15.4: Conventional paints consist of binder (orange) and cross-linking agents (red). Right figure: Nanopaints consist of organic binder with high elasticity (yellow) and inorganic nanoparticles with high strength (blue). The tightly packed nanoparticles make the paint scratch resistant [© Daimler AG].

Figure 15.5: Mercedes Benz passenger car painted with nanovarnish [© Daimler AG].

novel paint system has been developed by Daimler Chrysler. These nanopaints are being used in series production of Mercedes-Benz cars.

15.3.1.1 Maintenance and Sealing Products for Clean Surfaces

Maintenance and sealing products are being developed and offered for a series of exterior and interior applications in vehicles. Especially anti-mist and anti-dirt products are used to tune the surface with water- and fat-repellent nanoparticles. Applications include panes and mirrors but plastics and textile surfaces too, making use of the principle of self-organization where nanoscopic components arrange themselves independently. With the result that an invisible ultra-thin layer in the nanometer range prohibits bonding of oils, fats, water and dirt with the surface. Such surface seals consist of nanoparticles that tightly bond with the subsurface and other components, giving the coating the necessary hardness. The components responsible for hydrophobic and fat-repellent properties adjust themselves toward the surface. Most of the surface modifications or coatings against dirt known so far are hydrophobic, which means water repelling. The wettability is so low on these surfaces that the water just rolls off the surface when it is inclined in a certain way, thereby pulling the dirt with it. Many of those surfaces are smooth, others function on the lotus principle. However, practice has shown that the rain drops rolling off leave dirt stains on panes that are clearly recognizable as streaks when dried. Besides, hydrophobic substances tend to produce streaks when used in cleaning substances. If, however, surfaces are constructed such that water can wet it evenly and roll off faster, no runs can develop and the surface will stay clean longer. For this purpose, hydrophilic, which means "water-loving", nanoparticles can be used. The nanoparticles are chosen so that they can adhere to glass especially well. The particles are applied to the surface, where they autonomously form an even, invisible layer. Their negative charge keeps the neighboring particles at a distance and the surface becomes "waterloving."

15.3.1.2 Ultra-Thin Layers for Mirrors and Reflectors

Modern mirrors and headlights with superior optical quality and efficiency are based on glass and plastic components that are equipped with an ultra-thin reflecting layer of aluminum oxide. In the last few years, superior coating processes have been developed for ultra-reflecting layers with thicknesses below 100 nm.

An application field for glasses and mirrors in automotive engineering is being opened by the possibility to equip surfaces with water, fat and dirt-repellent features. These so-called hydrophobic and oleophobic layers are being produced using chemical vapor deposition processes.

15.3.2 Car Body

Car body	Lightweight/composite materials, carbon fiber-reinforced elements	Novel adhesives	Corrosion protection

In nanotechnologies, for forming high-strength steels in the manufacture of body parts, automotive producers had to make an enormous effort to fulfill increased requirements for crash and passenger safety. Forming and joining technologies for ultra-high-strength steel plates are being sought. The recasting of high-strength steel materials in cold state leads to problems relating to size accuracy and unwanted spring-back effects. An alternative to recasting special high-strength steel grades while avoiding these disadvantages is what is called hot forming, where steel sheets are heated to almost 1,000°C and formed in the forming tool and cooled discretely. Using this process, parts of the highest strength and of perfect fit can be produced, for example, for a safety cabin. For the hardening process, the steel sheet is heated up to a temperature of 950°C and pressed red hot into the form directly from the furnace until it reaches its final geometry. The hardening takes place by quick and selected cooling to temperatures below 200°C in the form. The coating of the plates, which has to fulfill several functions, is crucial for the control of such forming processes. The coating has to avoid the impurities of the unit surface caused by the heat treatment process through oxidation of the workpiece surface – called scaling. Furthermore, the friction between the workpiece and the tool has to be reduced and a sustainable corrosion protection assured. Based on a novel approach of the University of Kassel (Germany), a multifunctional coating system was developed that is capable of solving the problem of scaling at high temperatures. The multifunctional protective coating was realized through a combination of a nanotechnological approach with the principles of conventional paint technologies.

Since energy saving is a major goal, the implementation of lightweight materials is a must. Therefore, the search for advanced composite materials and the use of carbon fiber-reinforced elements will continue, and also improvements of conventional material systems, for example Al foam, show advantageous application potential.

15.3.3 Car Interior

Interior	Cover hoods, panels, reinforced plastic parts, movable sunshields	Nanofilters	Energy-independent sensor networks, low-cost sensors, heating elements, thermal management	Antiglare mirrors, touchscreens, antireflective instrument panel, thermal management, interior light designs, organic displays, flexible displays, panel lighting	Water- and soil-repellant coatings and textiles, antismell and antibacterial filters

15.3.3.1 Seat Protection

Car seats often come into contact with wet or dirty clothing. When the door is opened, rainwater or snow can get onto the seats causing water and dirt stains. These unwanted effects can be minimized or even avoided by using materials for the impregnation of fabric and leather coverings. Such substances have spread rapidly in the last few years and are being offered by many producers of accessories in spray cans or fluids. There are products for tissues, leather and leatherette alternatively that are commercially offered on inorganic–organic hybrid materials based on aqueous or alcoholic solutions. After a dipping or spraying application, a thin invisible film builds up covering the fibers. The impregnated layers have a significant hydrophobic and fat repellent effect. Thus the rate of humidity penetration in case of water and pollutants entering is reduced. A similar clever effect is exploited to produce specific fragrances on leather seats. Dispersions with microcapsules that can be sprayed contain scents. Aqueous microcapsule dispersions offer the possibility to furnish for instance leather with diverse fragrances. On the one hand, the capsules have to be small enough to penetrate the leather; on the other hand, they have to be big enough to adhere between the fibers. After spraying the capsules penetrate the leather reaching different depths. The shell of the capsule is only a few nanometers thick and made of polycarbamide. The scent is placed inside. Only when the capsule is stressed mechanically does the nanofilm burst and release the fragrance. If the seat is unused, the capsules stay intact and the scent stays enclosed. What is working on leather already can be applied on textiles as well.

15.3.3.2 Lighting and Displays

Lighting and displays are key factors for interior design. Transparent electrodes or organic light-emitting diode (OLED) electroluminescent foils offer a new degree of freedom for display geometries, and interior lights, but also for ambient light design of modern cars. They have already been introduced in production [10]. An automotive 3D effect instrument panel with transparent displays was reported recently as well [11]. As a possible future feature, haptic feedback displays are enabled by advanced ceramic substrates integrating electronic structures [12].

Quantum dot displays show higher efficiencies than LCD, and also a higher number of colors. They can be made out of flexible materials and are available on the consumer market in televisions since 2015 [13].

Cost-efficient displays may profit from carbon nanotube (CNT) inks for printed electronics or displays [14].

Rear mirrors are also a broad application field for new multifunctional displays. Using nanotechnology, it has already been shown to integrate rear view, navigation and communication functions are being developed [15]. It will also become possible to integrate the screen of the rear-view camera into the mirror.

Driver wellness improvement and driver health monitoring are enabled by nano-enhanced elements like antiallergic or smell filtering, sensors and local flexible heating elements.

The interior of the car is also the "hub" of the passengers' individual electronic devices, so it must provide compatible interfaces to the fast developing telecommunication market, but operating reliable under the harsh automotive conditions.

15.3.4 Chassis and Tires

Chassis and tires	Carbon black in tires, nanosteel	Magnetorheological damping, energy recuperating damping, electronic body control systems

An important role in the properties of tires is played by rubber mixtures. They determine the performance of the cover of tire that makes contact with the road. Usually, 30% of the cover consists of reinforcing filler which makes possible wanted properties such as grip, abrasion resistance, resistance against initial tear and tear propagation. For the optimization of tires, contradictory requirements sometimes have to be fulfilled. On the one hand, the tire needs good grip while, on the other hand, its rolling resistance has to be low. Furthermore, it needs to be resistant to abrasion while being slip-proof, prohibiting the vehicle to slide. The mechanisms behind these tire properties that partly contradict each other are many highly complex chemical and physical interactions between the rubber and the filler material. There are three products that significantly improve the properties of natural rubber: soot, silica and organosilane. Soot is added to the rubber mixture, whose exact composition is secret, as are silica, which need organosilica for the chemical bonding with the rubber. After being produced, the soot and the silica particles have dimensions in the nanometer range. The size and form of the particles as well as the cross-linking with the natural rubber molecules play a key role for tire properties.

Soot and silica, which are the two most important reinforcing chemicals in tires, originally had differing functions: while silica scores higher in passenger cars, especially on the cover of tire, soot prevails in utility vehicles not only because of its excellent abrasion resistance [9]. By using nanostructured soot as a filler in tires, prolonged durability and higher fuel efficiency can be achieved. These novel soot particles have a coarser surface than those that have been used so far. Due to the increased surface energy of the nanoparticles, interactions with the natural rubber molecules increase. This leads to a reduction of inner friction and, consequently, to better rolling resistance. At the same time, the strain vibrations that occur within the material at high speeds are reduced. The consequence is a superior traction, especially on wet roads.

With regard to driving comfort, nanotechnology enables advanced performance of damping systems which adjust their parameters depending on the road condition and driving situation. They make use of magneto- or electrorheological fluids, whose viscosity can be adjusted by electric or magnetic fields with time constants in the order of only a few milliseconds. One advantage compared to conventional solutions is that these systems can be built simple and compact. Since less mechanical parts are needed, they offer also a weight reduction potential.

15.3.5 Combustion Engine and Drive Train

Combustion engine and drivetrain	Low-friction components, wear-resistant coatings, advanced ceramics. Cylinders, pistons, valves, bearings	Catalysts, nanoporous exhaust filters	Piezoinjectors, energy harvesting	Catalysts, catalytic coatings, fuel additives

Saving fuel and reducing emissions are goals of highest importance. Therefore enormous activity is devoted to further improved combustion control (e.g., by exhaust sensing and multiple injection) better filtering, exhaust energy recuperation and reducing of friction.

For the latter, NANOSLIDE® is a technology for coating the cylinder surfaces in Al engines with an iron-carbon alloy by an arc-spray process allows to replace heavy cast iron cylinder bushes and to reduce friction losses by 50%. Since under partial load conditions up to 25% of the power losses in a combustion engine are caused by friction, fuel efficiency of several percent has been gained by this technology, which was first implemented in formula 1 racing cars. NANOSLIDE® was ranked under the top 100 high-tech products in 2014 [16].

Other techniques for tribological surface coatings of highly stressed motor components applied today are arc spraying of diamond like carbon (DLC) layers or physical vapor deposition (PVD) [17]. Also the use of graphene is in discussion, either as oil additive or as nanoparticle dope [18].

15.3.6 Electrics and Electronics

Electrics and electronics	Electronic packages, contacts, relays, connectors	New solder materials, nanosilver sintering or Cu sintering for die attach	GMR sensors, solar cells, memories, driver health monitoring, cooling media, quantum computing	Optoelectronic devices, lighting, optical computing and data transmission

Figure 15.6: ADAS systems in the new S-class Mercedes 2017.

Most innovations in cars today are enabled by electronics. An increasing number of sensors, actuators and information from the environment helps to improve energy consumption, safety and comfort. Some of these Advanced Driver Assistance System (ADAS) functions offered by the new S-class Mercedes are indicated in Figure 15.6. The growing complexity arising from merging these information needs high-speed data processing which will not become reality without increased packing density and speed of microprocessors and memories.

15.3.6.1 Electronic Devices and Functions

The eyes of modern cars consist of cameras, radar laser sensors, ultrasonic sensors and others. Shorter response times, for example, by faster radar microcontrollers, will enable 360° radio cocooning, better pedestrian recognition and thus higher safety on the roads [19].

LIDAR (light detection and ranging) is a remote-sensing method that uses pulsed laser light to measure distances. The first automotive adaptive cruise control systems used LIDAR sensors. Today, development goes to integrated robust and small systems, for example using multichannel laser for scanning LIDAR [20].

Silicon technology has shown a steady decrease of structure sizes throughout its existence, enabling higher integration densities, increased functionality and widespread distribution [21]. This development is generally referred as “Moore’s law.” The limit of this miniaturization will be limited by physical constraints, environmental requirements and production cost. As an example, optical photolithography sets minimum structure limits which to overcome needs patterning processes like extreme ultra violet (EUV) or electron beam processes. A fascinating approach to give new momentum to Moore’s law emerges from replacing silicon by germanium on insulator which could give chips a significant speed boost [22].

Higher microprocessor performance by shrinking feature size is a still unbroken. According to Intel, chips with 10 nm technology will enter the market in 2017, with 5 nm following in 2020 [23]. Nanosheet transistors with 5 nm have already been presented at the very-large-scale integration (VLSI) technology and circuits Symposium in Kyoto, 2017 by IBM [24]. They were fabricated with EUV lithography. As follow-up technology for FiNFETs, they will enable higher computing power and higher efficiency: compared to 10 nm technology, the computing power can be increased by 40%, or the energy consumption can be reduced by 75%.

The future development steps in data processing will be triggered by quantum computing and optical computing. Many research groups worldwide are active in this domain.

Electronic transistor structures below 2 nm have been achieved in the project "single nanometer manufacturing for beyond complementary metal–oxide–semiconductor (CMOS) devices" funded by the European Commission [22]. They mark the next step regarding performance and energy consumption.

A basic requirement for quantum computing is ballistic electron transport without loss of quantum information. Recently, successful ballistic transport of electrons in InAs nanowires on silicon has been reported by IBM [26].

Much research is also performed in the direction of optical computing. This will open the door to highest computing speeds. On-chip optical data transmission will completely avoid ohmic losses in Cu interconnections. One approach investigated at the Jülich Research Center FZI is to realize optical active elements on Si devices using Sn-doped SiGe layers [27].

In terms of data storage, magnetic devices like magnetoresistive random-access memory (MRAMs) are ramping up. They are nonvolatile memory devices, which base on the influence of magnetic fields on the electrical resistance of conductors. Magnetized stacked thin layers store information and change the resistivity of a sandwiched conductor layer. MRAMs allow fast reading and writing, high integration density and a quasi-unlimited number of read/write cycles [28]. The first 1 Gbit MRAM samples with a DDR4 compatible interface are already available as samples. They are produced in a 28 nm CMOS technology in a so-called perpendicular magnetic tunnel junction configuration [29].

Energy-independent sensors can be realized by energy harvesting or by on-chip integrated Li batteries [30]. They offer simple implementation inside the car and a new approach to fail-safe function.

Performance improvements of sensors, for example, of temperature sensors with fast response times, bending sensors or chemical sensors with extreme low-power consumption are addressed with developments in CNTs [31].

Last but not least, electronics offers advantages for exterior lighting [32], like for controllable LED headlights, which have already found their way to the market, or in holographic backlight.

15.3.6.2 Nanopackaging

Reduced feature size is accompanied by an increased sensitivity against stresses [33]. This implies that devices developed for the consumer market do not necessarily fulfill the requirements for automotive applications and care must be taken when using such devices in cars. The ongoing miniaturization trend in microelectronics therefore demands for improved materials and new packages. This includes plastic packages with tailored thermal expansion coefficient and higher thermal conductivity (e.g., by CNTs). A further challenge arises from integration of passive and active devices into printed wiring boards, which may be solved through nanotechnology.

Nanotechnology leads to novel technical challenges for system integration, through combination of multifunctionality/applications, multitechnologies/scales and multimaterials/interfaces. The assembly technology must be able to connect nanoscaled components with macroscopic structures in a reliable manner and to protect the devices at the same time against the outside stresses.

Actual progress and challenges of nanopackaging are reported in a recent special section by the IEEE Transactions on Components, Packaging, and Manufacturing technology [34]. This section includes an overview on nanowires [35] and on the status of graphene capacitors [36]. Also mechanically flexible interconnections to mechanically decouple package and chip are investigated [37].

Nanotechnology opens a variety of new opportunities for joining and attaching different materials. In electronics, assembly processes convert microelectronic and nonelectronic components into full systems. Assembly processes must not be forgotten since they have decisive impact on functionality, reliability and cost of the systems.

15.3.7 Electric Cars

Electric cars	Gearbox, housings, lightweight electric motors, high-speed motors	Super caps, fuel cells, Li-ion batteries	Magnetic materials, power devices, capacitors, battery cooling, battery heating, failure prediction

Energy storage and electric driving range are the most important fields of action in electric cars. Reducing the size and the cost of drive batteries is essential for a successful market development; the driving range for a given battery depends on the efficiency of the drivetrain components and on the overall weight. But also saving electric energy with auxiliary components, computation and air condition is decisive. So besides the electric drivetrain components, thermal isolation and reduction of mechanical losses may not be neglected.

15.3.7.1 Energy Storage

With the spreading of different hybrid drive concepts, systems for the reuse of braking energy, known as recuperation, are being enhanced. Thereby the moving energy is converted into electrical current via generator during braking and stored in accumulators or supercaps. This energy is available at the restart for acceleration. Principally, caps can buffer very quickly and almost without any loss compared to accumulators. Nevertheless, so far only caps with a low capacity have been available for the automotive sector. Using nanotechnology, supercapacitors – short supercaps, ultra-caps or scaps – with high-energy capacities are currently being developed and realized. Supercaps consist of a metallic contact foil and of highly porous layer electrodes with nanostructure, electrolytes and a separator foil. In superconductors, there are 10 to several 1,000 Farad in one package. If a 100-Farad supercap has the size of a match box, this would correspond to a capacity of 100 million standard capacitors connected in parallel with a capacity of 1 Farad. Therefore, the supercap is the link between the common capacitor and the battery. It combines the advantages of a capacitor, which is a quick current supplier, with those of the battery, which is an appreciable energy storage device. In comparison to conventional caps, this is a material that conducts ions but no electron is used instead of nonconducting electrolyte. The charge concentrates on both sides of the boundary surface on a very thin layer of approximately 1 nm, which leads to a high capacity. On the other hand, the application of highly porous layer electrodes allows for a large effective surface, which also contributes to the high capacity. Supercaps feature very high power densities and low internal resistances and are applicable for short-term maximum current supply. Therefore, they are flexibly usable high power storage elements with almost unlimited cycling durability and a solid energy balance and ecobalance.

Even with a 48 V hybrid system, considerable savings in the range of 10 g/km CO_2 are possible [38]. Here the use of supercapacitors can be highly beneficial.

Batteries and fuel cell systems can be further improved by carbon nanofiber materials and nanometer-sized catalysts for increasing energy volumetric density [39].

Li-ion batteries are smaller, less heavy and more powerful than lead-acid or Nickel-metal-hydride batteries. In these batteries, ion-permeable polymer membranes are used as separators between anode and cathode. Based on nanoscaled powders ceramic coatings could be developed, which provide chemical and thermal stability while retaining the flexibility of the polymer foils. Also increasing the surface of the electrodes or an increased ability to incorporate Li ions in the negative electrode will enable batteries with higher capacities [40].

15.3.7.2 Power Electronics

Power electronics is a fast growing market in automobiles due on the one hand to the higher extend of electric controllable actuators and control systems, and on the other hand on the exponential growth of electric-driven cars.

Nanotechnology in combination with wide bandgap materials such as GaN or diamond offers new degrees of freedom for electronic and optoelectronic applications including submillimeter wave generation in GaN-based quantum-well structures [41]. Recently, by optimization of the deposition process for nanocrystalline diamond, researchers succeeded to grow n-doped diamond films, which may open a new chapter in diamond electronics and sensor development [42].

SiC is already used in commercial on-board chargers (as diodes in the power factor correction circuit), because the reduced switching losses enable increased efficiencies. In development of DC–DC converters which supply the 12 V power net from the HV side, SiC metal oxide semiconductor field-effect transistors (MOSFETS) or GaN devices are under investigation, and there is also the potential to increase the efficiency and the power density of drive inverters [43, 44].

The possible increased junction temperatures in order to exploit the full benefit of these materials require robust die attach. Silver sintering as assembly process for power electronic devices shows much better lifetime performance than soldering [45, 46]. The use of nanosilver compositions allows to reduce processing temperature and pressure and to make this method compatible with many other applications. Also Cu nanosintering are under investigation.

15.3.7.3 Passive Devices and Electric Motors

Higher temperature and higher switching speed, and also size constraints lead to a growing need for improved passive devices like capacitors and inductors and for filtering and DC link energy storage.

Due to their high reliability and their self-healing properties, film capacitors are generally used as DC link capacitors in hybrid and electric vehicle. But they are bulky and restrict the maximum operation temperatures [47], so that new solutions are desired. Nanoparticle filled polymer films could offer increased specific capacitance and higher thermal stability [48]. In a project funded by the DoE, capacitors fabricated with a deposition process have been reported to achieve unprecedented compactness [49].

The fast switching properties of enhanced power devices arise new challenges in electromagnetic interference (EMI) filtering. New ferrite materials or filter systems are needed for an efficient use of these technologies.

In the area of electric motors, new magnet materials could reduce the dependence on rare earth metals. Lightweight rotor parts could enable higher rotational speeds needed to increase the power density. Lightweight housings for motors and gearboxes are further fields for employing nano-enhanced materials.

15.4 Outlook on Autonomous Driving

Autonomous driving is being developed worldwide to increase road safety and drivers' convenience. Figure 15.7 shows the development steps toward full automated driving. These functions rely on sensor systems and signal processing to identify the vehicle situation and its environment.

The data from different sensors, such as cameras, ultrasound and radar, are intelligently combined and analyzed. Already today, this allows autonomous driving functions, as realized in the new Mercedes E-Class with DRIVE PILOT (level 2 automation). The most important functions include semiautonomous driving on motorways, country roads and even in the city as well as assistance when changing lanes on multilane roads, for example when overtaking. Furthermore, it is capable in more and more situations of autonomously braking the vehicle if required and providing active assistance during evasive maneuvers.

The process from sensing to actuation in a driving situation is depicted in Figure 15.8. Acquired sensor data must be translated into recognition of the driving situation and the environment, combined with external information by vehicle-to-X systems; the local, behavioral and global drive planning has to be performed (Decision). Control units address the individual actuators. With increasing number of driving situations to be covered, the amount of data to be processed increases exponentially, whereas reaction times must be minimalized. At the same time the safety requirements grow, that is, high requirements on (functional) reliability. Interaction of the car with the environment and off-board data processing are also challenging in terms of data security and encryption. Managing the exponentially increasing complexity of the new features requires new components, architectures and communication systems. Fundamental work concerning nanotechnology in this aspect is performed in the EU project 3Ccar ("Integrated Components for Complexity Control in affordable electrified Cars," ECSEL JU Grant #662192) and transferred into application demonstrators of different automation levels in the project AutoDrive ("Advancing fail-aware, fail-safe, and fail-operational electronic components, systems, and architectures for highly and fully automated driving to make future mobility safer, more efficient, affordable, and end-user acceptable," ECSEL JU Grant #737469).

Making autonomous driving of higher levels become reality, it needs faster and more precise sensors, sophisticated systems for obstacle detection and faster decision-making and probable self-learning capabilities. The requirements are much higher than in conventional ADAS systems, so reliable high-speed devices, networked sensors, fail-safe system design and extended (predictive) self-diagnosis have to be developed. Especially the combination of automated driving with electric propulsion leads to new challenges concerning reliability and fail-safe functionality. In order to perform the automated real-time solutions properly, increasing computing performance on board or in the cloud are required, which calls for extreme

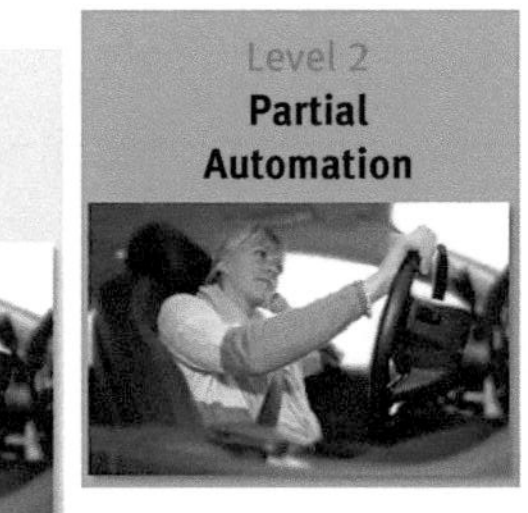
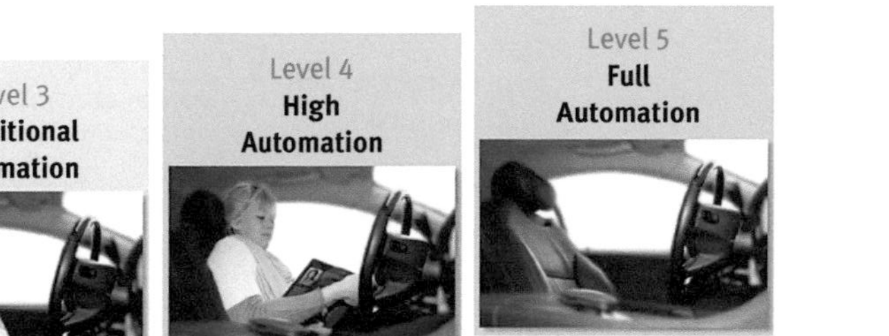

Figure 15.7: Development steps toward autonomous driving [50].

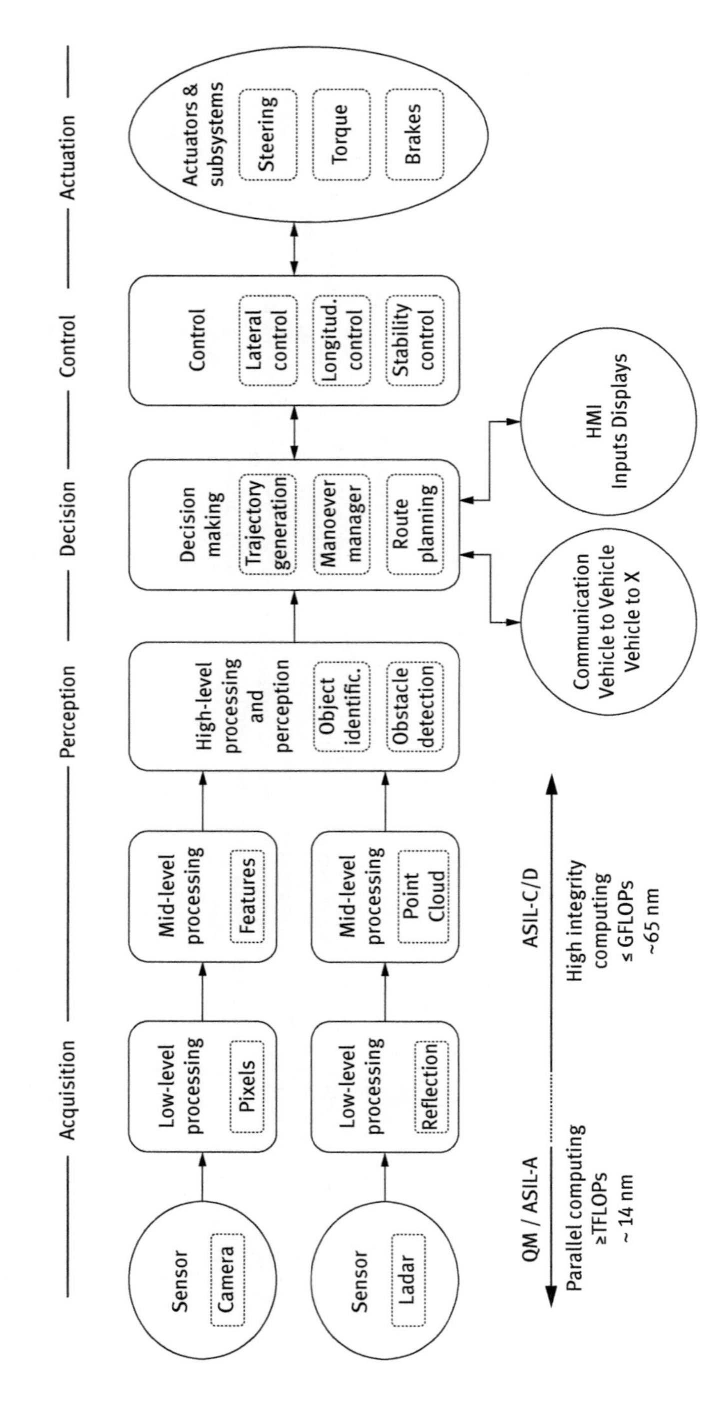

Figure 15.8: Decision-making and actuation process in automated driving (courtesy of M. Dendaluche, Tecnalia).

data security, or deep learning chip ICs, see for example, proceedings of the ISSCC 2016 [51]. In consequence, energy-efficient computing becomes an urgent issue, since computing power will affect the vehicles driving range. External information or data processing relies on car-to-car communication or other cloud functionalities (cf. Figure 15.3). This field is extremely dynamic and offers a huge platform for innovations.

15.5 Benefits Offered by Nanotechnology in the Automobile Industry

Nanotechnology will inevitably be a pacemaker for automobile industry on all levels of the supply chain. Besides enabling progressive designs, higher comfort and innovative communication systems in affordable manner, most applications will offer directly or indirectly environmental benefits and improve the ecologic footprint of transport systems in general. The main implications in that respect are:

a) Saving energy by better combustion processes, higher efficiency and higher level of control by more mechatronics solutions, and lightweight materials.
b) Reducing emissions by improved combustion engines, exhaust filtering and catalysts, vehicle control, energy harvesting and energy recuperation, and energy-optimized navigation including traffic situation as well as route and topographic data.
c) Eliminating local emissions by affordable and versatile electric vehicles with increased range. Not to forget the indirect impact on green electricity generation taking benefit from efficient systems developed for automotive.
d) Reducing the number of accidents by improved traffic surveillance and driver assistance systems, and enabling the performance needed for higher levels of automated driving.

Some of these enabling technologies and applications have been described in earlier chapters.

15.6 Outlook

The properties of nanomaterials, products and processes bear a multitude of opportunities to improve or replace existing solutions in automobiles and enable new functionalities giving higher customer values. In this chapter, some fundamental effects and possible applications are sketched highlighting the high dynamics of developments in transportation in general. Automotive industry is especially facing a major change toward electrification and automated driving. The great breakthroughs in these areas will not be possible without nanotechnology, for example, providing the necessary energy storage and electric drive systems, advanced environmental sensing, computing power and securing high-speed communication. Mass market introduction of networked vehicles may start in the early 2020s, of level 4 automated driving before 2025, of electric cars around 2025. Fully autonomous cars may gain maturity not before 2030. In the future, an increasing utilization of these functions will have also a substantial impact in the supplier industry. But, as usual, market introduction will depend strongly on the price–performance ratio, where nanotechnology will also play an important role.

Acknowledgments: The authors like to thank Mr. Martin Dendaluce-Jahnke from Tecnalia Research and Innovation (Spain) for discussion on requirements for autonomous driving and providing Figure 15.8.

References

1. Hogan B, Costantini F, Lacy E. Manipulating the mouse embryo: a laboratory manual. Cold Spring Harbor: Cold Spring Harbor Laboratory Press, 1986.
2. Werner M, et al., Nanotechnologies in automobiles, Wiesbaden, Germany, ministry of economics, energy, transport and regional development, Volume 3 of the Aktionslinie Hessen-Hessen-Nanotech Series of publications, 2007.
3. Presting H, Koenig U. Future nanotechnology developments for automotive applications. Mater Sci Eng C 2003;23:737–41.
4. TAB – Büro für Technikfolgenabschätzung beim Deutschen Bundestag (2003) TA-Projekt Nanotechnologie, Endbericht, Arbeitsbericht Nr. 92, 2003.
5. Mobility of the future-connected-autonomous-shared-electric, investor relations program, Mondial de l'Automobile, Paris, 30 September 2016 https://www.daimler.com/innovation/autonomous-driving/ Accessed: 30 July 2017.
6. Daimler corporate presentation https://www.daimler.com/dokumente/investoren/praesentationen/daimler-ir-corporatepresentation-september-2016.pdf. Accessed: 30 July 2017.
7. Malani AS, Chaudhari AD, Sambhe RU. A review on applications of nanotechnology in automotive industry. Int J Mech Aerosp Ind Mechatron Manuf Eng 2016;10(1);37–40.
8. Werner M, Wondrak W. Nanotechnology and nanoelectronics for automotive applications. In: Fecht Hans, Werner M, van de Voorde M, editors. The nano-micro interface: bridging the micro and nano worlds, 2nd ed. Vol. 2, Weinheim: WILEY-VCH Verlag GmbH & Co. KGaA, 2015;459–471.
9. Oberholz A. Mit Zwergentechnologie zum Automobil der Zukunft. Hessen-Nanotech 2006;4:6–12.
10. InovisCoat, Elektronik 22/2016, p.32. www.elektroniknet.de.
11. www.delphi.com , https://www.plasticstoday.com/automotive-and-mobility/delphi-automotive-s-3d-dashboard-display-delivers-wow-factor-along-driver-safety-features/123211369056980. Accessed 30 July 2017.

12. Kyocera, Elektronik 22/2016, p. 108.
13. Samsung TV since 2015, Elektronik 14/2017, p. 20.
14. www.brewerscience.com/research. Accessed 08 August 2017.
15. www.Gentex.com/fdm, Elektronik automotive, July 2017, p. 29.
16. http://media.daimler.com/marsMediaSite/en/instance/ko/NANOSLIDE-Mirror-smooth-surface-for-less-friction.xhtml?oid=14316637. Accessed 30 July 2017.
17. Kano M. Overview of DLC-coated engine components. In: Chul Cha S, Erdemir A, editors. Coating technology for vehicle applications. ISBN: 978-3-319-14770-3, 2015:37–62.
18. http://www.nature.com/articles/srep11579. Accessed: 30 July 2017.
19. l www.nxp.com, Elektronik 26/2016, p. 35.
20. www.osram.com, Elektronik 26/2016, p. 34.
21. Hellemans, Nanowire transistors could keep Moore´s law alive. IEEE Spectrum, http://spectrum.ieee.org/semiconductors/devices/nanowire-transistors-could-keep-moores-law-alive, 2013.
22. Ye PD. Switching channels. IEEE Spectrum Dec 2016:40.
23. Intel, IEEE Spectrum, Jan 2017:49.
24. Elektronik 14/2017, p. 23.
25. Elektronik 10/2017, p. 18.
26. SMT Magazine, June 2017, p. 40.
27. Elektronik 7/2017, pp. 38–41.
28. Xu Y, Thompson S (eds.) Spintronic materials and technology. Boca Raton: CRC Press, 2007.
29. https://www.electronicsweekly.com/news/business/first-gbit-mram-2017-08/. Accessed: 8 August 2017.
30. Wilkening M, Sternad M, Uni Graz, Elektronik 22/2016, p. 39.
31. www.brewerscience.com/research. Accessed: 8 August 2017.
32. www.hella.de. Elektronik 26/2016, p. 34.
33. Foucher B. Micro and nanoelectronics design: future challenges in R&D. ICT Workshop contributing to Framework Programme VII, 8–9 January 2008, Brussels, Belgium.
34. Raj PM, Mahajan R. Foreword special section on nanopackaging. IEEE Trans CPMT 2016;6(12):1731.
35. Jiu J, Suganuma K. Metallic nanowires and their application. IEEE Trans CPMT 2016;6(12):1733.
36. Song B, Moon K-S, Wong C-P. Recent developments in design and fabrication of Graphene-based interdigital micro-supercapacitors for miniaturized energy storage devices. IEEE Trans CPMT 2016;6(12):1752.
37. Zhang C, Yang H-S, Thacker HD, Shubin I, Cunningham JE, Bakir HS. Mechanically flexible interconnects with contact tip for rematable heterogeneous system integration. IEEE Trans CPMT 2016;6(11):1587.
38. Timmann M, Renz M, Vollrath O. Challenges and potential of 48 V starting systems. ATZ 2015;115(3):44–48.
39. https://www.vina.co.kr/eng/product/carbon_part.html. Accessed: 1 June 2017.
40. Sternad M, Forster M, Wilkening M, The microstructure matters: breaking down the barriers with single crystalline silicon as negative electrode in Li-ion batteries, https://www.nature.com/articles/srep31712. Accessed: 30 July 2017.
41. Iniewski K. (ed.) Nano-semiconductors: devices and technology. Boca Raton: CRC Press, 2012.
42. Wiora N, Mertens Mohr M, Brühne K, Fecht H-J N-type conductivity in nanocrystalline diamond films for MEMS applications. Microcar 2013, 25–26 February 2013, Leipzig, Germany.
43. http://newsroom.toyota.co.jp/en/detail/2656842.
44. http://www.mitsubishielectric.com/news/2017/0309-a.html.
45. Scheuermann U, Beckedahl P. The road to the next generation power module 100% solder-free design. Proceedings of the 5th International Conference Integrated Power Electronic Systems (CIPS 2008), 11–13 March 2008, Nuremberg, Germany: VDE Verlag Berlin und Offenbach.

46. Lei TG, Calata JN, Lu GQ, Chen X, Lou S. Large-area (>1 cm^2) Die-attach by low-temperature sintering of nanoscale silver paste. Proceedings of the PCIM Europe 2009 Conference, 12–14 May 2009, Nuremberg, Germany. VDE Verlag Berlin und Offenbach, ISBN: 978-3-8007-3158-9.
47. März M, Schimanek E, Schletz A, Poech M-H Mechatronic integration into the hybrid powertrain – the thermal challenge. Proceedings of the International Conference Automotive Power Electronics (APE), 21–22 June 2006, Paris, France. SIA, Suresnes.
48. Wondrak W, Nisch A, Pieger S, Rodewald A, Wagner M, Willikens A, Wurster P. Requirements on passive components for electric and hybrid vehicles. Proceedings CARTS-Europe 2010, 10–11 November 2010, Munich, Germany. ECA, Arlington (VA), USA.
49. Yializis A, High temperature dc-bus capacitor cost reduction and performance improvements. DoE Annual Merit Review Meeting, 06 July 2017, Project ID #: EDT059.
50. https://www.daimler.com/dokumente/investoren/praesentationen/daimler-ir-corporatepresentation-fy-2016.pdf
51. Proceedings of the 2016 IEEE International Solid-State Circuits Conference (ISSCC), IEEE Catalog Number CFP16ISS-PRT, January 2016.

Hans-Jörg Fecht

16 Nanotechnology and diamond: From basic properties to applications

16.1 Introduction

Diamond combines numerous extreme properties, such as high hardness, chemical inertness and exceptional mechanical and other physical properties, including electrical and thermal conductivity [1]. Table 16.1 provides an overview of several material properties in monocrystalline diamond.

This combination of different extreme properties lends itself to a series of applications, for example, in radiation detection, biosensor technology and chemical analysis [3] as well as heat management of high power transistors [4].

However, because of the difficulty in producing monocrystalline diamond in sufficient size and desired shapes it is not suited for a wider range of applications. In contrast, by using hot filament chemical vapor deposition (HFCVD), for example, it is possible to synthesize large-area polycrystalline diamond layers on various substrates to serve as the basis for varied applications.

The capability for depositing diamond on heat-resistant substrates (e.g., Si, Ti and WC [5]) also opens prospects for hybrid products, such as hybrid silicon-diamond technology. The latter leverages well-developed methods for processing silicon (lithography-based microstructuring, etc.) in combination with diamond technology for added or optimized functionalities [6, 7].

By varying the parameters of diamond layering (e.g., substrate temperature, pressure, process gas composition), diamond layers with differing micro- and nanostructure can be produced. Selecting specific process parameters chiefly enables changing the grain size, which allows producing both microcrystalline layers and nanocrystalline diamond layers.

As a result of conditions of growth, microcrystalline diamond layers generally exhibit a pronounced fiber texture. The typically large crystal facets consequently lead to a high roughness of the layer surface coupled with a high defect density between the individual cellular crystallites. Nano- and ultra-nanocrystalline diamond layers, on the other hand, consist of very small, unimodal and virtually equiaxed grains with grain diameters ranging from ca. 4 (ultra-nanocrystalline) to ca. 20 nm (nanocrystalline). This leads simultaneously to very smooth surfaces with a roughness in the grain size range and uniquely through this allows developing new applications and potential products [8, 9].

Nanocrystalline materials in general exhibit changed properties compared to their conventional large-crystal variants. This relates to the small grain sizes and consequent large grain boundary ratios with a disordered atomic structure of reduced density.

https://doi.org/10.1515/9783110547221-016

Table 16.1: Material properties of monocrystalline diamond [1, 10, 11].

	Single-crystalline diamond
Elasticity modulus (GPa)	1,050–1,210
Microhardness (GPa)	50–110
Mass density (g/cm^3)	3.52
Tensile strength (GPa)	93–97 (theoretical)
Thermal conductivity (W/mK)	1,000–2,500
Thermal expansion coefficient (K^{-1})	0.8×10^{-6}
Electron band gap (eV)	5.5
Disruptive strength E (V/cm)	5×10^5

These special physical and chemical features resulting from the material's nanostructure can be customized and further optimized by deliberately varying the grain size. This offers prospects for fine-tuning the material properties of nanocrystalline diamond layers to the requirements of the desired application.

16.2 Basics

The first reproducible process for making synthetic diamonds was developed in 1955 by General Electric, for which it employed high pressures and high temperatures analogous to natural conditions [12]. The first successful experiments to deposit diamond under low pressure had already been made in 1952 [13] and were patented by Union Carbide in 1962 [14–16].

Basically, to achieve this, methane under low gas pressure was piped over heated, already existing diamond crystals. The methane then would react with the hot diamond surface to grow diamond homoepitaxially on the crystallites [14]. In the process, growth is limited by the simultaneous production of graphitic carbon. Refinements made to this method today let us grow diamond layers continuously. Key to this is the targeted seeding with nanodiamond particles and the subsequent growth on these heterogeneous nucleation centers. To achieve the highest possible nucleation density, a nanodiamond suspension is used, which is decomposed by being exposed to ultrasound and an enhanced bias tension between filaments and substrates. This procedure achieves nucleation densities of ca. 10^{11} cm^2.

An appropriate suitable chemical environment created with monocarbon built up from the carbon transport gas (mostly methane) results in diamond layers growing on the nanocrystallites [17–19]. Substrate temperatures during diamond growth are kept between 700 and 7,500°C to create the optimum conditions for diamond growth and suppression of graphite layers [18].

This, therefore, makes the process suitable for scaling up a multiplicity of proffered substrates and other surfaces to industrial manufacturing standards. The

HFCVD process also makes it possible to produce three-dimensional layers, something that is only feasible to a limited extent using plasma-aided CVD processes.

16.3 Physical Properties of Nanocrystalline Diamond

Because diamond, unlike metallic alloys and other ceramic substances, combines extreme properties, it finds numerous potential applications in high energy physics, such as in heat spreaders or radiation detectors [20]. But diamond can provide an innovative further development of traditional silicon technology [21–23] even for mechanical [6, 7] and microelectromechanical components.

16.3.1 Mechanical Properties

16.3.1.1 Elasticity

The high elasticity modulus of diamond plays a role in many applications, for example, in manufacturing parts requiring a high rigidity combined with low weight. Accordingly, diamond films are also used for tips in atomic force microscopes (AFM) [24] in which high hardness and chemical stability yield advantages. Given the reduced bonding strength of grain boundary atoms, a progressive reduction in grain size and the concomitant increase in grain boundary volume lead to a reduction in the elasticity modulus. Figure 16.1 depicts the measurements of the elasticity modulus as a function of grain size. The values were obtained from measurements using nanoindentation and beam deflection using a nanoindentation setup [9, 25].

16.3.1.2 Hardness

Hardness measures the resistance of a material against intrusion by another body. Several hardness scales are available, depending on the measurement procedure used. In Figure 16.1, values measured with the so-called Martens hardness scale are indicated for diamond films by grain sizes. These values were also obtained using nanoindentation.

16.3.1.3 Tensile Strength

As grain size decreases, the critical tensile strength, the point at which catastrophic crack propagation occurs, markedly increases. The values shown in Figure 16.1 were

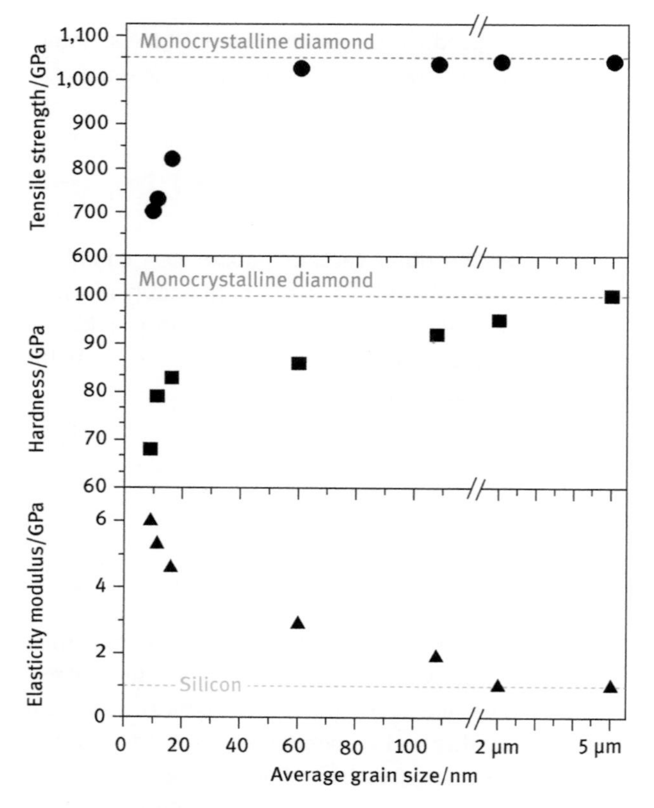

Figure 16.1: Elasticity modulus, hardness and tensile strength as a function of average grain size of different diamond films.

recorded from microbeam deflectors that were deflected in a nanoindenter setup until a break occurred. Clearly, because of the profusion of grain boundaries, nanocrystalline diamond layers exhibit strongly improved cracking characteristics. This is probably caused by the large number of grain boundaries permitting alternative deformation mechanisms – in the form of grain boundaries resembling superplasticity – and thereby improving ductility and cracking behavior that differs diametrically from catastrophic crack propagation in brittle ceramics.

This makes nanocrystalline diamond layers suitable candidates for applications in micromechanics, for example, for membranes in pressure sensors and in the microelectromechanical systems (MEMS) area. The heightened chemical inertness of (nanocrystalline) diamond layers across a wide temperature range offers an added advantage over silicon.

16.3.1.4 Coefficient of Friction

Diamond is known for its exceptional tribological properties, such as a low coefficient of friction and very high wear resistance [26, 27]. This is often explained chemically as resulting from the carbon atoms on the surface actually being terminated and therefore passivated by oxygen, hydrogen or gases in the environment [26].

Simulations of the abrasion process between two polycrystalline diamonds may point to the formation of an amorphous carbon layer on the surfaces [27, 28]. Analyses of the coefficient of friction under different environmental conditions revealed that it is higher in air than in water, which is explained by the liquid's passivating effect [29, 30]. Here, the coefficient of friction depends on the pH value of the liquid [29].

Examples of wear experiments on nanocrystalline diamond films are pictured below.

Figure 16.2 shows a "ball on disk" tribometer schematically. A diamond film-coated Si_3N_4 ball is pressed with a force of 1 N onto a silicon wafer coated with the same diamond film as the ball. With a ball diameter of 1 mm, the contact pressure can be estimated as roughly 0.64 GPa.

In Figure 16.3, "ball on disk" measurements on three different diamond films are indicative of different average grain sizes and the resultant differing surface roughnesses. At the start of the wear experiments, a slight abrasion is observed (coefficient of friction 0.06–0.1 – steel on steel ca. 0.3), which then progressively diminishes until it stops at a very low stationary value between 0.01 and 0.02. As shown in Figure 16.3, the break-in process duration varies, wherein it scales with the average grain size <D> or, rather, the surface roughness indicated by root mean square (RMS).

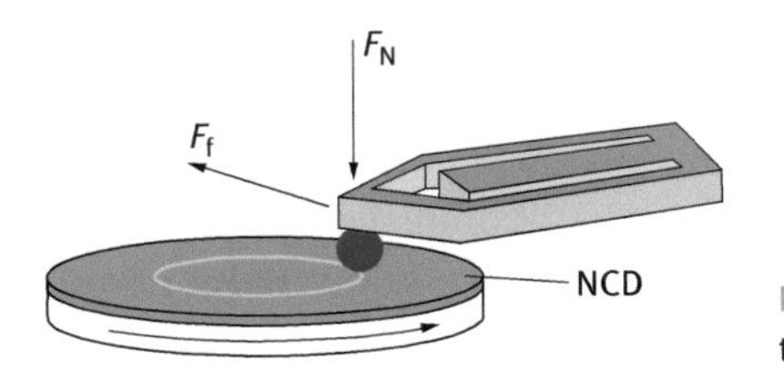

Figure 16.2: Schematic drawing of a "ball on disk" tribometer.

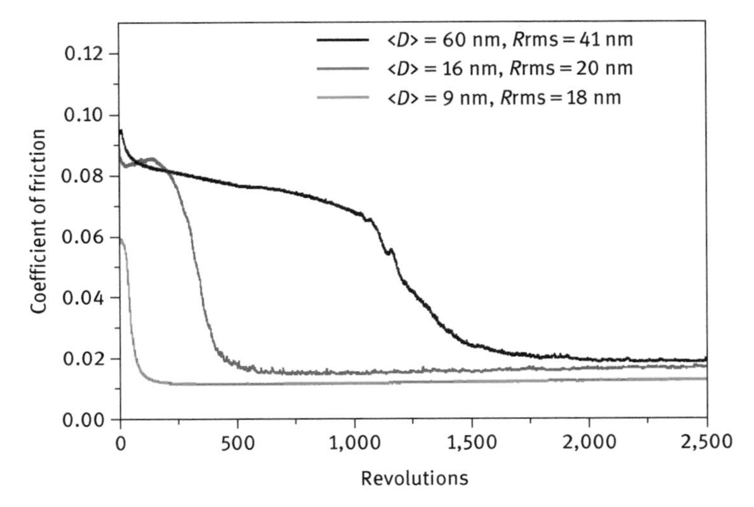

Figure 16.3: Coefficient of friction over time at different surface roughnesses.

The break-in process probably results from the erosion of the roughness of both surfaces. The smoother the surface at the start of the wear experiment, the faster the removal process ends and the steady state is reached [31].

The rate of wear for the experiments illustrated in Figure 16.3 ranges between 10^8 and 145×10^8 mm³/N/m and so is even smaller than the wear rates that appear for different diamond like coatings (DLC) layers [32].

As shown here, the material's mechanical properties (elasticity modulus, hardness, tensile strength, coefficient of friction) can be adapted to broad areas by varying the grain sizes.

16.3.2 Electrical Conductivity

Diamond has an electron gap band of 5.5 eV [10], making it a good insulator. For electronic components, theoretical calculations predict beneficial electrical properties through appropriate doping: high mobility, high disruptive strength and low leakage current [10]. However, doping diamond is relatively difficult [33].

Reproducible, controllable conductivities can be achieved with p-type boron doping. Nevertheless, the boron acceptor in diamond has an activation energy of 0.37 eV and therefore is not fully activated at room temperature. However, by increasing the doping concentration, the activation energy can be lowered [34].

By contrast, an n-type conductivity could be achieved in ultra-nanocrystalline diamond films [35–39]. This offers the simple possibility of creating electrically conductive diamond layers without precursor gases (such as diborane for boron doping) that are harmful to health. Because of the small grain sizes, here the charge carrier mobility is drastically reduced ($\mu = 0.3$–40 cm^2/V/s) [40].

However, conductivity ranging from insulating to "low metallic conductivity" can be set across a wide spectrum by varying the conditions of growth. Another important finding consists of the fact that a piezoresistive effect with a gauge factor up to ca. $k = 10$ could be demonstrated in this material [41]. The gauge factor is defined as the relative change in resistance $\Delta R/R$ to the mechanical strain $\varepsilon = \Delta l/l$:

$$k = \frac{\Delta R}{R}\frac{1}{\varepsilon} = 1 + 2\nu + \frac{1}{\varepsilon}\frac{\Delta\rho}{\rho}$$

Besides the geometrical effect resulting from the material's Poisson number, a change in the specific resistance ρ for semiconductor materials occurs, meaning that a gauge factor exceeding $k = 1 + 2\nu$ can be achieved because of it.

The piezoresistive effect in ultra-nanocrystalline diamond layers makes this material suitable for sensor technology, particularly at high temperatures and in environments for which other materials are not suited or where they are subject to excessive degradation due to environmental influences (e.g., hydrogen and strong acids/bases).

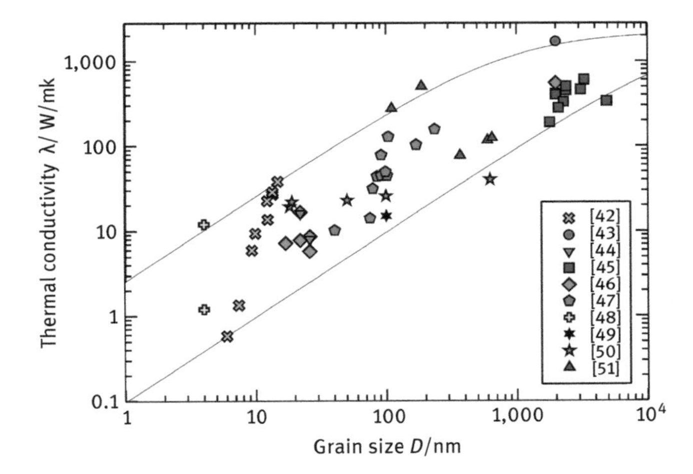

Figure 16.4: Thermal conductivity (at room temperature) as a function of grain size. Values from literature [43–51], plus own measurements [42] by the 3ω method.

16.3.3 Thermal Conductivity

Nanostructuring the diamond layers also changes other material properties. Because of decreased free paths of the phonons through scattering at the large number of grain boundaries and resistance to heat migration between the individual grains, the thermal conductivity properties of nanocrystalline diamond layers are significantly lower than those of monocrystalline diamond [42].

Figure 16.4 shows values from the literature and own measurements of thermal conductivity using the 3ω method with respect to grain size. It emerges clearly that the grain size is determinative for thermal conductivity, which is reduced in nanocrystalline diamond layers by two or three orders of magnitude compared with mono- or polycrystalline diamond.

16.4 Applications of Nanocrystalline Diamond and Hybrids

16.4.1 Microcomponents for Micromechanical Systems

The low coefficient of friction together with the high wear resistance of nanocrystalline diamond layers is an optimal prerequisite for being used in micromechanical components and systems. Of special interest here are, for example, mechanical watch movements in which precision, reliability and long-term stability are important features. Because of the low, approximately 1% coefficient of friction of diamond-coated

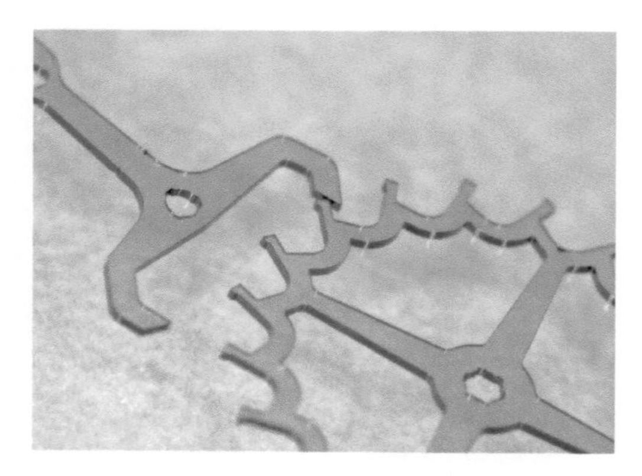

Figure 16.5: Partial view of an escapement consisting of escape wheel and pallet.

surfaces, they allow making lubricant-free watch movements that avoid the drawbacks of using oil-based lubricants (fouling, oil aging).

Figure 16.5, as an example, shows a partial view of a mechanical watch movement's escapement made of nanodiamond-coated silicon components. The nanodiamond layer has an average grain size of about 10 nm and a typical thickness of about 10 µm with an average surface roughness of 10 nm.

In order to investigate the long-term stability and performance of such nanocrystalline escapement parts, the escapement was assembled in an ETA 2892A2 watch movement and characterized repetitively by a dedicated chronoscope. The parts operate under the absence of any lubrication. The chronoscope measures the noise of the escapement by a microphone. The noise signal is then Fourier transformed. The accuracy and amplitude of the oscillator are calculated under the knowledge of the anchors lift angle. The observed run-in-phase at the beginning is induced by the residual nanometer-scale surface roughness of contact zones and can be directly correlated to the curves of Figure 16.3. As seen from Figure 16.6, the movement shows an excellent long-term stability meanwhile for more than 7 years. The test is ongoing. Intermediate standard electron microscopic investigations of the escapement contact zones do not show any indication of wear of the nanodiamond coating.

16.4.2 Nanodiamond-Based Cutting Tools

The high degree of hardness and high wear resistance of nanocrystalline diamond layers can also be utilized in the manufacture of durable, sharp cutting tools. Figure 16.7 pictures a diamond-coated carbide blade that was sharpened after diamond layering by an additional plasma sharpening process.

Using a hybrid diamond–silicon technology, structures suitable for use in cutting tools equipped with ultra-sharp diamond blades and rounded cutting edges with radii of only a few nanometers can be fabricated. Shown in Figure 16.7 is an example: a scalpel dedicated to medical use, for instance in eye (cataract) surgery.

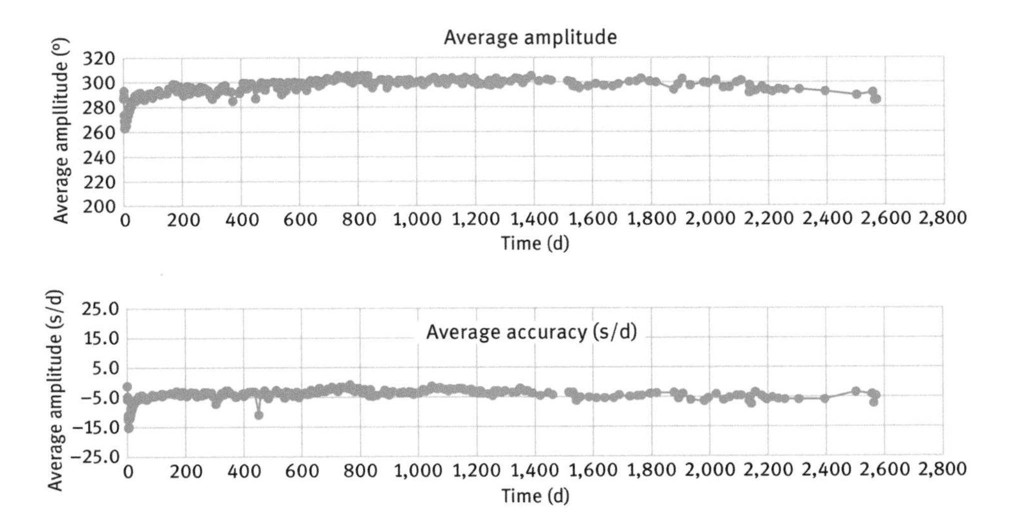

Figure 16.6: Long-term stability of a nonlubricated silicon-based escapement with a nanodiamond thin and ultrasmooth coating. The upper curve exhibits the measured average amplitude (degrees) while the curve below the corresponding average accuracy (seconds per day) over a time period of ca. 2,600 days. The curves show an excellent long-term stability meanwhile for more than 7 years without any visible degradation.

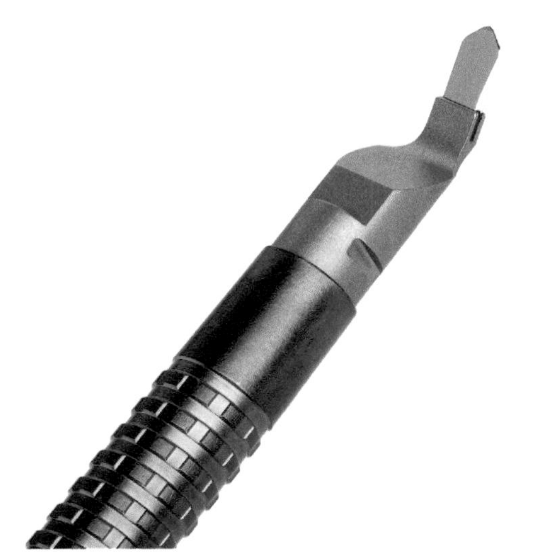

Figure 16.7: Diamond-silicon scalpel with an ultrasharp nanocrystalline diamond tip for medical use.

16.4.3 Piezoresistive Pressure Sensors

Sensors designed to withstand harsh environments pose new challenges for sensor materials, which silicon cannot always or durably meet [20]. One option is to rely on the established silicon sensor technology and shield instruments from the environment through an additional protective coating. In some cases, such as for membrane pressure sensors, this leads to reduced measurement accuracy and requires a complicated, expensive setup [20].

Diamond's chemical inertness and stability, by contrast, makes it an exceptional alternative. In the past, methods of fabricating diamond-based pressure sensors existed yet turned out to be unrewarding because of mechanical inadequacies, but, above all, due to the unreproducible properties of highly faceted microcrystalline diamond layers. This contrasts with the distinctly higher stability and reproducibility of nanocrystalline diamond, as well as its markedly elevated, critical tensile strength. Figure 16.8 shows a first prototype of a hybrid membrane pressure sensor made of silicon and nanocrystalline diamond [21].

The basic structure is shown in Figure 16.9. Located on a silicon substrate (Si) is a nanocrystalline diamond membrane (NCD), onto which the piezoresistive, ultra-nanocrystalline diamond (UNCD) resistors are introduced and fabricated using photolithographic processes.

This first prototype showed that a sensor of this type could, in principle, function durably [21]. Further efforts are being made toward developing and realizing a sensor combining optimized design and improved structural and connection technology to satisfy industrial norms also in the high-temperature regime.

Figure 16.8: Prototype of a hybrid silicon–diamond membrane pressure sensor.

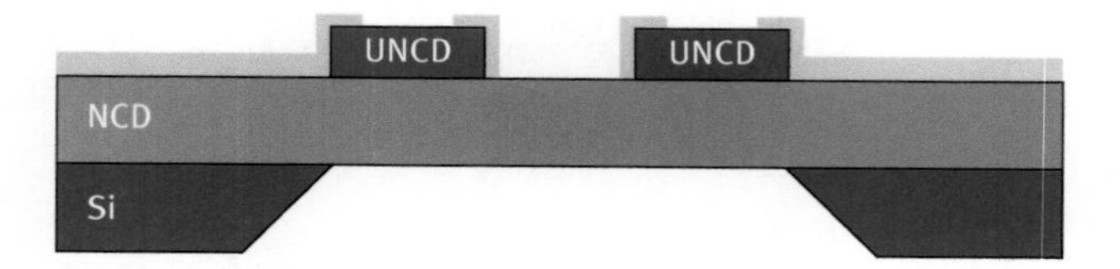

Figure 16.9: Schematic structure of the hybrid silicon–diamond membrane pressure sensor made of Si, nonconductive NCD, conductive UNCD and a gold metallization.

16.5 Summary and Perspectives

As demonstrated, application areas exist for which diamond (and, specifically, nanocrystalline diamond layers) is highly adapted for hybrid technologies, because of its outstanding, customizable properties. The advantages of nanocrystalline when compared with microcrystalline diamond above all reside in the distinctly lower roughness and five times higher tensile strength. These characteristics open the way to producing low-friction, low-wear microcomponents.

As further described, ultra-nanocrystalline diamond layers possess electron conductivity combined with a pronounced piezoresistive effect. This makes the material advantageous to use in sensor technology (e.g., in pressure-sensing equipment). Diamond's exceptional thermal conductivity predestines it for use in heat management of high-performance components. In this connection, important questions include how the diamond layer is deposited on the active structural element, since the synthesis of the diamond coating directly on the component is complicated by the high substrate temperatures during synthesis.

In summary, it can be stated that the physical material properties can be extensively specified and controlled over a broad spectrum through the growth environment and the resulting nanostructures. In the process, nanocrystalline diamond layers can be optimized to meet the requirements of a variety of applications.

Acknowledgments: High-level technical support and know-how by P. Gluche, K. Brühne, M. Mohr, M. Mertens and N. Wiora as well as financial support by BMBF are gratefully acknowledged.

References

1. Fecht H-J, Brühne K, Gluche P. Carbon-based nanomaterials and hybrids: synthesis, properties, and commercial applications. Pan Stanford Publishing, Singapore, 2014
2. Bachmair F. Diamond Sensors for future high energy experiments. Nucl Instrum Methods Phys Res Sect A 2016;831:370–7.
3. Luong JH, Male KB, Glennon JD. Boron-doped diamond electrode: synthesis, characterization, functionalization and analytical applications. Analyst 2009;134:1965–79.
4. Tadjer MJ, Anderson TJ, Feygelson TI, Hobart KD, Hite JK, Koehler AD, Wheeler VD, Pate BB, Eddy CR Jr, Kub FJ. Nanocrystalline diamond capped AlGaN/GaN high electron mobility transistors via a sacrificial gate process. Phys Status Solidi A Appl Mater 2006;213:893–7.
5. Liu H, Dandy DS. Studies on nucleation process in diamond CVD: an overview of recent developments. Diamond Relat Mater 1995;4:1173–88.
6. Gluche P, Flöter A, Ertl S, Fecht H-J. Commercial applications of diamond-based nano- and microtechnology. In: Fecht H-J, Werner M, editors. The nano-micro interface: bridging the micro and nano worlds. Wiley-VCH, Weinheim, 2004.

7. Wiora M, Gretzschel R, Strobel S, Gluche P. Industrial applications and commercial perspectives of nanocrystalline diamond. In: Fecht H-J, Brühne K, Gluche P, editors. Carbon-based nanomaterials and hybrids: synthesis, properties, and commercial applications. Pan Stanford Publishing, 2014.
8. Wiora M, Brühne K, Caron A, Flöter A, Gluche P, Fecht H-J. Synthesis, reliability and applications of nanocrystalline CVD-grown diamond and micro device fabrication, Technical Proceedings of the 2008 NSTI Nanotechnology Conference and Trade Show, NSTI-Nanotech, Nanotechnology 2008;1:206–9.
9. Wiora M, Brühne K, Flöter A, Gluche P, Willey TM, Kucheyev SO, Buuren AW van, Hamza AV, Biener J, Fecht H-J. Grain size dependent mechanical properties of nanocrystalline diamond films grown by hot-filament CVD. Diamond Relat Mater 2009;18:927–30.
10. Sauer R. Synthetic diamond – basic research and applications. Cryst Res Technol 1999;34:227–41.
11. Hess P. The mechanical properties of various chemical vapor deposition diamond structures compared to the ideal single crystal. J Appl Phys 2012;111:051101.
12. Bundy FP, Hall HT, Strong HM, Wentorf RH. Man-made diamonds. Nature 1955;176:51–5.
13. Angus JC. Diamond synthesis by chemical vapor deposition: the early years. Diamond Relat Mater 2014;49:77–86.
14. Eversole WG. Synthesis of Diamond, US Patent 3 030 187 (1962).
15. Eversole WG. Synthesis of Diamond, US Patent 3 030 188 (1962).
16. Eversole WG. Canadian Patent 628 567 (1961).
17. Spitsyn BV, Bouilov LL, Derjaguin BV: Vapor growth of diamond on diamond and other surfaces. J Cryst Growth 1981;52:219–26.
18. Matsumoto S, Sato Y, Tsutsumi M, Setaka N. Growth of diamond particles from methane-hydrogen gas. J Mater Sci 1982;17:3106–12.
19. Wiora M, Brühne K, Fecht H-J. Synthesis of nanodiamond. In: Fecht H-J, Brühne K, Gluche P, editors. Carbon-based nanomaterials and hybrids: synthesis, properties, and commercial applications. Pan Stanford Publishing, Singapore, 2014.
20. Bähr M. Synthetische Diamantschichten und Sensortechnologie. Mess- und Sensortechnik 2017:62–5.
21. Mohr M, Behroudj A, Wiora N, Mertens M, Brühne K, Fecht H.-J. Fabrication and characterization of a hybrid silicon and nanocrystalline diamond membrane pressure sensor. Quantum Matter 2017;6:41–4.
22. Kusterer J, Kohn E, Lüker A, Kirby P, O'Keefe MF. Diamond high speed and high power MEMS switches, 4th EMRS DTC Technical Conference – Edinburgh 2007.
23. Kusterer J, Kohn E. CVD Diamond MEMS. In: Sussmann RS, editors. CVD Diamond for Electronic Devices and Sensors. Wiley, 2009.
24. Malavé A, Oesterschulze E. All-diamond cantilever probes for scanning probe microscopy applications realized by a proximity lithography process. Rev Sci Instrum 2006;77:043708.
25. Mohr M, Caron A, Herbeck-Engel P, Bennewitz R, Gluche P, Brühne K, Fecht H.-J, Young's modulus, fracture strength, and Poisson's ratio of nanocrystalline diamond films. J Appl Phys 2014;116:124308.
26. Holmber K, Matthews A. Coatings technology, 2nd ed. Elsevier Science, 2009.
27. Pastewka L, Moser S, Gumbsch P, Moseler M. Anisotropic mechanical amorphization drives wear in diamond. Nat Mater 2011;10:34–8.
28. Fineberg J. Diamonds are forever – or are they? Nat Mater 2011;10:3–4.
29. Grillo SE, Field JE. The friction of CVD diamond at high Hertzian stresses: the effect of load, environment and sliding velocity. J Phys D: Appl Phys 2000;33:595–602.
30. Alahelisten A. Abrasion of hot flame-deposited diamond coatings. Wear 1995;185:213–24.

31. Wiora M, Sadrifar N, Brühne K, Gluche P, Fecht H-J. Correlation of microstructure and tribological properties of dry sliding nanocrystalline diamond coatings. Technical Proceedings of the NSTI Nanotechnology Conference and Expo NSTI-Nanotech 2011;2:164–7.
32. Jarratt M, Stallard J, Renevier NM, Teer DG. An improved diamond-like carbon coating with exceptional wear properties. Diamond Relat Mater 2003;12:1003–7.
33. Kalish R. The search for donors in diamond. Diamond Relat Mater 2001;10:1749–55.
34. Lagrange JP, Deneuville A, Gheeraert E. Activation energy in low compensated homoepitaxial boron-doped diamond films. Diamond Relat Mater 1998;7:1390–3.
35. Bhattacharyya S, Auciello O, Birrell J, Carlisle JA, Curtiss LA, Goyette AN, Gruen DM, Krauss AR, Schlueter J, Sumant A, Zapol P. Synthesis and characterization of highly-conducting nitrogen-doped ultra-nanocrystalline diamond films. Appl Phys Lett 2001;79:1441–3.
36. Williams OA, Curat S, Gerbi JE, Gruen DM, Jackman RB. n-type conductivity in ultrananocrystalline diamond films. Appl Phys Lett 2004;85:1680–2.
37. Wiora N, Mertens M, Mohr M, Brühne K, Fecht H-J. Synthesis and characterization of n-type nitrogenated nanocrystalline diamond. Micromater Nanomater 2013;15:1619–2486.
38. Zapol P, Sternberg M, Curtis LA, Frauenheim T, Gruen DM. Tight-binding molecular-dynamics simulation of impurities in ultrananocrystalline diamond grain boundaries. Phys Rev B – Condens Matter Mater Phys 2001;65:045403.
39. Birrell J, Carlisle JA, Auciello O, Gruen DM, Gibson JM. Morphology and electronic structure in nitrogen-doped ultrananocrystalline diamond. Appl Phys Lett 2002;81:2235–7.
40. Mertens M, Lin I.-N, Manoharan D, Moheinian A, Brühne K, Fecht H-J. Structural properties of highly conductive ultra-nanocrystalline diamond films grown by hot-filament CVD. AIP Adv 2017;7:015312.
41. Wiora N, Mertens M, Mohr M, Brühne K, Fecht H.-J: Piezoresistivity of n-type conductive ultrananocrystalline diamond. Diamond Relat Mater 2016;70:145–50.
42. Mohr M, Daccache L, Horvat S, Brühne K, Jacob T, Fecht H.-J. Influence of grain boundaries on elasticity and thermal conductivity of nanocrystalline diamond films. Acta Mater 2017;122:92–98.
43. Morelli DT, Beetz CP, Perry TA. Thermal conductivity of synthetic diamond films. J Appl Phys 1988;64:3063.
44. Liu WL, Shamsa M, Calizo I, Balandin AA, Ralchenko V, Popovich A, Saveliev A. Thermal conductivity in nanocrystalline diamond films: effects of the grain boundary scattering and nitrogen doping. Appl Phys Lett 2006;89:171915
45. Graebner JE, Mucha JA, Seibles L, Kammlott GW. The thermal conductivity of chemical-vapor-deposited diamond films on silicon. J Appl Phys 1992;71:3143.
46. Shamsa M, Ghosh S, Calizo I, Ralchenko V, Popovich A, Balandin AA. Thermal conductivity of nitrogenated ultrananocrystalline diamond films on silicon. J Appl Phys 2008;103:083538.
47. Plamann K, Fournier D, Anger E, Gicquel A. Photothermal examination of the heat diffusion inhomogeneity in diamond films of sub-micron thickness. Diamond Relat Mater 1994;3:752–6.
48. Angadi MA, Watanabe T, Bodapati A, Xiao X, Auciello O. Thermal transport and grain boundary conductance in ultrananocrystalline diamond thin films. J Appl Phys 2006;99:114301.
49. Sermeus J, Verstraeten B, Salenbien R, Pobedinskas P, Haenen K, Glorieux C. Determination of elastic and thermal properties of a thin nanocrystalline diamond coating using all-optical methods. Thin Solid Films 2015;590:284–92.
50. Engenhorst M, Fecher J, Notthoff C, Schierning G, Schmechel R, Rosiwal SM. Thermoelectric transport properties of boron-doped nanocrystalline diamond foils. Carbon 2015;81:650–62.
51. Rossi S, Alomari M, Zhang Y, Bychikhin S, Pogany D, Weaver JM, Kohn E. Thermal analysis of submicron nanocrystalline diamond films. Diamond Relat Mater 2013;40:69–74

P. Gluche

17 Nanocrystalline diamond films

17.1 Introduction

Diamond possesses outstanding material properties. It is well known to be the hardest material on earth. It has the highest thermal conductivity, which exceeds that of copper by a factor of 5. It is optically transparent from the near infrared to the far ultraviolet. Intrinsic diamond is highly insulating and shows a high electrical breakdown field strength, but it can also be doped and shows at very high doping concentrations also a metal-like electrical conductivity. Diamond is chemically inert and biocompatible [1]. It is also well known that diamond surfaces exhibit a very low coefficient of friction in the absence of any supplemental lubrication. No other material shows equivalent combinations of such extreme properties. Therefore, diamond is considered to be an ideal material for mechanical, thermal and electrical applications [2–4].

Unfortunately, diamond is expensive and due to its extreme material properties diamond cannot be machined in a conventional way. With the exploration of chemical vapour deposition (CVD) diamond during the last 30 years, big natural or synthetic diamond crystals can now be substituted by the deposition of thin diamond layers. This enables the utilization of diamond in industrial applications on a larger scale such as diamond cutting blades and tools, micromechanical parts, diamond spheres for metrology and scalpel blades for microsurgery. Especially the development of nanocrystalline diamond (NCD) films can be considered as a quantum jump for the industrialization of this innovative material [5–7].

17.2 Nanocrystalline Diamond

17.2.1 Diamond Deposition

In this section we are discussing the utilization of NCD films, which are deposited by the hot filament CVD technique [8, 9]. As thin films in the micrometre range are typically not self-sustaining, a suitable substrate has to be chosen carefully in order to provide the best possible support. Today, there are multiple suited substrates available, as cemented carbides (Co <10 vol%), silicon, silicon-based ceramics (e.g., Si_3N_4 and SiC), refractory metals such as W, Nb and Mo and most of their alloys and their carbides. Unfortunately, all the above-mentioned substrates are difficult to machine and in consequence expensive in contrast to steel. The most common substrates are cemented carbides for tools and silicon for micromechanical parts, electronics and scalpels.

https://doi.org/10.1515/9783110547221-017

This lack of cheap and easy to machine substrates is still a barrier for a larger commercial use of diamond films in industrial applications.

In case of cemented carbide substrates, a delicate pre-treatment procedure is necessary to deplete the cobalt concentration at the substrate surface, as diamond does not grow on cobalt [10]. After cleaning and nucleation, the substrates are placed in the reactor vacuum chamber close to the filaments. Methane (CH_4) serves as carbon source. By heating the filaments up to a temperature of 1,900–2,200°C the methane is dissociated and CH_x radicals are generated. At the substrate surface the carbon "condenses" as strong bound carbon (sp^3, diamond) and/or weak bound carbon (sp^2, "graphitic" carbon). The undesired sp^2 bound carbon fraction has to be removed. This is achieved simultaneously during the deposition process by adding a significant amount of hydrogen to the gas phase (typically 1–5% CH_4 in 99–95% H_2). The atomic hydrogen is dissociated accordingly when passing the filaments and is searching for a partner. It reacts with the sp^2-bound carbon molecule to a CH_x radical and returns back to the gas phase. Thus, there is a dynamic equilibrium of etching and deposition of sp^2 and sp^3 hybridized carbon and if the corresponding rates are adjusted correctly, a phase pure diamond film will grow. This equilibrium is controlled by the process parameters. In contrast to physical vapour deposition (PVD, sputtering), no ion bombardment is needed to form the strong sp^3 bond and the diamond film shows therefore a very homogeneous morphology. Furthermore, atomic hydrogen is not able to recombine in the absence of a collision partner. Thus, the atomic hydrogen is able to diffuse deeply into the deposition region and enables therefore highly isotropic diamond coatings even on complex-shaped substrate bodies.

In the following section, we are focussing on NCD films with a thickness of typically 6–30 μm and an average grain size of approximately 10 nm. These films show a very low surface roughness (approximately 10–20 nm) and a very high fracture toughness of up to 5 GPa [11].

The hot-filament diamond CVD process is scalable. Biggest commercial reactors exceed already a power consumption of 100 kW and a deposition area of more than 0.5 m^2 [12]. It is conceivable that the reactor design and upscaling processes have rather started and thus we expect for the near future larger and more efficient reactors, capable to coat surfaces in the mutli-m^2 range.

17.3.2 Properties and Morphology

Table 17.1 shows the material properties of diamond in comparison to other common materials. The diamond column shows in fact a range of values, which is correlated mainly to the morphology of the material. Since diamond has the smallest lattice constant of all materials, there's no foreign substrate suited for diamond epitaxy. In consequence, the CVD of diamond on a foreign substrate yields always to a polycrystalline

Table 17.1: Material properties of diamond versus other commonly used materials.

Property	Diamond	Cemented carbide	Si	Steel	Ti
Young's modulus E_m (GPa)	700–1,143	430–600	110–190	190–250	115
Fracture strength σ_B (GPa)	2–10.3	0.8–2.2	0.3–1.4	0.4–1.1	–
Indentation hardness HiT (GPa)	68–105	–	11–13	–	2–3
Density (g/cm^3)	3.2–3.5	6–15	2,321	6–8	4.5
Thermal expansion coeff. 10^{-6}/K	1	5–8	2.6	12	8.6

structure. As the CVD exhibits typically substrate temperatures of 600°C [10], a careful selection of the substrate is necessary to consider a low thermal expansion coefficient (typically below 5 ppm/K), the temperature stability and the unwanted solubility of carbon at elevated temperatures, as well as the desirable forming of a stable carbide at the interface for adhesion purposes. For mechanically stressed applications, it is furthermore advantageous to utilize substrates having an adapted Young's modulus in order to prevent interfacial stress and delamination of the diamond film.

The diamond column of Table 17.1 shows a rather wide range of values for each property, which corresponds to the fact that the diamond structure can be single crystal or polycrystalline (oriented/statistically oriented). Grain size and texture have a strong influence on the material properties of CVD diamond [13]. This fact, however, allows on the other hand for engineering the diamond film properties towards the application and the substrate characteristics. Especially NCD having grain sizes below 20 nm shows very interesting mechanical properties related to the substrate properties, for example, a decreased and "adapted" Young's modulus and a very high and isotropic fracture toughness [11].

A columnar grain boundary structure is, among others, responsible for a low residual fracture strength. It can be efficiently supressed by the reduction of the average grain size below approximately 20 nm as shown in Figure 17.1.

Figure 17.2A shows a cross-sectional scanning electron microscope (SEM) picture of the as-deposited NCD film on cemented carbide. Figure 17.2B shows the surface roughness of an NCD film deposited on a polished silicon wafer surface. The growth side surface roughness is around rms = 20 nm, measured by atomic force microscopy (Figure 17.2).

17.3 Applications

In this section, we will focus on three applications of great industrial relevance:

- Plasma sharpened diamond (PSD) blades for roll to roll conversion
- PSD tools for machining of carbon fibre-reinforced plastics (FRP)
- Diamond-coated spheres for metrology

Polycrystalline diamond (grain size >>100 nm)

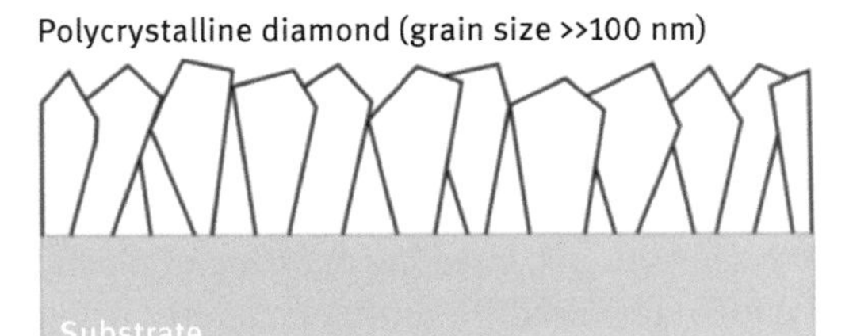

Nanocrystalline diamond (grain size 3–100 nm)

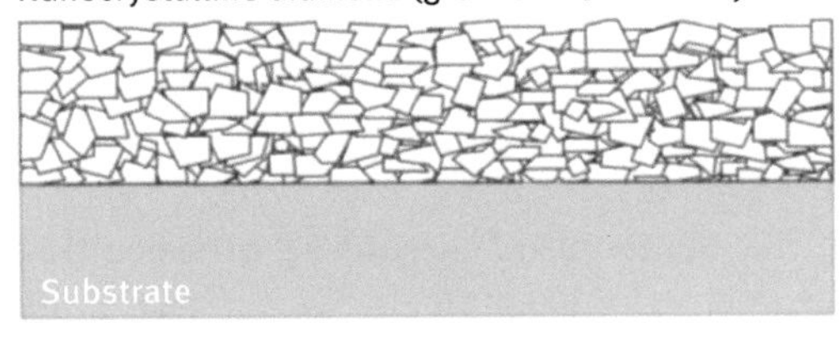

Figure 17.1: Schematic morphology of poly- and nanocrystalline diamond films (left). Scanning electron microscope cross sections of poly- and nanocrystalline diamond films (right).

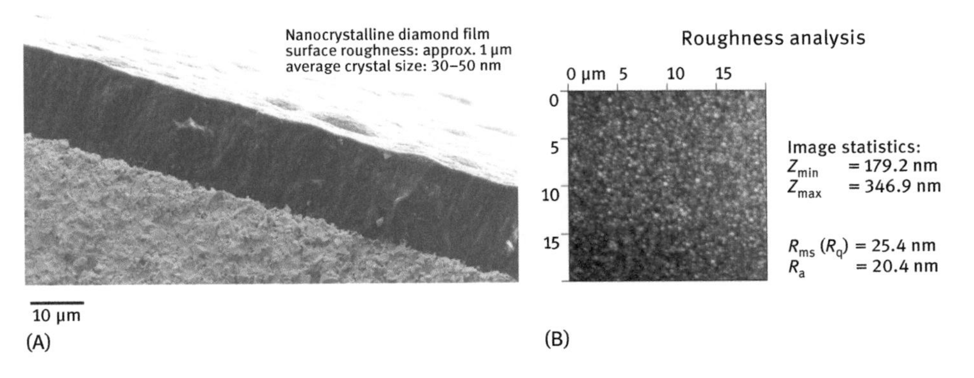

Figure 17.2: Cross section of a nanocrystalline CVD diamond film on a carbide blade (scanning electron microscopy picture) (A) and roughness analysis of a nanocrystalline CVD diamond film deposited on a very smooth surface by atomic force microscopy (B).

17.3.1 Plasma Sharpened Nanocrystalline Diamond Blades (PSD Blades)

Slitting is mainly seen as an insignificant part of web handling and changing blades permanently is a necessary evil. Blades with a coated finish are often regarded as being not sharp enough to fulfil the quality requirements. A new approach to overcome this dilemma is the use of CVD NCD coated, PSD blades. For cutting blades, this process is meanwhile well established and industrialized. First, prototypes entered in 2003 into the market. These types of blades typically exceed the lifetime of an uncoated steel (ceramic/carbide) blade by a factor of 800–1,000 (20–40) times [7]. This enables for the first time a very high and constant cutting quality in combination with a very low maintenance effort.

In almost all industrial applications, cutting is an important process in the production chain. For the converting industry, for example, sheeting or sizing of plastic foils, paper, fibre, cloth, the cutting procedure is realized by slitting machines using circular and slotted blades. Since there's a strong tendency towards tougher and more robust materials, the cutting performance and the lifetime of standard steel and ceramic/cemented carbide blades often reach their limit.

The most important parameters determining the performance of a cutting blade are:

a) The sharpness, better described as cutting ability
This property includes many parameters, like the radius of curvature r_c at the cutting edge, the friction properties, the blade geometry, for example, the wedge angle, the blade thickness, the surface roughness and many others. In the following, we concentrate on the most important parameter, the radius of curvature r_c, that determines the separation of the material at the cutting edge at a microscopic scale.

b) The durability or lifetime, better described as edge-holding property
This property includes also many parameters, like the hardness of the blade material or the microgeometry. But it is also influenced by the cutting process itself like the cutting speed, friction and temperature.

A blade fails, if the cutting quality falls below the pegged requirements. This corresponds to a cutting ability failure. Standard wear is the fundamental mechanism that "dulls" steel blades by increasing the radius of curvature r_c. The blade has to be exchanged if a certain maximum radius of curvature r_{cmax} is exceeded. The harder the blade material, the lower the wear propagation and therefore, the higher the edge-holding property. The hardness of the blade can be increased by applying a thin hard coating (e.g., by PVD) like TiN or CrN on the blade surface of a finished steel blade, for example. However, this approach reduces automatically the cutting ability, since the radius of curvature r_c is increased by the superposed hard coating layer. From many experiments in industrial environment, a maximum lifetime increase by a factor of 5 can be expected by the hard coating in contrast to an uncoated standard steel blade. Another approach is the substitution of the blade material, for example, by cemented carbide (also called solid carbide, tungsten carbide (TC)) or ceramics (typically ZrO). The production of these hard blade materials requires diamond grinding tools, and the fabrication process is therefore more expensive.

Such blades exhibit the lifetime of a standard steel blade typically by a factor of 20–40.

Here, we report on the utilization of the hardest possible coating – diamond. The NCD film thickness is in comparison to PVD hard coatings rather thick (12–18 µm) and needs therefore to be resharpened after coating.

The cutting ability (small r_c) is typically achieved by mechanical grinding and polishing. Even standard razor blades are fabricated this way. Another sophisticated approach is to utilize wet chemical etching in order to form a pointed cutting edge applied for instance for the fabrication of scalpel blades. This procedure is, however, not easy to control. In this chapter we will report on a nonmechanical procedure to sharpen blades based on the utilization of a plasma sharpening process.

17.3.1.1 Diamond Blade Fabrication

As substrate we use a cemented carbide blade. These blades are state of the art and available in different qualities on the market. Related to the granular structure of cemented carbide, the cutting edges of such blades are always less regular and are containing much more micro-chipouts than that of ductile standard steel blades. In consequence, the radius of curvature of an as-ground TC-blade exceeds significantly that of steel blades and is typically measured in the range of $r_c = 1–3$ µm. The TC grains having a typical size of 0.5–2 µm and in consequence the minimum radius of curvature is in the range of the grain size. Figure 17.3 shows a nicely machined cutting edge of a cemented carbide blade.

In contrast to thin hard PVD coatings, we are using a thick (12–18 µm) thick NCD film. Thus, the cutting ability is fully lost after the diamond deposition. Furthermore, as mentioned already in section 17.2.1, it is inevitable to pre-treat the cemented carbide in order to deplete the surficial Co-concentration prior to the diamond film deposition. This process will lead to a significant increase of r_c. After the pre-treatment, the nanocrystalline CVD diamond film forms a highly isotropic coating around the cutting edge. In consequence, the radius of curvature after diamond coating will significantly exceed the film thickness. Figure 17.3 shows SEM photographs of the as-ground cemented carbide blade and after diamond coating at the same magnification. The radius of curvature r_c is significantly increased after diamond coating. Such an as-coated blade has lost its cutting ability completely.

The high coating thickness fulfils several requirements:
- It stabilizes the substrate diamond interface mechanically and
- allows for a high residual wear volume after plasma sharpening.

In order to reduce the radius of curvature r_c we apply a plasma process. In contrast to the diamond deposition, this process takes advantage from a directed ion bombardment and thus, a high anisotropy. Activated ions and radicals are accelerated under a small angle towards the cutting edge. This results in an anisotropic removal of the NCD film. The angle of ion bombardment can be adjusted in such a way that approximately 75% of the initial diamond thickness at the cutting edge remain unaffected. On the bevel planes, however, the removed thickness is higher and can exceed 50% of the initial coating thickness. This process can be considered similar to “sand-

Figure 17.3: SEM micrograph of the cutting edge of the cemented carbide substrate. Left: as-ground; right: after CVD diamond coating (same magnification). The increase of the radius of curvature r_c is clearly visible.

blasting", however, supported by a reactive oxydizing component and having the smallest possible "blasting powder" with grain sizes in the range of atoms. The ion energy can be adjusted very precisely allowing for a controlled impact of the ions and therefore a highly reproducible removal rate in the range of several microns per hour. This slow but reproducible process allows for a precise fine adjustment of the micro-geometry of the cutting edge in the nanometre scale.

The cross-sectional SEM photographs of blades (Figure 17.4) show clearly the sharpening effect resulting in a visible reduction of the radius of curvature r_c from

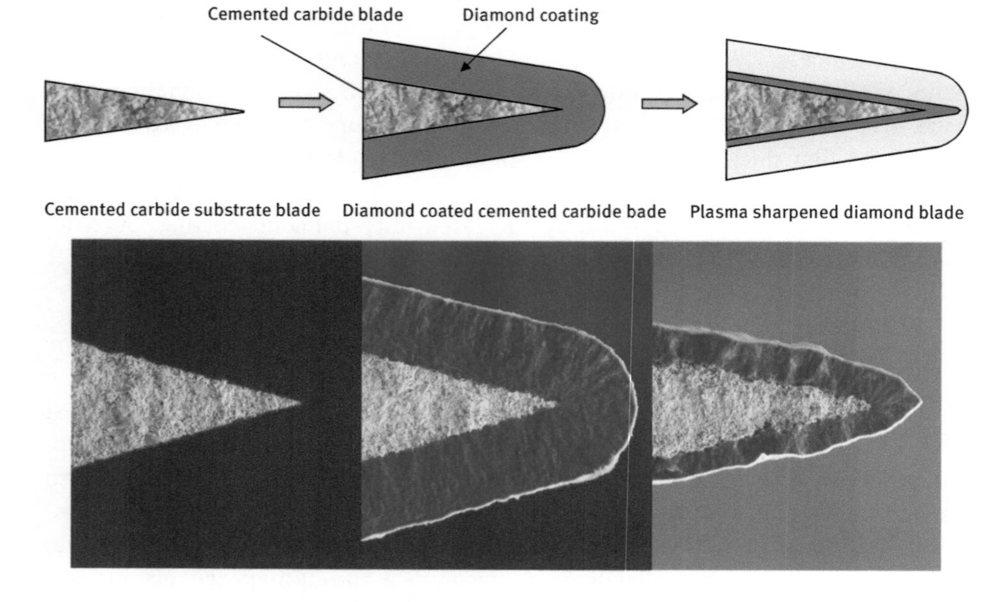

Figure 17.4: Cross sections taken by SEM of uncoated carbide blade (left); as-coated carbide blade (middle) and a diamond coated and plasma sharpened tungsten carbide blade (right).

several micrometres to less than 200 nm. Figure 17.5 shows a top view of the cutting edge before and after plasma sharpening.

The intensity of the plasma sharpening process and in consequence the resulting r_c is well controllable by the exposed time in the plasma. This means, the radius of curvature r_c can be adjusted to the application and thus the customer's needs.

Figure 17.6 shows the extreme case of a plasma sharpened diamond razor blade for shaving purposes at different magnifications. The radius of curvature could be successfully reduced to a level below 50 nm.

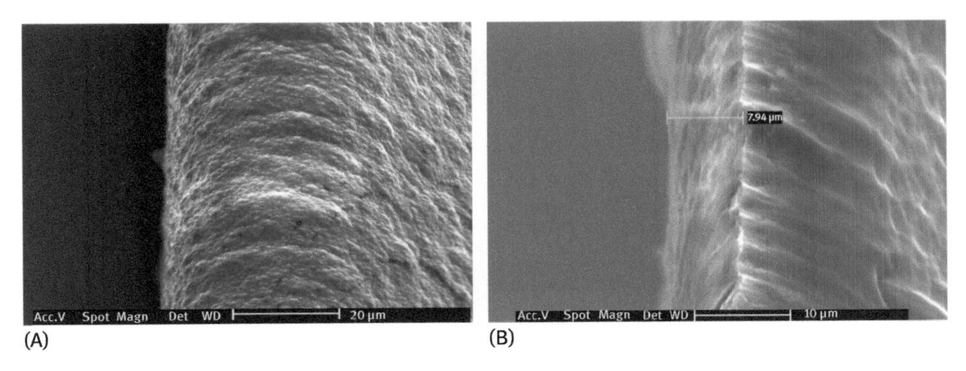

Figure 17.5: SEM top view of the cutting edge of as-coated (A) and plasma sharpened diamond blade (B). The right picture was taken at a higher magnification in order to see the microgeometry of the cutting edge.

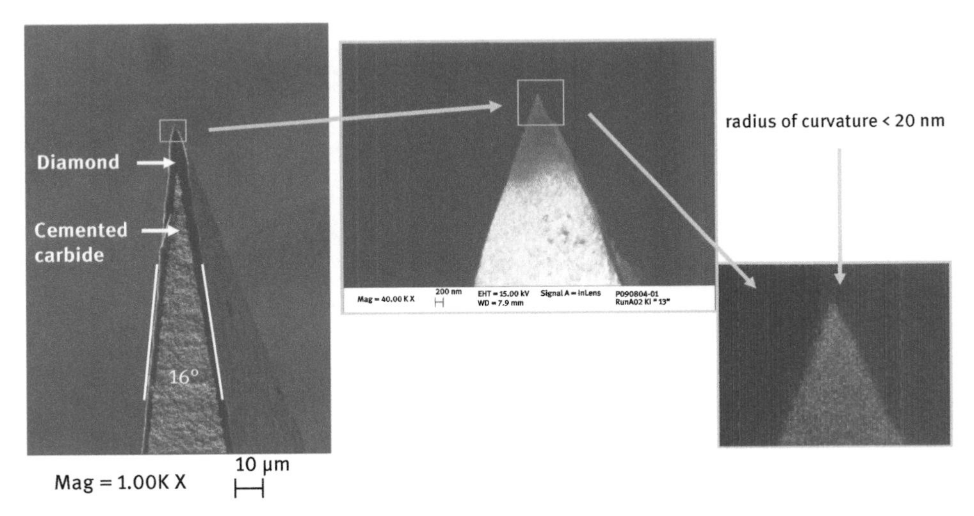

Figure 17.6: Cross-sectional SEM photograph of a diamond-coated plasma sharpened diamond razor blade. Middle and right: high-resolution SEM. The radius of curvature r_c could successfully be reduced below 50 nm.

17.3.1.2 Determining the Cutting Ability of a Blade

In order to characterize the cutting ability for cutting blades, a string cutting test has been established. Here, the cutting edge of a blade is driven perpendicular against a mechanically biased (F_y) polymer string. The cutting force F_x is measured during the displacement process resulting in a force versus dislocation curve $F_x = f(s_x)$. The maximum force is used to characterize the cutting ability (indirect characterization of r_c) of the blades. This test is not suited for large wedge angles as applied for milling or drilling tools. Figure 17.7 shows the experimental setup, Figure 17.8 shows schematically the measurement process.

Figure 17.7: Photo of the experimental setup of the string cutting test.

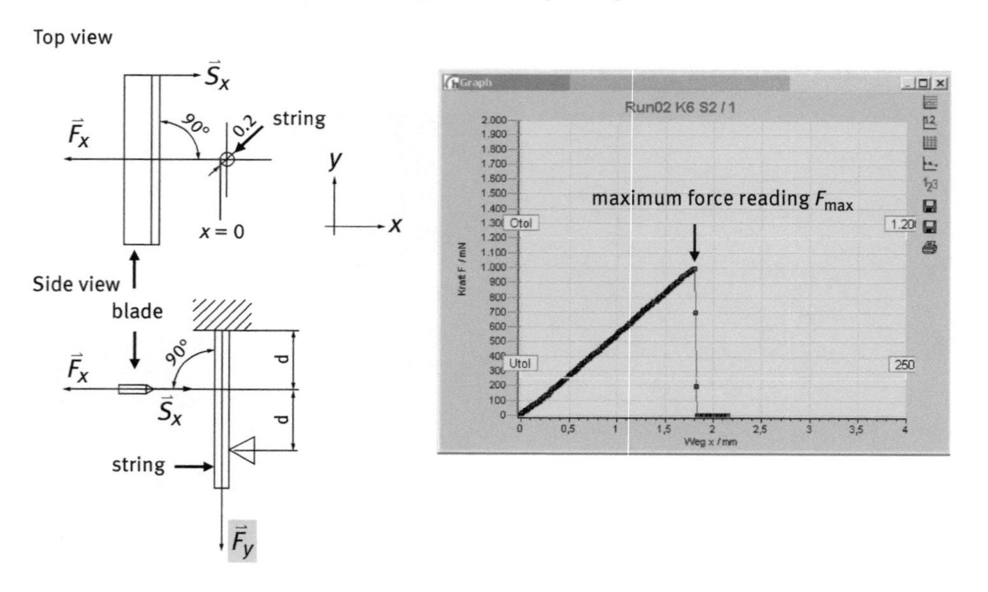

Figure 17.8: Measurement of the cutting ability. Schematic setup of the string cutting test and experimental measurement result.

17.3.1.3 Performance of PSD Blades and Comparison to Other Blades

In order to prove the influence of the process time on the cutting ability, a blade with 15° cutting angle was sharpened successively. Before depositing the NCD film the cutting ability of the uncoated blade was characterized to F_{xmax} = 100–800 mN. The strong variation of maximum force readings is originated from the inhomogeneous grinding process of the cutting edge (see also Figure 17.3). Due to the high isotropy of the diamond coating process and the high anisotropy of the plasma sharpening process, smaller chip outs and irregularities of the initial as-ground cutting edge can be smoothened out efficiently. This means in consequence that the cutting edge quality of the substrate blades is of minor importance as for uncoated cemented carbide blades.

The sharpening process was interrupted after certain process steps and the blade was characterized utilizing the string cutting test, respectively. Figure 17.9 shows the effect of the plasma sharpening process. This graph demonstrates that the sharpness can be adjusted towards the application and the customer's needs.

In order to compare the cutting ability of industrial available blades, the string cutting test was applied to different commercially available blades. Figure 17.10 shows the corresponding results.

Finally, different blade geometries were fabricated and tested in industrial environments. In order to compare the results, TC and ceramic blades were also added to the test. Statistically relevant data points were achieved by testing a total of 150 blades. The test material was a 0.2 mm thick plastic foil with TiO additive. Figure 17.11 shows the test results. The TC and ceramic knifes showed practically the same low cutting edge holding property of approx. 1.5–2 days. The PSD-blades, however, showed a significantly increased lifespan of approximately 36 days. In comparison to carbide blades, this corresponds to a lifetime increase of 24 times.

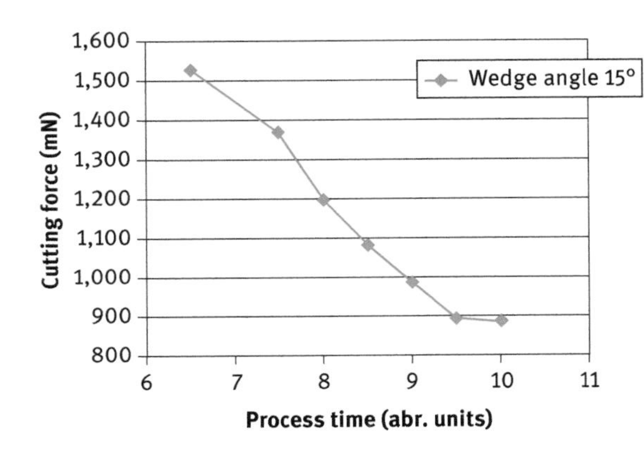

Figure 17.9: Influence of process time on the cutting ability. The cutting ability (sharpness) increases with increasing process time.

	Blade	Max. cutting force in mN
	Kitchen knife	Approx.2,000
Industrial blades	Industrial steel blade	469
Industrial blades	Industrial WC blade	446
Industrial blades	Diamaze PSD blade (normal sharpness)	427
Industrial blades	Industrial steel blade with TiN	178
Industrial blades	Diamaze PSD blade (extra sharp)	150
Industrial blades	Ceramic blade (best blade ever seen 20°)	126
Razor blades	3 whole razor blade (standard)	51
Razor blades	3 whole razor blade (best)	46
Razor blades	Razor blade from cartrige	38
Razor blades	Diamaze PSD razor blade	16

Increasing cutting ability

Figure 17.10: Maximum force readings of commercially available cutting blades. Different sharpness degrees of PSD blades (in green). The plasma sharpening process allows for a precise control of the microgeometry and therefore a sharpness adjustment.

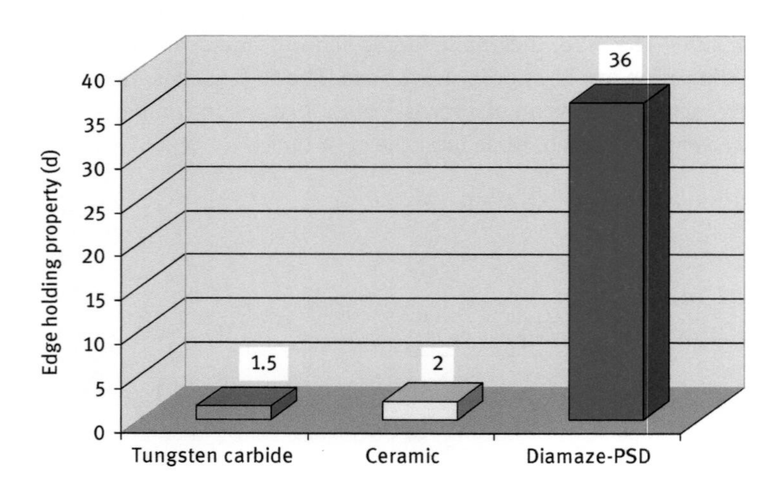

Figure 17.11: Comparison of the cutting edge holding property of tungsten carbide, ceramic and PSD blades.

17.3.2 Plasma Sharpened, Nanocrystalline Diamond Tools (PSD Tools)

Diamond-coated cemented carbide tools are state of the art for drills and milling operations for nonferrous materials such as graphite, aluminium/silicon alloys, FRP and ceramics. However, as explained above, the diamond coating increases the edge radius r_c of the tool significantly and limits thus the field of application. The microgeometry is especially for FRP materials of high importance, since the machining of FRP materials is not subject of chip formation but rather of fibre breaking. A key industry for these kinds of tools is the aircraft industry, utilizing carbon fibre reinforced plastics (CFRP) as a lightweight and mechanically stable alternative for aluminium. As an example for the economical relevance, the leading European Aircraft Company Airbus is expecting a need more than 50 Mio Drilling operations in CFRP for the year 2020 [14].

In order to overcome the existing limitations of the state-of-the-art tools, we recently developed a process to adapt the "quasi-planar" plasma sharpening process applied for blades to index tools to complex-shaped shaft tools, for example helix-shaped drilling and milling tools. This enables for the first time to combine the hardness of the NCD coating, the optimized substrate geometry of cemented carbide tools with a sharp and complex formed diamond cutting edge. These tools show a better cutting quality and a significantly improved life span in contrast to unsharpened CVD diamond tools.

17.3.2.1 PSD Tool Fabrication

As plasma sharpening is a slow but a very precise and highly reproducible process, it allows for a precise adjustment of the microgeometry also of tools. Furthermore, it is possible to apply the process from the cutting surface as well as from the clearance surface or both. Figure 17.12 shows the theoretical and practical results of diamond-coated tool cutting edges after different plasma sharpening treatments applied from the cutting surface. The cutting edge configuration is best described by referring to the remaining diamond thickness after the sharpening process for the two planes (S = cutting surface and F = clearance surface) instead of defining the removed diamond thickness. For example, S100F50 means that the cutting surface has not been affected at all and 50% of the diamond film remains on the clearance surface.

17.3.2.2 Performance of PSD Tools and Comparison to As-Coated Tools

The cutting ability and edge-holding property of plasma sharpened CVD diamond-coated tools was determined using a drilling tool with a shaft diameter of 4.76 mm for the machining of CFRP stacks. This composite material is well established in the

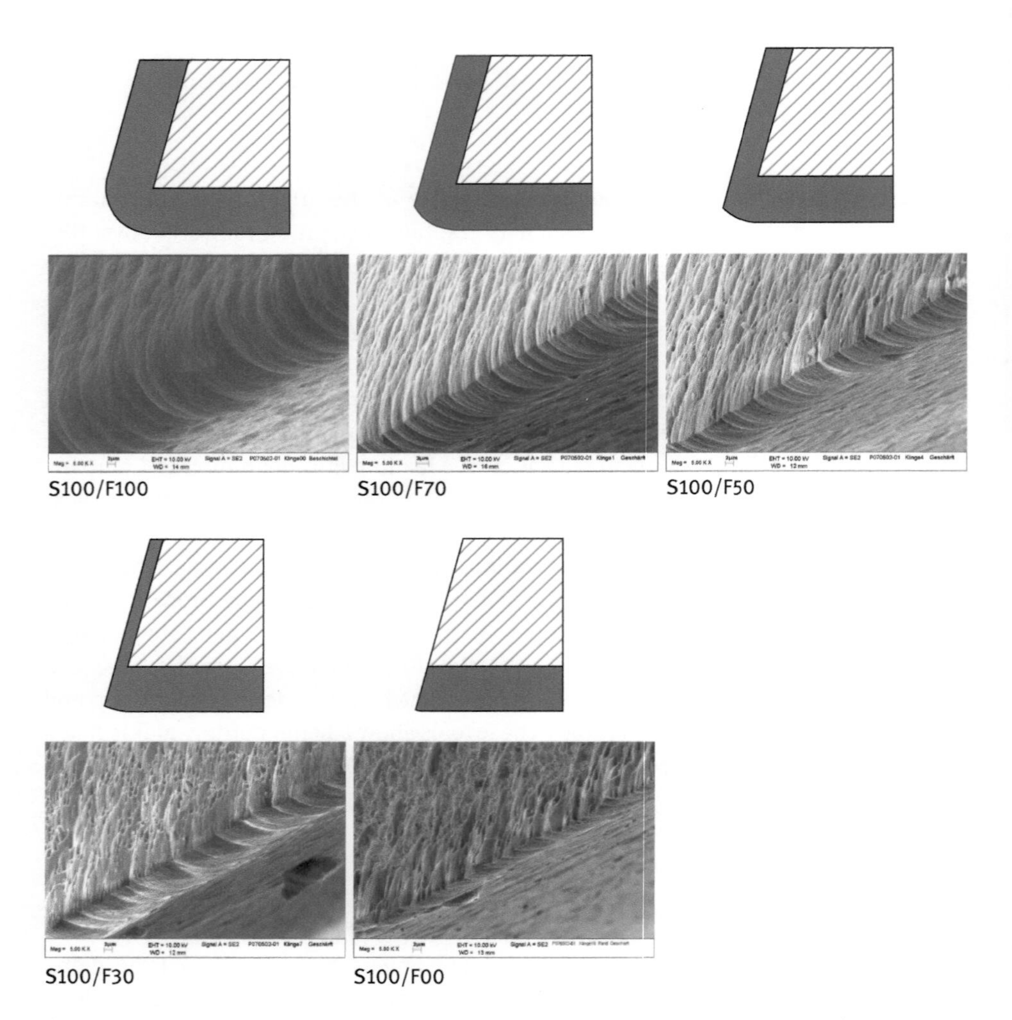

Figure 17.12: Comparison of the theory and results of plasma sharpened cutting edges (wedge angle 85°). Top: schematic cross section of the theoretical microgeometry (green: diamond coating). Bottom: SEM photographs at 5k magnification of the cutting edge. The sharpness degree is increased from the left to right picture.

aircraft industry and it is well known to be extremely abrasive. The material consists of a 5 mm thick layer of CFRP followed by a 5 mm thick layer of aluminium. Thus, only sharp solid cemented carbide and diamond tools can be used to machine this delicate and expensive material. Since the carbon fibres have a diameter of only approx. 6 μm, it is essential to utilize a tool having a sharp cutting edge. Otherwise, the composite material cannot be machined in an adequate quality. In order to compare the performance of an as-coated NCD tool with a plasma sharpened tool, drills of identical geometry were pre-treated, nucleated and CVD diamond coated in the same process run. The geometry of this tool is shown in Figure 17.13.

Figure 17.13: Drill for CFRP stack performance test. Tool diameter: 4.76 mm; number of cutting edges: 2; point angle: 100°; wedge angle: 70°; twist angle: 30° (courtesy: IFW Stuttgart).

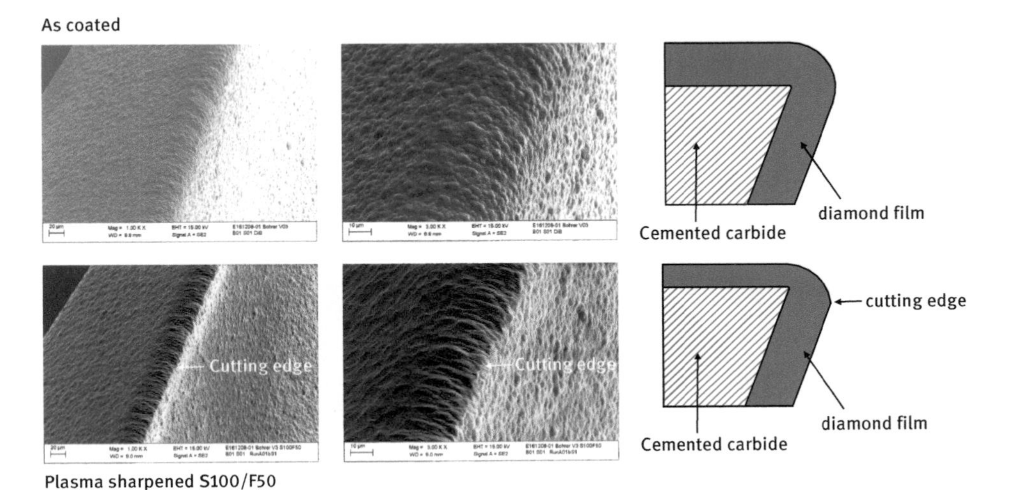

Figure 17.14: SEM photographs of the cutting edge of the as coated drills (top) and the plasma sharpened drills (bottom). A schematic cross section of the cutting edge configuration is shown on the right side. The cutting edge formation on the PSD tool is clearly visible.

Thus, the morphology, the thickness, the mechanical properties, etc. of the applied NCD film is identical for all tools. Here, we have chosen an initial coating thickness of 19.5 µm. Half of the tools were plasma sharpened respectively to a sharpness degree of S100F50 (see Figure 17.14), corresponding to a residual diamond film thickness of approx. 10 µm at the clearance face.

The industrial machining test was performed at the IFW Stuttgart, University of Stuttgart, utilizing a 5-axis CNC machine tool, type Maka PE 170. The drilling operations were performed at a cutting speed of $v_c = 36$ m/min and a feed velocity of $v_f = 289$ mm/min (see Figure 17.15).

The abrasive wear propagation was investigated using a SEM after 10, 600, 800, 1,200 and 1,600 operations. The diameter of the holes was measured regularly and the quality of the holes was determined by using an evaluation procedure developed by the

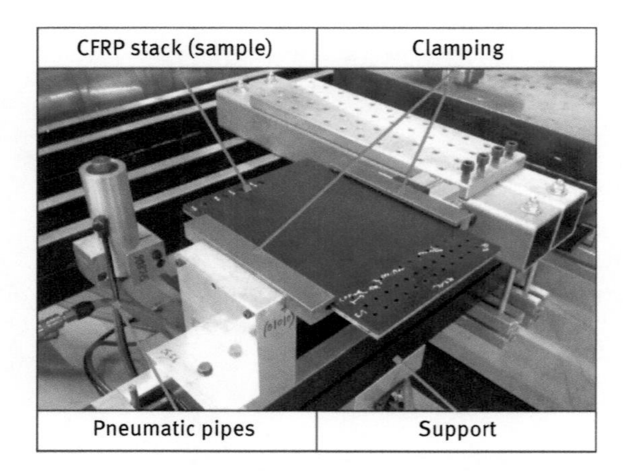

Figure 17.15: Experimental configuration of the PSD tool performance test at the IFW Stuttgart.

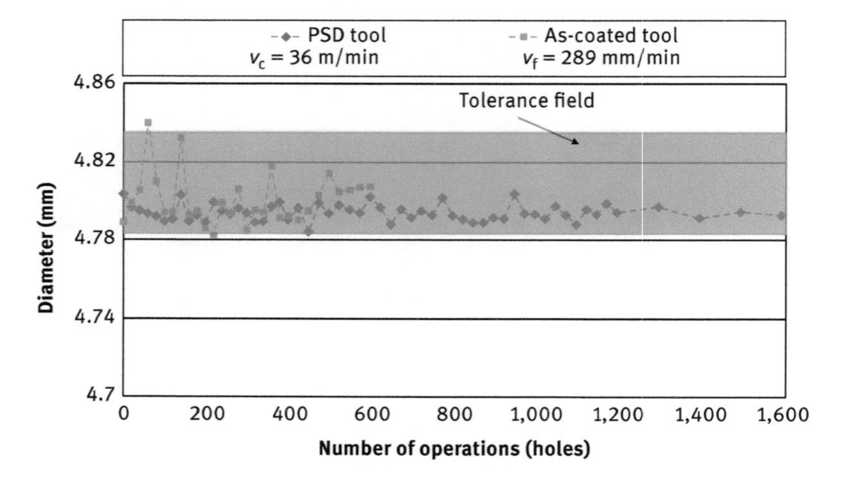

Figure 17.16: Tolerance field and measured bore diameters vs. the number of operations. Orange: as-coated tool; blue: PSD tool (courtesy: IFW Stuttgart).

Technical University Vienna, which allows to quantify the machining quality of CFRP [15]. In order to compare the influence of the plasma sharpening process, an as-coated tool from the same batch was added to the test as reference. This tool failed due to insufficient bore quality already after 600 operations when machining a less abrasive CFRP stack material (satin weave fabric) of identical thicknesses. From the subsequent SEM investigations of this reference tool, we measured a critical land-wear dimension of approximately 60 µm at the intersection of the main to the minor cutting edge. The same tool, however, plasma sharpened to S100F50, survived 1,600 operations at a land-wear dimension of still less than 41 µm. Furthermore, by analyzing the bore quality according to the evaluation procedure we could determine a significant improvement of the quality index by 25%. In addition, we discovered a highly improved tolerance field for the bore diameter (see Figure 17.16), thus, enabling a higher process stability.

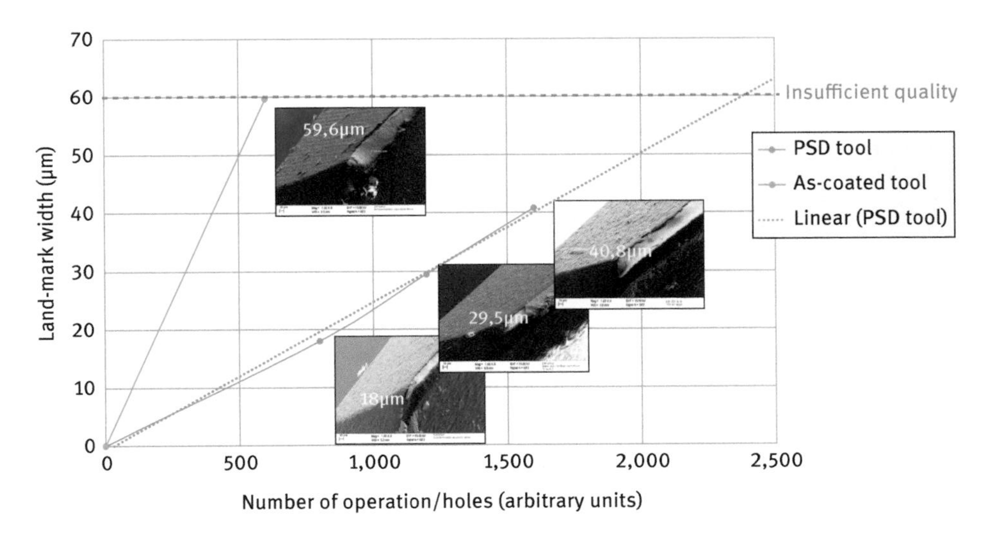

Figure 17.17: Land-wear propagation of the as-coated (orange) and PSD tool. The wear propagation of the PSD tool is approximately linear.

Figure 17.17 shows the propagation of land-mark dimension as a function of the number of holes. In contrast to the conventional square root law for an abrasive wear out of the cutting edge in woodworking as shown by Fischer for TC tools [16], an almost linear wear propagation was observed for the PSD tool. This can be explained, by considering the ultra-hard NCD coating as the predominant component concerning the wear. In this case, the wear rate of the diamond film is constant in time. The wear rate of the cemented carbide volume underneath is of minor importance. The function of the buried cemented carbide substrate serves therefore rather as shaping support for the diamond coating. However, the functionality and wear behaviour of the tool are fully determined by the thin NCD coating. A linear fit indicates that the lifetime end of this tool will be approximately 2,300 operations. This corresponds to a lifetime increase of approximately 4.

17.3.3 High Precision Diamond Spheres

Spheres from hard materials are used in many industrial applications such as ball bearings, ball valves and metrology.

Besides steel, spheres made out of ceramics like Si_3N_4, Si, ZrO, Ruby and cemented carbide are the state of the art. Since a few years, there are also spheres made out of single crystal diamond available. However, these products are extremely expensive and the maximum available diameter is limited up to approximately 3 mm due to the lack and the cost of large single crystal diamond substrates as well as the correlated machining costs of diamond. Although their performance is outstanding the factors of cost and size are the main entry barriers for these products preventing mass applications like ball bearings. In consequence, diamond spheres are rarely used in high price niche markets as for instance the metrology market. Here, diamond spheres, mounted on a cemented carbide shaft (so-called styli) are used as measurement tips for coordinate measurement machine (CMM) systems. These tips are enabling the CMM to perform scanning operations in contrast to point to point measurements. The hardness of diamond serves here as enabling factor to prevent an inacceptable wear propagation or material built-up correlated with a loss of accuracy e.g. when scanning soft materials like aluminium.

In order to overcome these limitations we present in this section a novel approach by utilizing thin NCD coatings on foreign substrate spheres for styli applications.

17.3.3.1 Fabrication of Nanocrystalline Diamond Coated Spheres

As substrate material we are recommending a Si-based ceramic such as Si_3N_4 or SiC. As already explained in Section 17.3.1.3, the diamond deposition is highly isotropic and therefore it smoothens out smaller irregularities of the substrate surface. Figure 17.18 shows the schematic structure and a SEM image of a coated sphere.

The substrate surface quality is therefore also of minor importance. Ceramic spheres of high-quality grades are commercially available. As they are used in mass markets manly for ball bearing applications, these spheres are not expensive. Although the CVD diamond coating process is highly isotropic, it will never end up in a perfect homogeneous film thickness all over the substrate sphere. Thus, a post treatment is necessary in order to obtain the desired sphericity and surface roughness.

According to DIN [17], there is a classification of sphere qualities in grades (G) according to Table 17.2. Precision spheres for CMM applications typically exceed G10, thus, a tolerance of sphericity of less than 250 nm and maximum peak to valley surface roughness r_t of less than 20 nm are required.

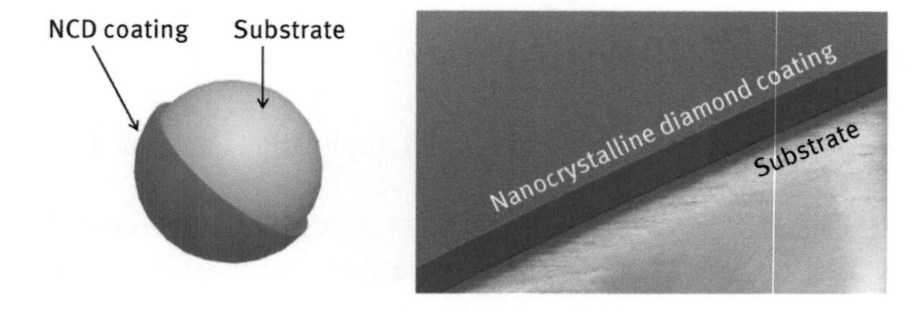

Figure 17.18: Schematic structure (left) and SEM photograph with partially removed diamond film (right).

Table 17.2: Grade of spheres and the correlated specification according to DIN [17].

Grade	Deviation of sphere radius (nm)	Max. surface roughness r_t (nm)
G3	<80	10
G5	<130	14
G10	<250	20
G16	<400	25
G20	<500	32
G200	<5,000	150

Figure 17.19: Diamond spheres of different diameters after CMP processing. The diameter ranges actually from 0.6–8 mm. It is expected that larger and smaller diameters can be produced in the near future.

The thickness of the diamond film is of minor importance and should costwise be kept as low as possible. In order to guarantee a sufficient mechanical support for the post treatment process, we have chosen a diamond film thickness of approximately 25 µm. After coating, the relative uniformity of the coating thickness measured over the entire sphere surface is less than 10%. In consequence a minimum film thickness removal of 2.5 µm is required in order to achieve the desired sphericity. Such a shaping process, a chemical mechanical polishing (CMP), is adapted from metal and ceramic spheres.

Figure 17.20 shows SEM pictures of the as-grown diamond surface resulting in a typical surface roughness of approximately rms = 20 nm. After the CMP the surface roughness is significantly reduced. The surface roughness is measured by AFM to be less than rms < 2 nm (measured value rms = 1.4 nm) and a maximum peak to valley roughness of r_t = 10 nm. The sphericity of the sphere is determined to be 60 nm (class G3), which corresponds to the resolution limit of the tactile roundness measurement set up. For CMP of diamond spheres, it is necessary to enhance the removal rate by chemical additives and to adjust the polishing speed in order to keep the polishing times reasonably low. The utilization of a batch treatment is a necessary evil

Figure 17.20: High-resolution SEM micrographs (magnification 100kx) of the diamond surface as grown rms = 20 nm (left) and after CMP rms <2 nm (right).

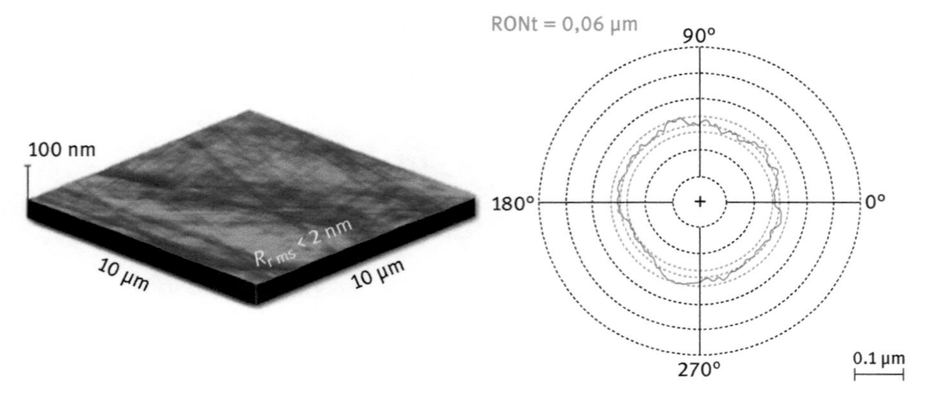

Figure 17.21: Surface roughness and sphericity analysis of a diamond-coated sphere after CMP. The grade according to Table 17.2 of this sphere is G3.

for the CMP process in order to keep the production costs reasonably low. This batch process is highly efficient, but there is on the other hand a latently risk of losing the entire batch in case the diamond film of one single sphere even partially peels off the substrate. In this case, the delaminated diamond shell will cause a catastrophic chain reaction and destroy all other spheres of the batch. Thus, the diamond coating process and the material properties of the NCD film have to be chosen very carefully in order to enable the best possible adhesion of the diamond film on the substrate and a sufficient mechanical stability to survive the applied forces of CMP. In contrast to polycrystalline diamond films, the reduced surface roughness of as-coated spheres, the lower Young's modulus and the higher fracture toughness of the NCD film are supporting the efficient CMP of spheres and allow therefore for a significantly reduced diamond coating thickness and process time.

Finally, as the desired grade is achieved, the precision sphere is brazed on a cemented carbide shaft (see Figure 17.22) and is ready to be used in the CCM.

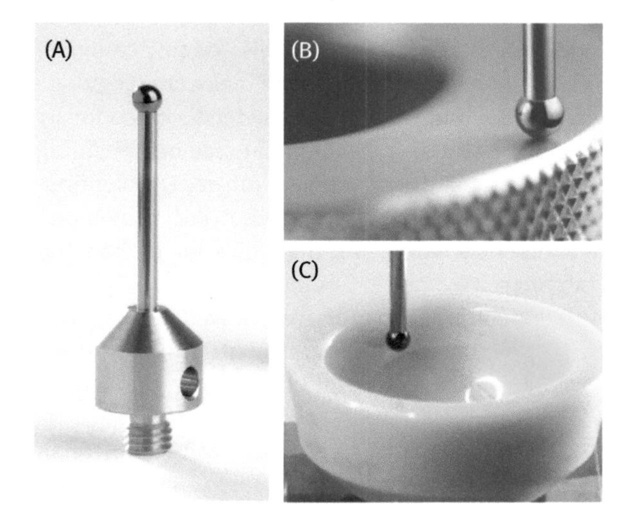

Figure 17.22: Diamond stylus for CMM application (A). Diamond stylus scanning an aluminium part (B) and an abrasive ceramic hip implant (C).

Due to their hardnesses, diamond spheres are enabling the scanning operation mode of CMM, thus enhancing the possibilities of the machine significantly. Single crystal spheres are, however, extremely expensive and very limited in size. The NCD-coated sphere overcomes both limitations and may yield in a change of the marketing model by merging the CMM and the stylus market towards a service measurement sales model according to smart manufacturing. The actual production process was designed for mass fabrication and the development is currently focussed on upscaling of the above-described fabrication methods. In consequence, it will be feasible that mass markets like diamond ball bearings can be addressed in the near future. This, however, requires further optimization towards thinner diamond coating thicknesses and CMP processes of higher efficiency.

17.4 Summary

Diamond, especially NCD films, can be utilized in many industrial applications. In contrast to single crystal diamond or coarse-grained polycrystalline diamond material, the mechanical properties of NCD films can be engineered towards the substrates and the applications needs, thus enabling an improved adhesion on foreign substrates. If applied correctly, the buried substrate serves rather as shaping support for the functional coating. Functionality and wear propagation are fully determined by the thin and cost-effective NCD coating.

We have demonstrated this working principle in different applications such as diamond-coated industrial cutting blades, innovative and complex-shaped cutting tools for CFRP machining and finally for precision spheres in CMM applications. For the sake of completeness, it should be mentioned that there are many other applications such as micromechanical watch parts or scalpels for microsurgery, showing the industrial capability of this intersectoral cross-sectional technology.

During the last 20 years, the CVD diamond technology emerged tremendously and it is conceivable that the availability of large-scale diamond reactors, enabling coating surfaces in the m^2 – range, will be probable in the near future. This continuous scale-up process will lead to a further reduction of the production cost of diamond films and will lead to decrease the main entry barriers for the utilization of diamond in mass applications such as ball bearings.

The parallel development and industrialization of innovative and efficient diamond machining technologies, like plasma sharpening, is contributing towards an efficient and cost-effective precision post-processing and finishing of the hardest material on earth.

The combination of this innovative toolbox allows for the successful application of this technology in emerging markets such as CFRP machining, for example for aerospace industry, and contributes therefore indirectly in the energy and resource-saving progress of key industrial sectors. It can be considered to be an enabling factor for the industrialization of lightweight construction and the application of novel materials such as CFRP stacks.

We have shown that one single PSD diamond blade replaces approximately 2–40 TC blades, accompanied by a significant reduced production yield loss and increased productivity. Thus, a thin NCD film enables a remarkable conservation potential of resources besides the substrate materials such as tungsten and cobalt.

We therefore expect a rapid growth of this technology in different markets and applications.

Acknowledgments: This work was supported by the BMBF framework “Mai Carbon.” Many thanks to all contributions and to the collaboration partners of “Mai ProCut”. The fruitful collaboration with the University of Ulm, Institute of Nano and Micromaterials, namely Dr. K. Brühne, Dr. M. Mohr, M. Mertens, Dr. A. Minkow and Prof. H.J. Fecht, is greatly acknowledged.

References

1. Tang L, Tsai C, Gerberich W, Kruckebeug L, Kania D. Biocompatibility of chemical vapour deposited diamond. Biomaterials 1995;16:483–488.
2. Williams AO. n-type conductivity in nanocrystalline diamond films. Appl Phys Lett 2004;85:1680–1682.
3. Mertens, M, Lin I-N, Manoharan D, Moheinian A, Brühne K, Fecht H-J. Structural properties of highly conductive ultra-nanocrystalline diamond films grown by hot-filament CVD. AIP Adv 2017;7:015312-1-7, Article ID 015312:1–7.

4. Shenderova A, Guen M. Ultrananocrystalline diamond. 2nd ed. Great Britain: Elsevier Inc., 2012.
5. Williams AO. Nanocrystalline diamond. Diamond Relat Mater 2011;20:621–640.
6. Williams AO, Nesladek M, Daenen M, Michaelson S, Hoffman A, Osawa E, Haenen K, Jackman BR. Growth, electronic properties and applications of nanodiamond. Diamond Relat Mater 2008;17:1080–1088.
7. Fecht H-J, Brühne K, Gluche P. Carbon-based nanomaterials and hybrids. Singapore: Pan Stanford Publishing, 2014:155–167.
8. Wiora M, Brühne K, Flöter A, Gluche P, Willey TM, Kucheyev SO, Van Buuren AW, Hamza AV, Biener J, Fecht H-J. Grain size dependent mechanical properties of nanocrystalline diamond films grown by hot-filament CVD. Diamond Relat Mater 2009;18:927–930.
9. Hess P. The mechanical properties of various chemical vapor deposition diamond structures compared to the ideal single crystal. J Appl Phys 2012;111:051101-1-15, Article ID 051101:1–15.
10. Cappelli E, Pinzari F, Ascarelli P, Righini G. Diamond nucleation and growth on different cutting tool materials: influence of substrate pre-treatments. Diamond Relat Mater 1996;5:292–298.
11. Mohr M, Caron A, Herbeck-Engel P, Bennewitz R, Gluche P, Brühne K, Fecht H-J. Young's modulus, fracture strength and Poisson's ratio of nanocrystalline diamond films. J Appl Phys 2014;116:124308-1-9, Article ID 124308:1–9.
12. http://www.cemecon.de/en/coating-systems, 2018.
13. Mertens M, Mohr M, Wiora N, Brühne K, Fecht H-J. N-type conductive ultrananocrystalline diamond films grown by hot filament CVD. J Nanomater 2015;2015:527025-1-6, Article ID 527025:1–6.
14. Stuhrmann J., Challenges when drilling CFRP and CFRP sandwich compounds in the Airbus A350XWB. In: Proceedings of the 6th IfW-Conference – Machining of Composites, Stuttgart, 20 October 2016:51–64.
15. DIN SPEC 25713: Beurteilung der Bauteilqualität nach der trennenden Bearbeitung von faserverstärkten Kunststoffen. Berlin: Deutsches Institut für Normung e.V., 2017.
16. Fischer R. Berechnung der Schneidenabstumpfung beim Fräsen (Teil 2). HOB die Holzbearbeitung. Ausgabe 1997;6:S. 71–77.
17. DIN 5401 ISO 3290-1:2008
18. Komanduri R, Hou BZ, Umehara N, Raghunandan M, Jiong M, Bhagavatula RS, Noori-Khajavi A, Wood ON. A ‚gentle' method for finishing Si3N4 balls for hybrid bearing applications. Springer Link. Tribol Lett 1999;7(1):39–49.

Part IV: Nanomaterials in Occupational Health and Safety

Thomas H. Brock

18 Nanomaterials in occupational health and safety

18.1 Introduction

Nanomaterials have had a firm place at the workplace for many years and are used in manufacturing, processing and disposal or recycling. The volume and type of nanomaterials used has increased continuously during the course of the years and in many cases they are not even recognized as nanomaterials. They are deployed in workplaces encompassing laboratories, the production and processing of the nanomaterials, the user and finally the disposal firm.

There is a general consensus that nanotechnologies are technologies that use structures within the range of approximately 1–100 nm. There are many attempts to establish a clear definition for this complex field, all of which more or less aim to narrow down the topic [1]. In order to determine and assess risks and derive protective measures, it has been proven useful to generously expand the upper limit towards the micrometre range (many measuring devices detect the range up to 1,000 nm) as the properties of the materials do not abruptly alter at 100 nm. In addition, larger secondary particles such as agglomerates and aggregates also play a significant role at the workplace. Although primary particles in the range of 1–100 nm will seldom be encountered at many workplaces, secondary particles made up of these primary particles do occur, which may disintegrate into the primary particles again in the body. The application of strict limits would simply mask a significant part of this problem.

Nanomaterials occur more rarely than free, unbonded nano-objects (nanofilms and nanoplates, nanotubes, nanorods or nanowires, as well as spherical nanoparticles). Alongside agglomerates and aggregates, composite materials are often used, for example nanoparticles or nanotubes, which are integrated in a polymer matrix. This categorization in accordance with the definition of the ISO Technical Committee 229 is helpful for occupational health and safety [2].

Furthermore, two types of material with nanoscale structures occur at the workplace (a third type would be ultrafine dusts from natural sources). Ultrafine aerosols are differentiated from the nanomaterials that are generally understood to be nanomaterials and are consciously and intentionally manufactured or used as such, as although they can share the same properties in principle, they can also be created unintentionally and cause exposure. Both types are to be considered equally for occupational health and safety measures. The ultrafine aerosols can be released by nanomaterials, although these are often created through process and processing steps on coarser materials. Although we will mainly discuss nanomaterials in the following, in most cases the statements made can be applied to ultrafine aerosols, too, although

https://doi.org/10.1515/9783110547221-018

the characterization of the dusts for risk assessment can be difficult due to the considerable time and effort involved. Assumptions about the nature and concentration of ultrafine dusts emitted during working procedures like milling, grinding or laser ablation can be made but may be wrong. If no detailed guidelines for the safe processing are available, expert judgement is needed. In most cases available technologies for an efficient exposure control will minimize the exposure to ultrafine dusts also, as long as filtered air is not blown back into the working area or this air is filtered properly (very efficient filters are available). Processing materials on the benchtop without further exposure control measures are possible according to expert judgement or measurement at least of the particle number concentration in comparison with the background concentration.

The assessment and evaluation of the risks are hindered by the large number of nanomaterials and remaining gaps in our knowledge of the effects. The emergent properties of many nanomaterials also make this task difficult, as conclusions drawn on the characteristics of the nanomaterial on the basis of the properties of the coarser material can be misleading. For instance, nanoscale dusts from metals can be far more ignitable – in some cases even auto-ignitable – than their dusts in the micrometre range and of course the compact material. If this is not taken into account, explosions or even detonations can lead to serious consequences for people and the environment as well as severe material damage.

The European Commission explains how nanoscale fractions in coarser materials are to be taken into consideration: “A natural, incidental or manufactured material containing particles, in an unbound state or as an aggregate or as an agglomerate and where, for 50% or more of the particles in the number size distribution, one or more external dimensions is in the size range 1–100 nm. In specific cases and where warranted by concerns for the environment, health, safety or competitiveness the number size distribution threshold of 50% may be replaced by a threshold between 1 and 50%. By derogation from the above, fullerenes, graphene flakes and single wall carbon nanotubes with one or more external dimensions below 1 nm should be considered as nanomaterials” [3]. It will certainly be necessary to also consider other large molecules within this sub-1 nm range, C_{60} e.g.

The primary particles have a tendency to congregate quickly to the less tightly bound aggregates or the polyvalent and very tightly bound aggregates. These may disintegrate under physiological conditions releasing smaller or primary particles again, depending on the forces between the primary particles and the mechanisms of energy transfer (mechanical energy on gloves or the skin e.g.) or biological and chemical effects (separating and coating the smaller particles with proteins e.g.). So it is necessary to take these aggregates and agglomerates into account too. They may also function as carriers for molecules transporting them into biological structures.

Although many studies on effects of nanomaterials in humans have been published, it is absolutely necessary to keep in mind that there is no common behaviour of all nanomaterials, that there is no phenotypical nanoparticle or nanomaterial. So these results have to be interpreted with expert knowledge to yield a useful picture near to the truth.

18.2 Strategy for Accident and Illness Prevention

The focus of occupational health and safety is to minimize risks but not necessarily to eliminate them entirely, as this is often not possible. In this context, the risk is often described as the product of the probability of occurrence of damage and its severity. The probability of occurrence is directly dependent on the exposure, which can now be determined – albeit with some effort. In order to determine risks, however, the severity of the damage resulting from the respective scenario also needs to be quantified. This is considerably more problematic, as we are not fully aware of the mechanisms of possible nanospecific effects (insofar that these are relevant) and their effects on the organism. As such, if no sufficiently sound data are available on the effect, the risk level can only be controlled by limiting the extent of exposure. The categorization in four in its respective current version can provide assistance here.

There is a harmonized European Union law for occupational health and safety with regard to substances and materials. This is supplemented by national implementations and additions, which can contain further details on working safely with nanomaterials and on protection from ultrafine aerosols. For instance, specific information is provided by the Announcement on Hazardous Substances 527 [4] or the laboratory guidelines "Working Safely in Laboratories" [5] with the complementary publication on "Nanomaterials in the Laboratory" [6]. Further publications are available in various countries [7–17]. The procedures described here are similar, but differ occasionally in detail.

Performing a risk assessment is a common and proven strategy for handling hazardous substances. This requires the compilation of various kinds of information in order to at least make a qualitative assessment of the risks or, better still, perform a reliable estimation of the risks. Appropriate measures can then be derived from this and put in place.

The currently used nanomaterials are – with regard to the used quantities – predominantly substances that have been known for a long time, for example carbon, silicon dioxide or titanium dioxide. However, because the number of available types of nanomaterial is much larger, and the number of those that can be manufactured in theory so enormous, there will be applications involving the handling of nanomaterials about which we know very little.

For this purpose, information will be required on various parameters. Some, but unfortunately not all, of this information can be found in the safety data sheets and technical data sheets (product specifications) if these are informative and complete.

With regard to nanomaterials or solid materials that contain a relevant proportion of nanomaterials, some gaps sadly remain. When handling these substances, the raw and auxiliary materials, solvents and reagents also have to be assessed. This includes in particular information and data on the following:

- Size distribution and shape
- Fire and explosion behaviour, auto-ignitability
- Additions, coating, soiling, enclosures, adsorbates, “foreign atoms” incorporated through chemical bonds (e.g. metal atoms from the growth process)
- Chemical and physical–chemical properties, such as catalytic effects, reactivities, radical formation, surface activity, specific surface, ζ potential, absorptivity, kinetics of the agglomeration or aggregation, dustiness behaviour, solubilities, ageing and stability
- Toxicology
- Exposure possibilities and levels (inhalation, dermal and oral)

At times it can be difficult to recognize whether a product contains nanomaterials and whether such nanomaterials are released or created during (intended) use. Not all safety data sheets contain information on this; in case of doubt, the manufacturer or supplier should be consulted. The use of nanotechnology is occasionally advertised in products, in particular for the downstream user and consumer segment, even though it is of no significant relevance in the product itself. This is done simply for the reason that the label “nano” carries positive connotations, especially in some technically savvy markets. However, there are also cases where the contrary applies. This can make it much more difficult to gather information.

18.3 Controlling Risks Posed by Nanomaterials

The toxicology of nanomaterials is not always analysed sufficiently: in some cases there are considerable data gaps that need to be circumvented as a precautionary measure. If a precautionary renunciation of certain materials or applications is not possible in a socioeconomic context, because this would deprive society of significant advantages, for example in the therapy of illnesses, or the prevention of improvements in drinking water supply in water-scarce countries would be hard to justify in an ethical sense; prudent compromises must be made on the basis of a risk assessment with sensible and justifiable assumptions always keeping in mind that our knowledge is limited and there may be risks being not properly addressed, especially when it comes to new materials with new properties emerging from the (self) organization on nanomaterials. At the moment such materials are used only in small quantities or are still in research and development stage.

If no usable and plausible data are available from the manufacturer or supplier or, if applicable, no dedicated examinations are available, the categorization as per

Table 18.1: Categories of nanomaterials according to BekGS 527.

Cat. I	Soluble nanomaterials without specific toxic properties (solubility at least 100 mg/L water at room temperature): e.g. many inorganic salts like sodium chloride
Cat. II	Soluble nanomaterials with specific toxic properties, e.g. quantum dots, inorganic salts like heavy metal compounds, dendrimers and lipid shells carrying toxic atoms or molecules inside
Cat. III	Granular biopersistent dusts without specific toxic properties, e.g. silicon dioxide
Cat. IV	Fibrous nanomaterials, e.g. carbon nanotubes

the Announcement on Hazardous Substances 527 (BekGS 527) 4 represents a possible procedure for taking toxicology into account. This divides nanomaterials into four categories with regard to the toxic properties (Table 18.1).

If no conclusive data about a material are provided by a manufacturer or supplier, it can be allocated to one of the categories with some expert knowledge. The associated measures are categorized according to time and effort involved. Category I materials require no special measures above and beyond the general protective and hygienic measures. Materials of categories II and III often have threshold values which must be adhered to. Category IV materials can be handled in the same manner as materials from one of the three previous categories if it has been proven – for instance by the manufacturer – that these do not possess any asbestos-like properties. HARN belong to category IV*.

The physical and chemical properties are often not sufficiently perceived and taken into consideration as causes of possible risks, with the focus instead predominantly being on the toxic properties. These deficits in the assessment of risks can nevertheless cause immense damage to people, the environment and assets. The large specific surface areas of nanomaterials often cause higher reactivities and reaction speeds, meaning that flammable materials in the nanoscale range can have significantly lower minimum ignition energies than coarser material or can even take on self-igniting properties (Figure 18.1) [18].

18.4 Occupational Health and Safety with Nanomaterials in Practice

18.4.1 Risk Assessment

One problem with the manufacturing and use of nanomaterials as well as the production of ultrafine aerosols is the often insufficient data situation for assessing

* HARN: high aspect ratio nanomaterial: WHO fibre with at least one dimension between 1 and 100 nm.

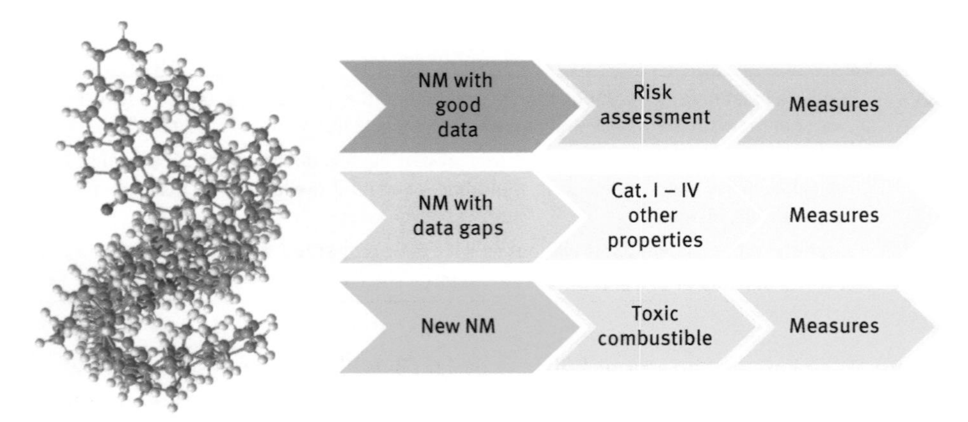

Figure 18.1: Strategy for nanomaterials (NM) with good data (material safety data sheet, plausible and complete), NM with data gaps, that can be matched to one of the categories I–IV (Table 18.1), and new NM, to be handled like all other possibly toxic and combustible substances (in laboratories). Measures to be taken are described in Section 18.4.

the risks. An important factor is the early acquisition or generation of data, at the latest when materials outside of the field of research are to enter the market. Extensive data often already exists for nanomaterials that are manufactured and used on a larger scale. Measurements can be performed easily for some properties, but can be difficult for others.

The source of knowledge for operational use is primarily the product information material provided by the manufacturer or supplier, i.e. the safety data sheets and technical information sheets. Enquiries to the manufacturer or supplier can help to close information gaps in the documentation. Sometimes it is necessary to show some perseverance when making enquiries to this end. All available information is not always provided immediately, but the reminder of a possible partial legal responsibility or at least a voluntary obligation within the scope of responsible care can be very helpful here. Manufacturers have to gather information elsewhere, as of course they cannot simply refer to safety data sheets, but rather have to draw these up in full themselves. After all, it also has to be possible to determine the level of risk for internal use (Figure 18.2). Classification into one of the categories by four must be possible, otherwise the most dangerous properties must be assumed in each instance. This is often the case in research and development work, as there is frequently no data available. Here, however, is where the protective measures in laboratories come into play, which are also effective for other potentially very dangerous substances and have proven their worth over many years. These also include strategies for assessing exposure, for example measuring options.

The risk assessment must be performed prior to commencing the work, as otherwise risks may be recognized too late. Even if no damage occurs, costly retrofitting or a delay to pending work is often necessary. When setting up production facilities and, if

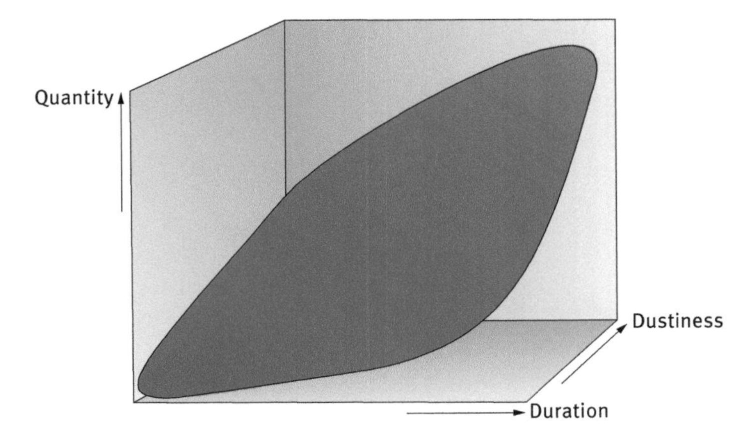

Figure 18.2: Risk correlating to quantity of the nanomaterial, the duration of handling and the dustiness of the material, reaching from green for a lower risk level to red for comparably higher risks.

applicable, also laboratory and technical rooms, it is necessary to provide the technical and organizational prerequisites for safe working. When setting up a production facility, such planning errors can even lead to the cancellation of a project. In a laboratory that was set up for a specific purpose and where, for example, supply and exhaust air capacities were reduced for cost-related or other operational reasons (e.g. as the implementation of such reductions was deemed possible following an assessment of risks and specification of measures sufficient in this specific case), work on manufacturing nanomaterials will often not be possible without significant retrofitting of the ventilation technology. It may even be discerned that it is not possible to perform this retrofitting as, for example, no further supply or exhaust air capacity is available in the building and cannot be expanded, and procedures cannot be adapted to the extent that supply and exhaust air technology can be replaced by alternative exposure minimization methods. As a result, this would mean that the planned work – at least in this room or building – cannot be performed. If the work is still carried out, for example due to economic considerations or in order to avoid endangering research results, serious – and even legal – consequences may be the result. The careful and thorough compilation of the risk assessment – or the prospective risk assessment, depending on the planning stage – and, where required, the immediate adaptation to changed operational conditions such as the use of new chemicals with a risk potential that was previously not existent in this form and therefore not taken into account, therefore sit at the heart of occupational health and safety. Of course, regulatory provisions also have to be observed.

All exposure routes are to be taken into account for determining the exposure situation and checking the effectiveness of the measures taken. The process of assessing the risks (Figure 18.3) may seem to be difficult and arduous, and this actually may be true in some cases. However, in many countries this is not only a legal requirement but also a prudent strategy to avoid damage and woe.

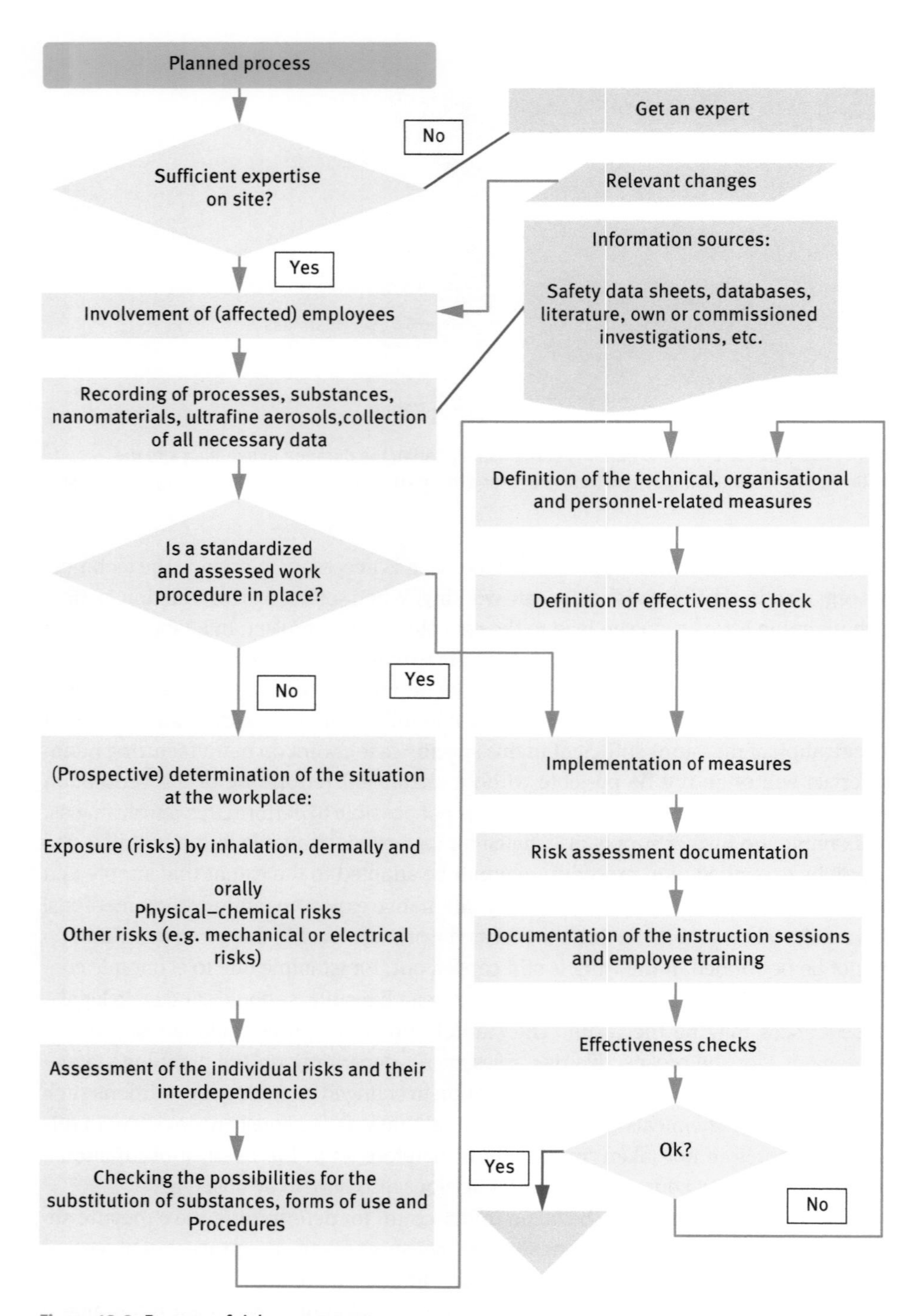

Figure 18.3: Process of risk assessment.

All possible exposure routes – inhalation, dermal, oral – must be taken into account for determining the exposure situation and checking the effectiveness of the measures taken. In particular, oral exposure to hazardous substances is often underestimated, as it is generally correctly assumed that hazardous substances are not intentionally or accidentally eaten or drunk. Due to hygienic deficiencies at the workplace, however, dermal and oral exposure can still occur, for instance if substances are transported to the face by touching hand rails that have been contaminated by soiled gloves with the bare hand and from there get into the eyes or mouth, for example by sweating.

The work procedure and the dustiness behaviour of the material are significant for assessing skin contact. Nanomaterials that are chemically very similar can differ greatly in terms of their dustiness behaviour. For example, in the case of two iron oxides with similarly sized primary particles, one of them released large amounts of nanoscale particles when decanting and the other very little. If the documentation contains no details on the dustiness behaviour and enquiries to the manufacturer or supplier do not yield the desired information, it is recommended to identify this data prior to their introduction into production at the latest. Standardized procedures are available for this.

With the data obtained in this manner and the exact knowledge of the work procedure and equipment to be used, it can be determined whether a relevant release of nano-objects will occur at all during the (intended) processes. Even if a nanomaterial does not release relevant amounts of nano-objects, this can still take place as a result of the work procedure, for example through mechanical or thermal influences that destroy the matrix or split aggregates and agglomerates again. Even the drying of a solution or suspension can lead to nano-objects being released into the air from the dry residues. For some work procedures, further investigation is unnecessary: if a high-performance mill, which creates nano-objects out of larger material through intensive milling within the scope of a top-down process, is opened in a room after the milling process without any additional protective measures having been taken, one would generally determine that enormous quantities of nano-objects have been released.

As well as defining the technical (and other) measures, knowledge of the dust formation also allows the contamination risk to the skin presented by the deposit of released nano-objects to be assessed. Although in line with current knowledge it is believed that human skin provides a reliable barrier to nanomaterials when in good condition, this must not result in work being performed with unprotected hands, arms or even unprotected facial skin due to the existing data gaps in this area, unless this has already been proven for the specific nanomaterial processed. It is necessary to wear protective gloves (in the case of disposable gloves it is generally recommended to wear two pairs on top of each other as a precaution), regularly check them for contamination and damage and change them regularly.

The collection of dust can also lead to deposits on the surfaces of work equipment and devices, surfaces in the room and on clothing. These deposits can then be released back into the air in an uncontrolled manner at a later stage and cause exposure by inhalation, which can occur at a different place and time to the release of material that caused it. Because it is particularly difficult to detect such exposure, it is essential that this dispersal of contamination is avoided by taking technical measures. Swipe samples or direct reading measurements of manual resuspensions can provide clarity here [19].

The risk assessment process also involves documenting the effectiveness of the measures taken. This can be the ventilation test of the supply and exhaust air unit or regular checks by the supervisors that the employees have understood the specified measures and implement them reliably. The risk assessment also documents scopes of responsibility.

Quantifying the risks can be difficult for risks caused by toxicological effects due to a lack of data. Minimization of the exposure will minimized the risks too (for self-organizing or self-reproducing materials this may be different). Risks of fires and explosions are addressed as for other combustible substances. The costs of the materials that can be very high or the amount that can be collected can be a problem for some standard procedures in need of large amounts of sample material, using a 20 L or a 1 m^3 explosion pressure test system, for example. The safety of the personnel carrying out these works has to be guaranteed also, cleaning the testing apparatus and its surrounding area from rests of the material and its reaction products.

18.4.2 Measuring Strategies for the Workplace

Measurements of nanomaterials can be very costly and time-consuming, which impedes broader application. A staggered approach has therefore proven useful for assessing exposure by inhalation. In a first step it should be determined whether relevant exposure has occurred or can even be hypothesized. In an enclosed system operated in a vacuum, releases would not be expected as long as the system is sealed and monitored. The critical operating states in this case would be the extraction of the products, depending on how the filling process takes place, and in particular leaks and accidents as well as cleaning and maintenance work. For these, the exposure would have to be assessed separately across all exposure routes.

The measuring strategy contains a basic exposure assessment, which can be performed at all workplaces with limited effort, expenditure and basic skills for handling the measuring equipment. The results of this assessment must then be compared with the background concentration that is to be determined in parallel. Of course, comparability must be ensured and it must be known whether the same nanomaterials in the air are being compared. Due to factory traffic, engine emissions can release large quantities of nanomaterials into the air in which comparative measurements are made, resulting in a background level that is very high but is not comparable to the nanomaterials to be measured at the workplace. Hand-held devices are used

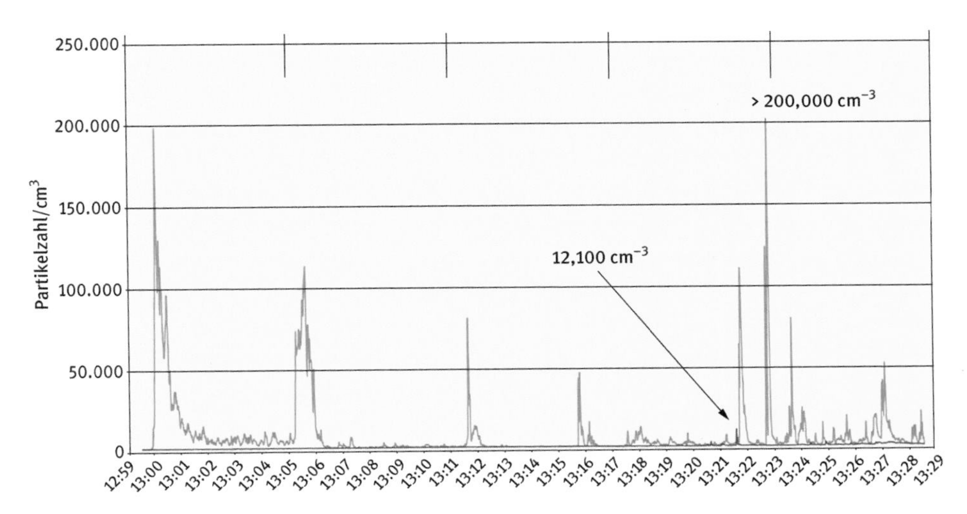

Figure 18.4: Particle concentrations measured inside and outside of a fume hood while handling nanoscaled TiO_2.

as measuring equipment here. If the nano-objects deviate too significantly from the spherical shape, no useful measuring results are generally obtained. Further information can be found in [20, 21].

An example for measuring the emission from a fume hood is shown in Figure 18.4. While handling nanoscale TiO_2 in a fume hood (filling several g from one bottle to another and back) the particle concentration in the air inside of the fume hood and outside at the researchers position a high amount of TiO_2 nanoparticles was released, but no increase of the background concentration in the air of the laboratory is detected. The particle size was detected in a range between about 1 nm up to about 1,000 nm covering agglomerates and aggregates too. To test the experimental set-up air from the inside of the fume hood was blown out waving a sheet of paper to give a short and small peak [22]. Precondition is the proper use of a well-maintained and tested fume hood (in this case according to EN 14175 [23]) without air draughts in the laboratory atmosphere disturbing the air flows in the fume hood.

The technical expenditure is disproportionately higher within the scope of the expert exposure judgement, and is therefore accompanied by high requirements on the people carrying out the measurements. This approach should be employed if it was not possible to achieve a sufficient level of certainty in the first stage. One scale for such measurements is provided by the Organization for Economic Cooperation and Development's tiered approach [24]. The mostly limited resources for carrying out such measurements can be better adapted to the level of difficulty of the enquiry and the scale of the problem in this way. These measurements allow reliable documentation of conditions at the workplace in accordance with present knowledge, and it is possible to define measures appropriate to the scale of the problem.

There are few limit values for exposure to nanomaterials, and none based on health. However, several assessment values can be used as a basis for the assessment [25, 26]: although mass concentrations are traditionally used for assessment in the context of protection from hazardous substances, number concentrations and surface concentration are probably more meaningful for nanomaterials. For criteria for assessment of the effectiveness of protective measures, see [27].

As it often cannot be ruled out that in particular the less tightly bound aggregates of primary particles or fibrous nano-objects can be split open again in the body, the scale of dimensions to be considered should not be limited to 1–100 nm, but rather also extend into the higher range, i.e. up to the micrometre range. Many measuring devices automatically detect these sizes. In normal breathable air, nano-objects usually occur in close proximity to the place from which they were released and are almost exclusively gathered in aggregates or agglomerates having been attracted towards other coarser dust particles or droplets [28]. In a low-particle atmosphere, on the other hand, the nano-objects primarily agglomerate together due to a lack of other bodies to collide with.

18.4.3 Measures at the Workplace

Typical work procedures that can lead to exposure to nanomaterials are:
- Manufacturing of primary particles and their agglomerates and aggregates
- Processing into products for the market
- Processing of coarser nanomaterials while releasing nano-objects (clogged nano-objects or nano-objects that have been removed from the matrix during processing, e.g. nanodroplets from polymers) or compound structures made up of nano-objects and matrix (e.g. carbon nanotubes that protrude from such polymer nanodroplets)
- Waste disposal
- Cleaning of systems, devices and rooms (e.g. to repair a reactor)
- Accidents and leaks

It is advantageous to take the protective measures in the following order (Figure 18.5):

Figure 18.6 demonstrates the differences between the preparations of a nanomaterial, sun blocking TiO_2 in this case. TiO_2 embedded in a matrix of water and oils as a sunscreen does not release any particles, the TiO_2 isolated from the sunscreen by incineration in a crucible is agglomerated and does not release particles, as long it is not milled. Nanomaterials embedded in matrices are relatively safe to handle, since exposure via inhalation is not possible as long the material does not become dry or is not sprayed. Although exposure can occur through the skin or oral (hygiene), agglomerates and aggregates of primary nanoparticles may pose a danger when they are separated into smaller particles.

When handling other (hazardous) substances, field-tested and effective protective measures are available that also offer a high level of protection when working

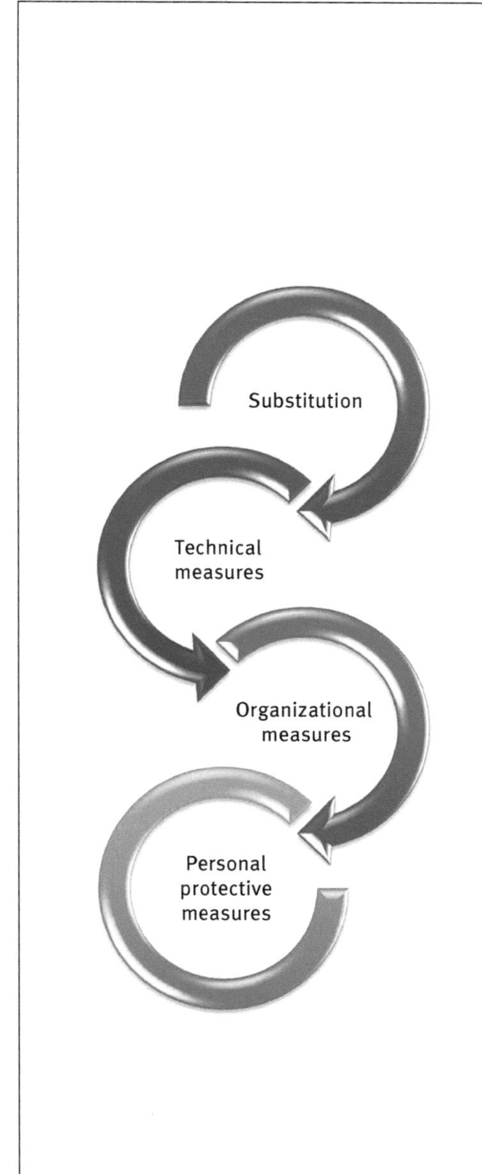

Substitution of hazardous substances, use of applications or processes that lead to lower exposure, e.g. different nanomaterials that result in similar product properties chemical modifications of the nanomaterial that lead to lower toxicity or ignitability use of master batches or suspensions instead of pure nanomaterials use of modifications of the nanomaterial resulting in less formation of dust (e.g. through a coating)
Technical measures, e.g. using closed apparatuses and systems, working in the vacuum range, instead of (partially) opened systems cleaning systems before opening them by fitting cleaning devices ventilation measures, measuring at the point of emission so that the breathing zone is not contaminated and no deposits occur at the workplace and in the neighbouring rooms
Organizational measures, e.g. staff training (see e. g.) spending as little time as possible in contaminated areas minimize the number of people in these areas taking particular groups of people into account, e.g. young people hygiene policy for avoiding dispersal of contamination
Personal protective measures, e.g. single-use protective suits, closed work clothes or laboratory coats that cover the body well protective gloves (with long cuffs if necessary) sleeve guards respiratory protection (not a permanent measure)

Figure 18.5: Strategy of minimizing the exposure; starting point of the process is the replacement of the material as far as it is reasonable.

with nanomaterials. A possible risk can thus also be reduced by minimizing exposure. Therefore, it is usually not necessary to develop completely new protective measures, but rather to search the existing toolbox for the appropriate measures and apply these as intended. As such, technical ventilation measures not only help against toxic gases

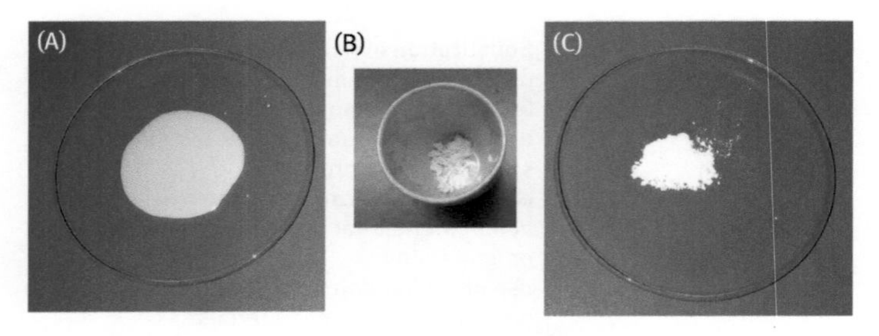

Figure 18.6: (A) Typical sun screen (factor 50) with nano-TiO_2; (B) residue in the crucible after incineration; and (C) milled TiO_2.

and dusts, but also with nanomaterials, and their effectiveness always depends on whether they have been designed, built and applied correctly. Sufficient protective measures can be specified using the available information from safety data sheets, other manufacturers' information, existing literature or by contacting the manufacturer to make specific enquiries. The categories in four can provide assistance here in the event of data gaps, but if necessary experts have to be consulted.

On the basis of several points it is possible to gain a first impression as to whether it is necessary to go into more detail in the risk assessment (see also [29, 30]).

18.4.3.1 Structural Prerequisites

Appropriate construction, infrastructure and fixtures and fittings of buildings and working areas with thorough planning documents, systems and apparatuses built and operated in accordance with the state of the art.

18.4.3.2 Operational Organization

Operation of buildings and systems with precise definition of responsibilities, in particular in the case of facilities where different (legally separate) parties are responsible for ownership, technical operation and utilization, supporting communication and control structures.

18.4.3.3 Occupational Health and Safety

Specification and awareness of responsibilities across all levels of the operational organization, expert safety and occupational medical advice available internally or externally, inclusion of employees in matters relating to occupational health and safety (e.g. in the risk assessment or in the selection and suitability test of personal protective equipment).

18.4.3.4 Agency Workers and Persons from Outside the Company or Laboratory

Include cleaning staff, service technicians, manual workers and others in the risk assessment, instruction and monitoring of these persons by internal staff.

18.4.3.5 Gathering Data for the Nanomaterial

Plausible and complete information not only on toxicity but also on burning and explosion behaviour, dangerous reactions (is it apparent that information is missing?), at least to a sufficient level to categorize the material in accordance with four. This includes the up-to-date safety data sheet for the nanomaterial in question, not for similar products from other manufacturers or suppliers.

18.4.3.6 Informative Labelling of the System Parts and Containers

In particular, exact designation of the nanomaterial, pictograms, H-phrases and P-phrases as per the GHS are necessary, and simplified labelling for own storage bottles in the laboratory is possible (Figure 18.7) [31].

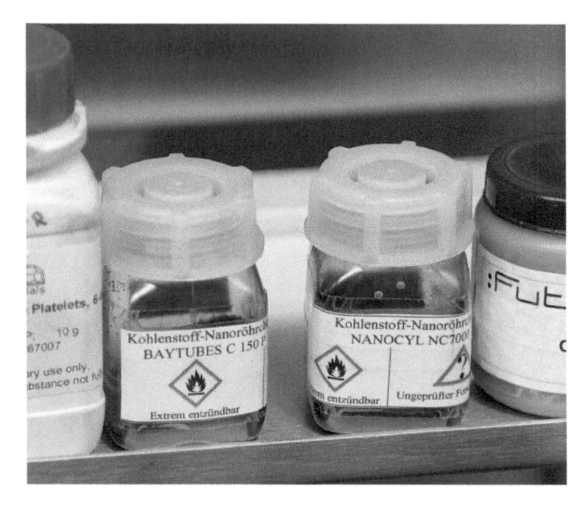

Figure 18.7: Example for a simplified labelling of a storage bottle with pictogram and short phrase.

18.4.3.7 Measures Against Fires and Explosions

Inertization, vacuum, constructional explosion protection or avoidance of ignition sources, fire alarm systems and firefighting (even small quantities can lead to incipient fires or cause severe explosions when mixed with air).

18.4.3.8 Chemical Properties

For example, catalytic activity or auto-ignitability of the nanomaterial (when mixed with other substances there can be a risk of a severe reaction or explosion; large surfaces can lead to accelerated reaction or self-ignition).

18.4.3.9 Protection Against Dust Formation

If possible, enquire about or investigate (or commission an investigation of) the dustiness behaviour and specify protective measures in accordance with this (significant quantities of dust can be formed, particularly when filling and decanting the nanomaterial, but this is not necessarily the case). If nanomaterials are handled openly during the activity, substance release and contamination of the working area are possible and there may be a risk of ignition. If suspensions nebulize there is a risk of inhalation of the aerosol as well as of deposits on skin and mucous membranes and on the surfaces in the working area. This may also present a risk of ignition. Powders can give off dust more or less easily and can be blown away easily. Filling processes can take place in a practically exposure-free manner in (semi-)automatic weighing chambers for sacks, barrels and other containers while extracting and separating escaping air contaminated with nanomaterials. If necessary, manual weighing processes can be performed in specially vacuumed and low-vortex exhaust booths, in sealed automatic scales or through reweighing of sealed containers that are only opened in the fume hood (glove box, etc.). A multitude of further, in some cases, very simple technical solutions or minor but effective modifications to the workflow can be applied here. Dust can escape from unstable containers such as bags during handling. When compressed, air escapes with dust (this may even escape in the vacuum lock). You should therefore use solid, sealable containers, which should not be too large. Machines that can release nanomaterials (e.g. mills) must have a closed design or they must be enclosed and vacuumed. The extracted air can be cleaned well using high-performance filters.

18.4.3.10 Apparatuses and Systems

These have to be engineered or designed in such a way that they have to be opened rarely and as little as possible (e.g. this can be achieved using automatic metering, filling or cleaning devices).

18.4.3.11 Avoiding Dust During Work

Dust formation, nebulization or dried-up leaks can be the result when performing work (dried-up substances may result in dust-generating solids); depending on the risk level, additional technical measures are to be taken. Contaminated auxiliary material for removing leaks can in turn release nanomaterials again, e.g. damp cloths when drying. Contaminated auxiliary material must therefore be packed in a sealed container prior to this.

18.4.3.12 Air Flows in the Working Area

For example from supply or exhaust air units, or unintentional flows caused by draughts in the room. These must not carry the nanomaterials away and distribute or deposit them in suction channels or in the room. Some nanomaterials give off dust so easily that they need to be protected from air flows (e.g. in a glove box or with a decanting pipe from inert gas laboratory technique [32]) during handling. Flow conditions can be determined and assessed using battery-operated fog generators or larger mains-operated equipment for larger rooms, but test setups in fume hoods or systems in rooms with targeted ventilation can also be employed. Apparatuses and systems can block air flows or cause vortexes, and as a result unexpected exposure can occur.

18.4.3.13 Protection from Dermal Exposure

Wear protective gloves if there is a risk of contact with the solid or suspended nanomaterial, or a danger of suspensions dripping down, splashing and nebulizing.

18.4.3.14 Processing Materials and Nanomaterials

Nanoscale aerosols can be created out of coarser or composite materials through milling, grinding, cutting, laser ablation, etc. These can release free nanomaterials or nanomaterials completely or partially enclosed in a matrix. Nanomaterials can be formed and released from materials without nanostructures too.

18.4.3.15 Cleaning

Preclean contaminated equipment so that it can be cleaned by auxiliary staff without any danger. Systems such as reactors must be cleaned sufficiently that staff can also access them from the inside without any risk. Adhering nanomaterials can contaminate the surroundings and put people at risk. A release of dusts from contaminated equipment must be avoided until it is cleaned. The cleaning should therefore take place without delay. Until then, the containers must be sealed or placed in a clean enclosure.

18.4.3.16 Hygienic Measures

These are personal measures for avoiding dispersal of contamination. Contaminated equipment and work tools, e.g. writing equipment or contaminated gloves, must not be removed from the possibly contaminated area (e.g. the fume hood) before being cleaned (or being packed ready for disposal). Hand rails should be used but not touched with gloves. Sleeve guards can prevent nanomaterials clinging to the arms when working in the fume hood. Contaminated protective, work and everyday clothes are to be packed in dust-free containers and given to the specialist cleaning company or disposed of. Used protective gloves must be disposed of if they cannot be cleaned. The dispersal of contamination can also be prevented by the appropriate design of workplaces, processes and experiments. Do not interrupt work in the contaminated area unnecessarily, do not remove contaminated objects (e.g. gloves) from the area, ensure that special attention is paid to hygiene when leaving the area.

18.4.3.17 Other Substances

The other reagents, products, solvents and auxiliary materials are to be taken into account in the risk assessment. The risks associated with these materials must also be taken into account, e.g. handling in particle-filtered circulation operation is critical if gases or vapours also occur, or when an extraction box with absorption or adsorption filters is used and an opening cannot be detected in time.

18.4.3.18 Disposal

Collect and seal waste in suitable and labelled containers and dispose of it at appropriate intervals. The larger the container, the more dust can be released from it. The potential risk increases with the amount of waste stored, and in addition dangerous mixtures or ageing processes can lead to gases and vapours being released, which in turn can cause fires or explosions.

18.4.3.19 Personal Protective Equipment

Alongside work clothes or laboratory coats, protective goggles and suitable shoes are to be worn. Protective gloves with low porosity – which also have to be resistant to the other hazardous substances they may come into contact with – protect against contamination. If there is a risk that the face, head or body can become contaminated with nanomaterials, a protective screen, hood or disposable suit (preferably made of closed textiles without seams or hollows) must be worn (define hygiene measures and disposal procedures).

18.4.3.20 Briefing and Instruction

The properties of and risks posed by nanomaterials are not universally known and must be explained; staff must be familiarized with and trained in safe work processes and protective measures.

Further properties are to be taken into account in order to assess the risk, in particular the ignition and burning behaviour, ageing and decay and the catalytic effect. Please take into consideration that people with disabilities at workplaces may need special equipment and assistance. An example: Contamination may be of greater concern for a person in a wheelchair sitting at a bench or fume hood when traces of nanomaterials may be collected on the sleeves of the lab coat touching the benchtop.

The following guidelines assign protective measures to the nanomaterials for orientation purposes; the risk assessment can provide a different result in specific cases (Table 18.2).

The measures required in the individual case must be specified precisely within the scope of the risk assessment. Measures for avoiding unwanted contamination of work areas may also be necessary without a direct risk being established. Measures

Table 18.2: Categories of nanomaterials according to four and assignment of typical protective measures.

Cat.	Measures
I	Basic protective and hygiene measures, avoidance of all unnecessary releases and exposure
II	See measures of cat. IV.
III	In addition to the measures for category I, in particular protective measures against exposure by inhalation are to be taken (effectively vacuumed enclosures, dust-free handling in fume hood, safety cabinet*; glove box or closed apparatus or systems also possible)
IV	– In addition to the measures for category I, in particular protective measures for avoiding inhalative, dermal and oral exposure are to be taken (closed systems, effectively vacuumed housings, dust-free handling in the tested fume hood, safety cabinet*, glove box, closed apparatus or system, preferably *clean-in-place* or *wash-in-place* in the case of a closed apparatus or system), personal protective equipment must be used, but respiratory protection is only necessary in exceptional cases – If it is proven that there are no asbestos-like properties, measures of the other categories are possible depending on the material

* If no protection from gases and vapours or from fragments or splashes propelled around the room is required.

are also related to other hazardous substances used in the process (e.g. reactants, solvents, auxiliary materials), equipment or systems.

Particle-filtering respiratory masks offer high filtration efficiency for nano-objects and are particularly effective in the range below approximately 200 nm. However, respiratory protection measures should be used with care. The wearing of additional equipment can be a burden for employees, and it should be considered that, in contrast to volatile gases and vapours, a long-lasting contamination in the area of the workplace is possible, even at greater distances depending on the air flow conditions. Deposits can occur along the entire length of exhaust channels, particularly in places with little flow or where vortexes can form. This can be prevented through special structural design and expert construction.

Special care should be taken when handling nanotubes. They too can have a carcinogenic effect on humans if they have certain geometries, length–diameter ratios and a rigid structure [33]. They often form tangled structures with individual nanotube ends protruding from them and can also cause inflammatory reactions. When handled, these sometimes release no nanotubes at all, although it is unclear whether this is also the case under physiological conditions. In this case, safe work can be performed using closed apparatuses, tested enclosures and extractions, tested laboratory fume cupboards or glove boxes. The exhaust air can only be released into the work area after special cleaning. Overviews of this can be found in Refs. [34, 35], for example.

Until now, the measures and effectiveness checks focused on nanomaterials of the first two generations, and also on the third generation to a limited extent [36]. Many nanotechnological applications in the pharmaceutical industry are to be classified as passive (first-generation) or active (second-generation) structures. These are characterized by the fact that they behave like normal substances with regard to quantities, concentration and dose. The systems begin to resemble biological systems to the extent that a multiplication that is actively controlled from the outside to an ever smaller extent can occur. Up to a point this is already the case with the self-organizing processes of the third generation. However, whether multiplication mechanisms of the fourth generation can establish themselves outside of real biological systems is the subject of much debate, but it is not unthinkable. As NY City College's and NY City University's famous physicist Michio Kaku states: "it is very dangerous to bet against the future" [37]. In these cases one would expect an independent increase in particle numbers, quantities, concentrations and doses, as is usual in nature. However, concepts for protective measures against such substances already exist; indeed, the methods are also available for safe handling of organisms and DNA. And like every self-reproducing system, unlimited multiplication is inconceivable as all systems of this kind reach the boundaries regarding available space, nourishment (reagents) and room for waste. With technological adjustments, this should therefore also make it possible to safely control nanomaterials of future developments [19].

Following these prudent practices and proven measures it can be assumed that risks from handling nanomaterials are minimized.

18.5 Further Needs

Continued safety research is required in the future, not only in toxicology but also for combustible materials and testing methods for smaller quantities. The efficiency of ventilation systems should be improved to reduce the risk of wrong usage (very often the wrong positioning), the costs for installing such systems and also the need of energy for running the systems (electric energy for the electric motors or heating and cooling for the treatment of the fresh air).

One of the key components are materials for the information of workers and managers in small and medium enterprises, researchers and all people who have to start handling nanomaterials without sufficient experience and knowledge about the safety procedures. It is also necessary to create the needed level of awareness that there might be risks associated with the planned work caused by nanomaterials (and ultrafine dusts).

This kind of information has to be in practical and concrete terms detailed enough to enable the reader to choose safety measures that are known to be effective and appropriate for the concrete work to be carried out. At the moment most publications deal with the theory of nanomaterials and general principles of prevention, but the gap between this theoretical level and the needs of most users is too wide for them to be bridged by themselves. Here a lot of practical and actionable information is needed. This includes print material with concentrated information (otherwise people will not read it due to the common lack of time), useable also for training purposes. It has been shown that a combination of printed background information and a separate short concentrate from it in key phrases easily to be memorized are quite effective. The concentrated paper can be a folded leaflet to be carried in a pocket of the working clothes or the lab coat. An example is shown in Figure 18.8.

Especially people in research and development and younger people are addressed much easier by modern information technologies, tablet and personal computers, apps and Internet portals. This is quite costly but catches the interest and transports the right kind and level of information. An example is shown in Figure 18.9 [38].

18.6 Conclusion

Since there is no such thing as a one-for-all type “nanomaterial”, the typical nanoeffect or a specific nanotoxicology one has to assess the hazards of the specific material using data from safety data sheets, publications, guidelines and analogies. This looks more complicated than it actually is, at least in the majority of cases, since the materials widely used very often have properties that can be estimated by interpolation from data of coarser fractions and of single molecules or crystal units. For example,

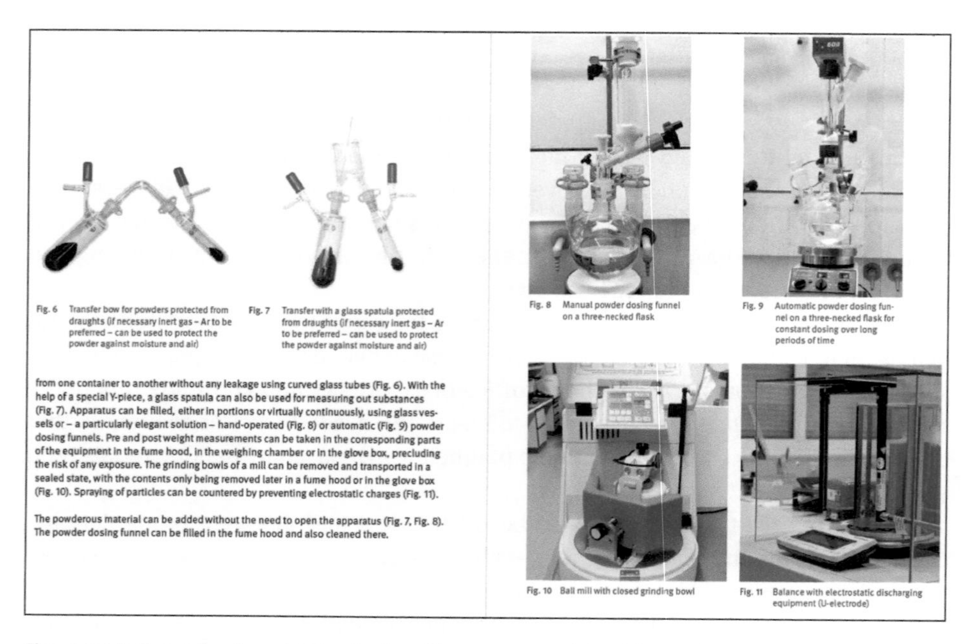

Fig. 6 Transfer bow for powders protected from draughts (if necessary inert gas – Ar to be preferred – can be used to protect the powder against moisture and air)

Fig. 7 Transfer with a glass spatula protected from draughts (if necessary inert gas – Ar to be preferred – can be used to protect the powder against moisture and air)

from one container to another without any leakage using curved glass tubes (Fig. 6). With the help of a special Y-piece, a glass spatula can also be used for measuring out substances (Fig. 7). Apparatus can be filled, either in portions or virtually continuously, using glass vessels or – a particularly elegant solution – hand-operated (Fig. 8) or automatic (Fig. 9) powder dosing funnels. Pre and post weight measurements can be taken in the corresponding parts of the equipment in the fume hood, in the weighing chamber or in the glove box, precluding the risk of any exposure. The grinding bowls of a mill can be removed and transported in a sealed state, with the contents only being removed later in a fume hood or in the glove box (Fig. 10). Spraying of particles can be countered by preventing electrostatic charges (Fig. 11).

The powderous material can be added without the need to open the apparatus (Fig. 7, Fig. 8). The powder dosing funnel can be filled in the fume hood and also cleaned there.

Fig. 8 Manual powder dosing funnel on a three-necked flask

Fig. 9 Automatic powder dosing funnel on a three-necked flask for constant dosing over long periods of time

Fig. 10 Ball mill with closed grinding bowl

Fig. 11 Balance with electrostatic discharging equipment (U-electrode)

Figure 18.8: Example of a guideline for handling nanomaterials with practical information to be adapted easily to the needs of the user.

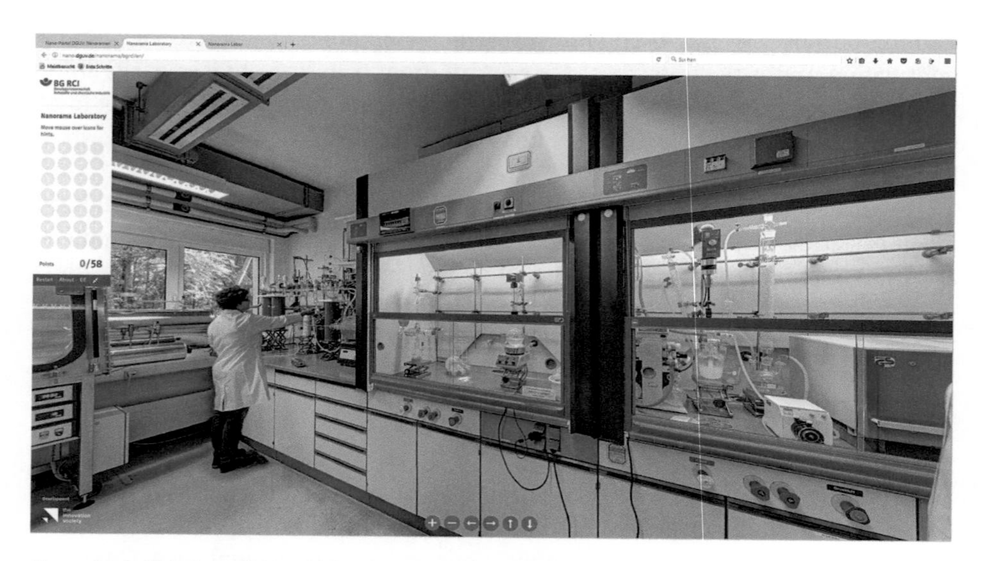

Figure 18.9: Web-based virtual laboratory for nanomaterials running in standard browsers which allows the user to walk around and investigate the different positions with information windows popping up and links to further reading material.

the increase of the surface area per gram ratio is no surprise but well known for all kinds of solid matter. Also the penetration of particles being small enough through biological structures is not a new finding. Adverse effects in the human body may be caused by such particles, but not as a general principle. Problems correlated with the toxicology of the substance per se or the problems caused by overload effects with dusts are known. However, in some cases one detects new properties emerging from the nanoscale, quantum effects in small portions of matter, for example. Whether these may be a problem is not always known. So toxic quantum dots (many are based on cadmium compounds) will show toxic effects due to the cadmium, and quantum effects could be responsible for the catalysis of chemical reactions. On the other hand, iron oxide nanoparticles ("nanomagnets") are useful tools in cancer therapy, even if there might be still also some unknown effect on the body. But it would be not reasonable, maybe even ethically unjustifiable to withhold a cancer treatment due to this (distant) possibility.

Were we to face nanomaterials with self-reproducing properties in the future, we would have to take into account these "biological properties" too. But exposure control for microorganisms is well-established practice already and should not be too difficult to adapt to such nanomaterials.

Minimizing the risks by exposure control apparently is the right strategy in most cases. The toolbox for exposure control is well equipped with strategies and technical solutions for controlling the risks with all kinds of substances, even the ones to be well known for their very dangerous properties. Laboratories in research and development have taught us to cope with very toxic chemicals and new substances without data at all by using proper exposure control strategies – from the appropriate awareness of possible risks to the correct use of all protective measures. Decades of experience have shown that following these strategies the risks of accidents are comparably low to other workplaces. This leads to the conclusion that based on the actual state of knowledge it is possible to work safely with nanomaterials and does not have to be no more complicated than working with other substances. But the implementation of the appropriate measures in all relevant businesses and workplaces is condition sine qua non.

18.7 Summary

While the great expectations of a revolution through nanotechnologies have receded and have been replaced in many areas by a more realistic assessment of an evolutionary development, a large number of nanomaterials can still be found at workplaces. The established protective measures are generally well suited to reducing the possible risks to a responsibly low degree by minimizing exposure. Following these prudent practices and bestowing the appropriate care one would not expect risks correlated with nanomaterials to be higher compared to other risks at workplaces. However, there is no such thing as zero risk. So further research is needed, but also a suitable measure of attention and conscientiousness on the part of all involved.

References

1. Shatkin JA. Nanotechnology – health and environmental risks, 2nd ed. Boca Raton: CRC Press, 2013.
2. ISO. Nanotechnologies – vocabulary – part 1: Core terms, ISO/TS 80004-1:2015.
3. COMMISSION RECOMMENDATION of 18 October 2011 on the definition of nanomaterial. Available at: http://eur-lex.europa.eu/legal-content/EN/TXT/?uri=CELEX:32011H0696. Accessed: 20 November 2015.
4. Committee on hazardous substances of the German Federal Ministry of Labour and Social Affairs: Announcement 527 "Manufactured nanomaterials". Available at: http://www.baua.de/en/Topics-from-A-to-Z/Hazardous-Substances/TRGS/Announcement-527.html;jsessionid=C4F5BF7EE5E373563E80E35FA8461BC0.1_cid323. Accessed: 20 November 2015.
5. German Social Accident Insurance (DGUV). DGUV Information 213-851 Working safely in laboratories – basic principles and guidelines. Available at: http:www.guidelinesforlaboratories.de. Accessed: 20 November 2015.
6. German Social Accident Insurance (DGUV). DGUV information 213-854 nanomaterials in the laboratory. Available at: http://www.dguv.de/medien/fb-rci/dokumente/info_zu_213_854.pdf. Accessed: 20 November 2015.
7. Centers for disease control and prevention – national institute for occupational safety and health: safe nanotechnology in the workplace, 2008. Available at: http://www.cdc.gov/niosh/docs/2008-112/pdfs/2006-112.pdf. Accessed: 20 November 2015.
8. Centers for disease control and prevention – national institute for occupational safety and health: general safe practices for working with engineered nanomaterials in research laboratories, 2012. Available at: http://www.cdc.gov/niosh/docs/2012-147/pdfs/2012-147.pdf. Accessed: 20 November 2015.
9. US Food and Drug Administration. Considering whether an FDA-regulated product involves the application of nanotechnology, 2014. Available at: http://www.fda.gov/RegulatoryInformation/Guidances/ucm257698.htm#intro. Accessed: 5 December 2015.
10. Italian Workers' Compensation Authority (INAIL). White book, exposure to engineered nanomaterials and occupational health and safety effects. Rome: INAIL, 2011.
11. Australian Government Department of Health: Nanotechnology and therapeutic products, https://www.tga.gov.au/nanotechnology-and-therapeutic-products. Accessed: 5 December 2015.
12. REACHnano consortium. Guidance on available methods for risk assessment of nanomaterials. Valencia 2015. Available at: http://www.invassat.gva.es/documents/161660384/162311778/01+Guidance+on+available+methods+for+risk+assesment+of+nanomaterials/8cae41ad-d38a-42f7-90f3-9549a9c13fa0. Accessed: 2 December 2015.
13. Austrian Workers' Compensation Board (AUVA). Nanotechnologien. Vienna, 2010. Available at: http://www.auva.at/portal27/portal/auvaportal/content/contentWindow?viewmode=content&action=2&contentid=10007.672853. Accessed: 2 December 2015. [German language only]
14. Ministry of Health, Labour and Welfare (Japan). Notification on precautionary measures for prevention of exposure etc. to nanomaterials, 2009. Available at: https://www.jniosh.go.jp/publication/doc/houkoku/nano/files/mhlw/Notification_0331013_en.pdf. Accessed: 5 December 2015.

15. Institut national de recherche et de sécurité (INSR): *Les nanomatériaux.* Définitions, risques toxicologiques, caractérisation de l'exposition professionnelle et mesures de prevention, Paris, 2012. Available at: http://www.inrs.fr/dms/inrs/CataloguePapier/ED/TI-ED-6050/ed6050.pdf. Accessed: 2 December 2015.
16. Health and Safety Executive (HSE). Using nanomaterials at work, 2013. Available at: http://www.hse.gov.uk/pubns/books/hsg272.pdf. Accessed: 2 December 2015.
17. European Medicines Agency: Scientific Guidelines: Nanomedicines. Available at: http://www.ema.europa.eu/ema/index.jsp?curl=pages/regulation/general/general_content_000564.jsp&mid=WC0b01ac05806403e0. Accessed: 5 December 2015
18. Bouillard JX. Fire and explosion of nanopowders. In: Dolez PI, editor. Nanoengineering – global approaches to health and safety issues. Amsterdam: Elsevier, 2015.
19. Brock TH. Occupational safety and health. In: Cornier J, Owen A, Arno K, Van de Voorde M, editors. Pharmaceutical Nanotechnology: Innovation and Production, Weinheim, 2017:331.
20. DGUV: Ultrafeine Aerosole und Nanopartikel am Arbeitsplatz. Available at: http://www.dguv.de/ifa/fachinfos/nanopartikel-am-arbeitsplatz/index.jsp page visited 2017-07-04
21. European Chemicals Agency: REACH Guidance for nanomaterials published. Available at: https://echa.europa.eu/de/-/reach-guidance-for-nanomaterials-published (in particular https://echa.europa.eu/documents/10162/13643/appendix_r14_05-2012_en.pdf/7b2ee1ff-3dc7-4eab-bdc8-6afd8ddf5c8d). Accessed 4 July 2017.
22. Brock TH, Schulze B, Timm K. Unpublished.
23. EN 14175: Fume cupboards – Part 1 & 2:2003, Part 3 & 4: 2004.
24. Organisation for Economic Co-operation and Development: Harmonized tiered approach to measure and assess the potential exposure to airborne emissions of engineered nano-objects and their agglomerates and aggregates at workplaces. ENV/JM/MONO (2015) 19.
25. Institute for Occupational Safety and Health of the German Social Accident Insurance organisation (DGUV). Criteria for assessment of the effectiveness of protective measures. Available at: http://www.dguv.de/ifa/Fachinfos/Nanopartikel-am-Arbeitsplatz/Beurteilung-von-Schutzma%C3%9Fnahmen/index.jsp. Accessed: 2 December 2015.
26. van Broekhuizen P, van Veelen W, Streekstra W-H, Schulte P, Reunderss L. Exposure limits for nanoparticles: report of an international workshop on nano reference values. Ann Occup Hyg 2012;56:515.
27. Schumacher C, Pallapies D. Criteria for assessment of the effectiveness of protective measures. Available at: http://www.dguv.de/ifa/fachinfos/nanopartikel-am-arbeitsplatz/beurteilung-von-schutzmassnahmen/index-2.jsp. Available at: 14 August 2017.
28. Seipenbusch M, Binder A, Kasper G. Temporal evolution of nanoparticle aerosols in workplace exposure. Ann Occup Hyg 2008;52:707.
29. Brock TH. Safe handling of nanomaterials in the laboratory. In: Vogel U, Savolainen K, Wu Q, van Tongeren M, Brouwer D, Berges M, editors. Handbook of nanosafety. Amsterdam: Elsevier, 2014:296.
30. German Social Accident Insurance (DGUV). Sicheres Arbeiten in der pharmazeutischen Industrie [German language only], Heidelberg: DGUV, 2012.
31. DGUV Laboratory Safety Expert Committee: Simplified labelling of laboratory containers. Available at: https://www.bgrci.de/fachwissen-portal/topic-list/laboratories/guidelines-for-laboratories/simplified-labelling-of-laboratory-containers/. Accessed: 28 June 2017.
32. Shriver DF, Drezdzon MA The manipulation of air-sensitive compounds, 2nd ed. Hoboken: Wiley-Interscience, 2017.
33. Donaldson K, Poland CA, Murphy FA, MacFarlane M, Chernova T, Schinwald A. Pulmonary toxicity of carbon nanotubes and asbestos – similarities and differences. Adv Drug Deliv Rev 2013;65:2078.

34. Centers for disease control and prevention – national institute for occupational safety and health: current intelligence bulletin 65: occupational exposure to carbon nanotubes and nanofibers. Available at: http://www.cdc.gov/niosh/docs/2013-145/. Accessed: 29 November 2015.
35. Lotz G. Arbeitsmedizinisch-toxikologische Beratung bei Tätigkeiten mit Kohlenstoffnanoröhren (CNT), Dortmund 2015 [German language only]
36. Roco MC. Nanoscale science and engineering: unifying and transforming tools. AIChE J 2004;50:890.
37. Michio Kaku. Physics of the future. New York: Knopf Doubleday Publishing, 2011.
38. DGUV Expert committee for dangerous substances. Available at: http://nano.dguv.de/en/nanoramas/, Accessed: 24 July 2017.

Part V: **Outlook**

Marcel Van de Voorde

19 Conclusions and outlook

The book gives an overview of the breakthroughs in nanotechnology in multiple fields for the coming decades. It spans topics from health care to electronics to advanced textiles.

Great changes will take place in human health care, with nanotechnology-based techniques allowing for the early detection of diseases followed by efficient treatment. It will become possible to keep people much more active into old age.

Information and communication technologies will still see revolutionary developments in nanoelectronics, nanophotonics, and nanomagnetics, resulting in spintronic applications and quantum computers for a wide range of usage.

Industry will be transformed through products with multiple functions, such as smart textiles with embedded computers. Energy supply will become cheaper and environmentally friendly. Transport will become autonomous, with improvements in safety.

The toxicity aspects that concern consumers will be solved, and confidence of the public will be secured.

The following sections provide more details on these aspects.

19.1 Nanotechnology-Based Medicine

Nanotechnology has enabled medical tools such as ultrasensitive diagnostics and 3D molecular bioimaging, which help in earlier and more precise detection of diseases. In addition, more focused treatment will be possible by use of nanotechnology-based drugs and related treatments. Targeted delivery of innovative drugs, formulated into nanoparticles, to the disease site (such as cancer) will enhance efficacy of treatment and reduce undesired side effects in healthy organs. “Theragnostics” combine the option of diagnostic bioimaging with direct subsequent targeted treatment. Nanomedicines also comprise therapeutic agents stored in nanostructured implanted matrices for controlled release in tissue regeneration, or long-term chronic drug medications. Biochemically or physically programming nanomedicines, optionally combined with microelectronics, enables controlled continuous or pulsed release of active substances.

Currently, available nanotechnology has demonstrated impressive proof of concept for severe, previously incurable diseases. After more than two decades, the first effective gene therapies have reached approval as medical drugs. Further gene therapy products (including anticancer DNA vaccines) are close to registration.

https://doi.org/10.1515/9783110547221-019

In 2017, six oligonucleotide therapeutics are already on the market and several RNA interference-based drugs are in late-stage clinical trials.

For wide success and application in medicine beyond life-threatening acute or genetic diseases, the applied delivery technologies need further refinement, such as sequence-defined biomimetic carriers optimized by chemical evolution. Importantly, the steady increase in knowledge on the interaction of the human body with nanoparticles, providing additional tools and markers for nanomedicine safety and efficacy, will present a critical mass of experience for the application of nanomedicines to severe diseases. This includes cardiovascular and lung diseases; and neurodegeneration, where incidence is growing rapidly in our population as a consequence of rising life expectancy. Standardization of the required high-end technologies and production, and harmonization of clinical trial procedures and safety assessment at the European and international level will be important. Biodata and genomic data banks combined with bioinformatics data banks, collecting the acquired preclinical, clinical and postclinical medical experience in correlation with individual genetic patient profiles, will contribute to a new approach to medicine which handles currently incurable severe diseases and provide general improvements in health and quality of life.

19.2 Nanodentistry – Nanotechnology in Oral Health

The branches of dentistry where nanotechnology has become popular, apart from implants, are tooth regeneration, periodontal therapy, soft and hard tissue reconstruction, jawbone repair, plaque control, caries (tooth decay) diagnostics, and caries treatments. One should note that dental caries is the most common disease worldwide and the related costs are immense.

Prostheses and dental implants have become standard, using pure titanium and titanium alloys. Current research focuses on improving the mechanical performance and biocompatibility of metal-based systems, toward anisotropic biomimetic implants, changes in alloy composition, modification of internal and external micro- and nanostructures, and dedicated surface treatments to improve biocompatibility properties, especially with soft tissue components. Nanostructured materials exhibit enhanced mechanical, biological, and chemical properties compared to their conventional counterparts.

The improvement of both short- and long-term tissue integration of implants can be achieved by modification of the surface roughness at the nanoscale level for increasing protein adsorption, and cell adhesion for enhancing osteoconduction; and by the addition of drugs for accelerating the bone healing process in the implant area.

The detailed understanding of the interactions between proteins, cells, tissues, and the artificial implant material is the key factor for developing strategies for oral health. The local release of stimulating drugs will help in many situations, and acceleration of osseointegration time will offer patients shorter and safer rehabilitation.

19.3 Nanotechnology in Communication Technologies

There is no doubt that nanoelectronics has been a game-changer, leading to a complete revolution not only in manufacturing and business but also in our social life. Nanoelectronics has also opened the way to technology sectors by making available the tools to develop and mass-produce complex structures, and to detect and process weak signals, enabling the use of legacy information carriers for high-speed data transmission.

Nanophotonics uses photons for material analysis, data transmission, lighting, industrial and medical applications; while spintronics exploits the magnetic moment (spin) of electrons for signal storage and transmission, with the aim to achieve significant power advantages. Initial practical applications have already emerged in both fields, but much more can be expected in the future with the foreseen progress in materials development and basic physical understanding. A further step into the future is quantum computing, which aims at exploiting the fact that a quantum variable can exist in a combination of states until measured to revolutionize the field of computing. A machine that can operate on such variables (qubits), instead of using variables frozen to a single value (bits), could check all possible variable combinations in one pass, instead than through lengthy recursive calculation. Practical implementation of the concept is complex, and many approaches are being pursued, with no clear winner. However, also here, the first practical applications are starting to appear, mainly in secure communications.

19.4 Nanotechnology in Catalysis

The successful synthesis of mesoporous catalysts in the last decades of the twentieth century represents an important breakthrough in chemistry, nanosciences, and nanotechnology. Mesoporous oxide catalysts play a crucial role in manufacturing processes for the synthesis of chemicals, pharmaceuticals, energy resources, fuels, biofuels, and other products. Nanoscale catalysts can improve the performance of industrial catalytic processes, increasing not only the reaction rate but also the selectivity to the desired reaction. In addition to the well-known particle size effect (smaller is better, as this means a higher surface-to-volume ratio, with more active sites exposed and a higher reaction rate), nanotechnology enables designing and producing catalyst particles with a specific morphology, exposing the most active crystalline facets. This achievement is of crucial importance, because differences in reaction rate between various crystalline faces of a catalyst particle can be several orders of magnitude. Such a nanocatalyst allows producing a higher volume of target product at lower temperature (energy saving) with better selectivity (fewer by-products, more environmentally friendly processes). Industrial nanocatalysis will respond to global challenges such as efficient water splitting to produce pure hydrogen, or carbon dioxide transformation to useful products.

19.5 Porous Materials

Porous materials are of scientific and technological importance due to the presence of controllable dimensions down to the nanometer scale. Research efforts in this field have been driven by rapidly emerging applications such as biosensors, drug delivery, gas separation, energy storage, and fuel cell technology. These research activities offer exciting opportunities for developing strategies and techniques for the synthesis and applications of innovative materials. The perfect control of the structural parameters of porous materials is of fundamental importance in order to tailor and verify their properties. Dedicated three-dimensional imaging techniques are necessary to determine not only the average porosity but also further parameters including the accessible pores, the pore, and channel geometry as well as the nanostructures on the pores' surfaces.

19.6 Nanotechnology Applications in the Automotive Industry

The current key topics in the automotive industry are electric mobility, autonomous driving, communication-based services, and shared mobility. Nanotechnology has cross-sectorial impact and therefore benefits a very broad range of applications in transport systems and the automotive industry. Improvements in mechanical, electrical, magnetic, thermal, optical, and chemical properties have opened the doors to innovations in almost every domain of cars. Although several of them are already well established on the market, there are many more to come: reducing weight and emissions of automobiles, increasing efficiency and range of electric vehicles, and enabling reliable and secure sensing, communication and data processing still require higher performance. Nanotechnology gives huge potential in all of these fields and can be regarded as one of the pacemakers for future automobiles and transport systems to sustain competitiveness.

19.7 Nanodiamond: From Basic Properties to Potential Applications

Due to their outstanding, customizable physical and chemical properties, nanocrystalline diamond layers and components are well suited for hybrid technologies and industrial usage. The advantages of nanocrystalline diamond with a grain size typically around 10 nm are found in their extreme hardness and wear resistance as well as low surface roughness and consequently low coefficient of friction of about 1%. Together with advanced microprocessing technology based on photolithography and reactive-plasma ion-etching, complex-shaped micrometer-sized parts and components are being produced on an industrial scale for a number of micromechanical and biomedical applications. Highly conductive nanocrystalline diamond with tailored grain size and boundary width together with the strong piezoelectric effect makes nanodiamond membranes suitable for sensing applications.

19.8 Nanotechnology in Industrial Applications – Nanocrystalline Diamond Films

Nanocrystalline diamond films show a combination of outstanding material properties, thus making them very attractive for multiple industrial applications as wear-resistive coatings on tools, blades, and bearings. Although the manufacturing scale-up process has made remarkable progress during the last 20 years, the deposition of diamond is still expensive. Advantage can be taken from the engineering of the material properties of nanocrystalline diamond films, with a substrate serving as shaping support for the functional, wear-resistant, thin, and cost-effective nanocrystalline diamond coating.

Due to their outstanding properties, nanocrystalline diamond films have quit niche markets and serve as enabling factors in industrial applications such as machining of innovative lightweight construction carbon fiber reinforced plastic materials for the aircraft industry. This intersectoral technology contributes therefore to the achievement of economic objectives such as energy saving and conservation of resources.

19.9 Conclusions

The overall conclusion of the book is that the nanomaterials revolution will cause great changes in industry, the economy, and all our lives. The book provides guidelines to prepare for these changes. It is encouraged that all society from students to industrialists, consumers to investors become familiar with the topics in this book, to have a full and informed understanding of the benefits and challenges to come.

Index

https://doi.org/10.1515/9783110547221-020